THE HISTORY OF THE STUDY OF LANDFORMS
OR THE DEVELOPMENT OF GEOMORPHOLOGY

VOLUME THREE

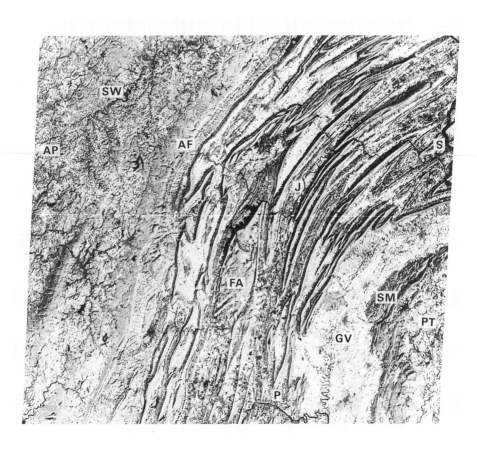

Frontispiece: A LANDSAT image of the archetypal region of the Middle Appalachians (see Figure 7.1), studies of the denudation chronology of which provided one of the most potent influences in geomorphology during the first half of the twentieth century (Courtesy NASA). Key: AF – Allegheny Front; P – Potomac River; AP – Allegheny Plateau; PT – Piedmont Province; FA – Folded Appalachians; S – Susquehanna River; GV – Great Valley; SM – South Mt. (The northern tip of the Blue Ridge Province); J – Juniata River; SW – Susquehanna River (West Branch).

THE HISTORY OF THE STUDY OF LANDFORMS

OR
THE DEVELOPMENT OF GEOMORPHOLOGY

VOLUME 3: HISTORICAL AND REGIONAL GEOMORPHOLOGY 1890–1950

ROBERT P. BECKINSALE

AND

RICHARD J. CHORLEY

LONDON AND NEW YORK

First published 1991
by Routledge
11 New Fetter Lane, London EC4P 4EE
Simultaneously published in the USA and Canada
by Routledge
a division of Routledge, Chapman and Hall, Inc.
29 West 35th Street, New York, NY 10001
© 1991 Robert P. Beckinsale and Richard J. Chorley
Typeset by J&L Composition Ltd, Filey, North Yorkshire
Printed and bound in Great Britain
by Biddles, Guildford and King's Lynn

British Library Cataloguing in Publication Data
The History of the Study of Landforms.
or
The Development of Geomorphology
Vol. 3, Historical & regional geomorphology, 1890–1950.
1. Geomorphology history
I. Beckinsale, R. P. (Robert Percy) II. Chorley, Richard J.
(Richard John)
551.4109

ISBN 0–415–05626–8

Library of Congress Cataloging in Publication Data
has been applied for

TO MONICA AND ROSEMARY

Contents

PART IV: REGIONAL GEOMORPHOLOGY

INDEXES

List of figures

List of plates

Preface

In volume 1 of our *History of the Study of Landforms* (1964) we dealt with major contributions in this field up to the later years of the nineteenth century, and in volume 2 (1973) we dealt with the life and work of William Morris Davis, whose concepts dominated so much of world geomorphological thought until long after his death in 1934. Davis' work was primarily concerned with regional and historical interpretations of landforms, and the present volume 3 explores these themes in some detail on a worldwide basis during the period from about 1890 until the middle of the twentieth century. Volume 4, on which we are presently engaged, will concentrate on the large number of studies in process geomorphology and in Quaternary landforms during the same period, but will carry these themes forward into the second half of the twentieth century, which has been so dominated by them.

History seldom allows itself to be disarticulated into timespans which are convenient for students of history but, in the study of landforms, the years 1890–1950 do provide such a reasonably distinct timespan. The last few years of each century seem impatient to cast off the attitudes of preceding decades and to reach forward towards the new ideas of the new century. Just as the publication of James Hutton's *Theory of the Earth* (1788, 1795) (see volume 1, chapter 4) set the stage for much of nineteenth-century geomorphology, so the years 1888–90 must surely rank as one of the major turning points in the earth sciences. Within this short span were formally stated for the first time four of the concepts which have given twentieth-century geomorphology much of its distinctive character. In 1888 the second volume of Eduard Suess' *Das Antlitz der Erde* set out the eustatic theory; in 1889 Clarence Dutton named the concept of isostasy and William Morris Davis gave the first complete statement of the cycle of erosion and of polycyclic denudation chronology; and in 1890 G. K. Gilbert formulated the concept of epeirogeny. The Second World War provided a similar hiatus in geomorphology. By this time regional description and the speculative basis of much of historical geomorphology were losing their appeal, and both were soon to be overwhelmed by studies attempting to relate landforms to the operation of detailed processes, commonly climatically dominated. Of course, such process studies had always been present in geomorphology but, despite their inhibition by the influence of W. M. Davis and his supporters, they grew apace during the first half of the present century, as our volume 4 will demonstrate.

The present volume is divided into four parts. Part I deals with global influences concerned with crustal, climatic and eustatic changes which formed a backcloth to geomorphic studies during the period 1890–1950. Part II provides a natural link with our volume 2 by examining Davisian influences in the USA and Germany, in France and Britain, and worldwide. Part III is concerned with historical geomorphology through a consideration of four national themes – American polycyclic studies, French eustatic planation, British subaerial–marine synthesis, and the German tectonic approach. Part IV is devoted to regional geomorphology, with particular reference to regional classification and climatic geomorphology. In preparing the present volume we have been concerned to link it as closely as possible with its two predecessors and in this connection we have referred to the works of W. M. Davis according to the format of his complete bibliography given in volume 2, pp. 793–825.

In embarking on volume 4, we find ourselves once again facing a formidable mountain of literature and a vast ocean of geomorphological ideas. The outcome, as with the present volume, will depend heavily on the continued willingness of kind helpers to keep us on the right track and, we hope, to favour us with their own recollections and thoughts; and their knowledge of the existence of personal photographs, influential scholarly researches and of what may loosely be termed significant archival sources. For the present, we are reminded of the comment of Dr Spooner, the sometime Warden of New College, Oxford, and allegedly the orginator of the 'spoonerism', to a young Don – 'I have just read your recent book. It fills a much-needed gap in the literature'. In contrast, we trust that our volume 3 is a much-needed book which fills a gap in the literature which has existed far too long.

R. P. BECKINSALE   R. J. CHORLEY
University of Oxford   *University of Cambridge*

Acknowledgements

During the 16 years which have elapsed since the publication of volume 2 we have lost the help of several valued friends including Professor André Cailleux of Université Laval, Quebec; Professor T. Walter Freeman, for many years editor of *Geographers: Biobibliographical Studies*, sponsored by the International Geographical Union; Professor Preston E. James, formerly of Syracuse University; Professor J. A. Steers and Mr Bruce Sparks of Cambridge University; and Professor George W. White of the University of Illinois at Urbana. Other old friends, and many new ones, have supported us with their assistance and we would especially like to express our thanks to Professor Geoffrey J. Martin of Southern Connecticut State University, the new editor of *Geographers: Biobibliographical Studies*, for allowing us to share his wide and detailed knowledge of the history of geography; to Professor David Stoddart of the University of California, Berkeley, for his important insights; to Professor Chauncy D. Harris of Chicago University, Professor Mikhail Piotrovskiy of Moscow State University, Ms Rosemary Graham of the Scott Polar Research Institute, Cambridge, and Dr Theodore Shabad, editor of *Soviet Geography: Review and Translation* (New York) for help with Russian geomorphology; to Professor Marton Pécsi, Director of the Geographical Research Institute, Budapest, for help with Hungarian studies; to Professor Olavi Granö of the University of Turku and Professor Sten Rudberg of Gothenburg University for help with the works of Scandinavian geomorphologists; to Professor Frank Cunningham of Simon Fraser University for his studies in the earlier history of geomorphology; to Professor William A. Koelsch of Clark University, Massachusetts, for help with the W. W. Atwood archives; to Professor Philippe Pinchemel of the Institut de Géographie, the University of Paris, Professor Numa Broc of the University of Perpignan, Professor J.–P. Nardy of the University of Besançon, Professor Michel Florin of the Société de Géographie, Paris, and Mlle Pelletier of the Département des Cartes et Plans, Bibliothèque Nationale, Paris, for help with geomorphological writings in France; to Professor Otto Fränzle of the University of Kiel for help with German geomorphology; to Professors Ionitá Ichim, Vintila Mikhailescu, Tiberiu Morariu and Victor Tufescu for help with geomorphology in Romania; to Dr J. Babicz of the University of Warsaw, Dr Adam Malicki and Dr Jan Trembaczowski of Marie Curie-Sklodowska University, Lublin, Professor S. Kozarski of Adam Mickiewicz

University, Poznań and Professor A. Jahn of the University of Wrocław for similar help in Poland; to Professor M. Maska of the University of Prague and Dr Tadeáš Czudek of the Czech Academy of Sciences, Brno; to Professor Diniy Kanev of the University of Sofia, Bulgaria; and to Mrs L. M. Richmond, Deputy Archivist of Glasgow University. Thanks are also due to the Leverhulme Trust for the Emeritus Fellowship awarded to one of us to assist in the preparation of this volume.

We have been much assisted by the publication of the *Dictionary of Scientific Biography* (1970–1980) edited by C. C. Gillispie and published by Charles Scribners' Sons, New York, in sixteen volumes. We also note with pleasure and gratitude the publication of other important works relating to the history of geomorphology which have appeared since we began our labours almost 30 years ago. These include:

Davies, G. L. (1969) *The Earth in Decay: A history of British geomorphology 1578–1878*, London; Macdonald.

Cunningham, F. F. (1973) *The Revolution in Landscape Science*, Vancouver: Tantalus Research.

Greene, M. T. (1982) *Nineteenth Century Theories of Geology*, Ithaca, NY: Cornell University Press.

Tinkler, K. J. (1985) *A Short History of Geomorphology*, Beckenham: Croom Helm.

Ellenberger, F. (1986) *Histoire de la Géologie: Des anciens à la première moitié due XVIIe siècle*, vol. 1, Paris: Lavoisier.

The authors and publishers would like to record their gratitude to the following editors, publishers, organizations and individuals for permission to reproduce illustrations and other material:

Editors

Abhandlungen der sächsischen Akademie der Wissenschaften (Leipzig), Math.– Phys. Klasse for figures 10.2 and 10.4.

American Geographical Society, Occasional Publications for plates 6.2 and 11.3.

American Journal of Science for figures 1.3, 7.8 and 7.13.

Annals of the Association of American Geographers for figures 12.10, 12.11 and 12.12

Bulletin, Academie des Sciences Mathématiques et Naturelles, Belgrade, for figure 2.4.

Bulletin of the American Association of Petroleum Geologists for figure 7.2(A).

Bulletin of Denison University for figure 7.12.

Bulletin of the Geological Society of America for figure 7.14.

Connecticut Geological and Natural History Bulletin for figure 7.9.

Geographical Journal for figures 9.14, 9.15 and 9.16.

Geographical Review for figure 7.5.

Geographische Abhandlungen for figure 10.9.

Institute of British Geographers, Baulig (1935) quotes for figures 8.3, 9.2, 9.4 and 9.5.

International Geological Review for figure 2.9.

Journal of Geology for plate 3.1.

Meteorologische Zeitschrift for plate 1.1.

National Geographical Society, Monographs, for figure 7.4.

Natural Science for figure 9.8.

Proceedings of the Geological Society of America for plates 3.2 and 9.2.

Proceedings of the Geologist's Association for plate 9.1.

Proceedings of the Yorkshire Geological Society for figure 9.10.

Quarterly Journal of the Geological Society of London for figures 9.14 and 9.17.

Science Progress for figure 9.9.

Scottish Geographical Magazine for figure 9.12.

Transactions of the Geological Society of Glasgow for figure 1.7.

Zeitschrift für Geomorphologie for figures 12.5 and 12.14.

Zeitschrift der Gesellschaft für Erdkunde zu Berlin for figures 10.5 and 10.7.

Publishers

Armand Colin, Paris for figure 8.2 and 8.4 from H. Baulig (1928) *Le Plateau Central de la France* and H. Baulig (1928) quotes.

Cambridge University Press for figure 9.11 from F. R. C. Reed (1901) *The Geological History of the Rivers of East Yorkshire*.

Columbia University Press, New York for figures 7.2B, 7.10 and 7.11 from D. W. Johnson (1931) *Stream Sculpture on the Atlantic Slope*.

Gauthier-Villars, Paris for plate 5.3 from A. Lacroix (1932) *Figures de Savants*.

Gebrüder Borntraeger, Berlin for figure 1.1 from L. Kober (1928) *Der Bau der Erde*; for figure 1.6 from W. Köppen and A. Wegener (1924) *Die Klimate der geologischen Vorzeit*; for plate 2.1 from J. Büdel (1982) *Climatic Geomorphology*.

Hutchinson, London for figure 2.3 from F. E. Zeuner (1959) *The Pleistocene Period*.

McGraw-Hill, New York for figures 7.1, 7.6 and 7.7 from N. M. Fenneman (1938) *Physiography of Eastern United States*; for figure 12.1 from G. T. Trewartha (1954) *An Introduction to Climate*.

Macmillan, London and Basingstoke for plate 10.4 and figures 10.3, 10.6, 10.10 and 10.11 from W. Penck (1953) *Morphological Analysis of Landforms*; for figures 12.4, 12.9, 12.15 and 12.16 from E. Derbyshire (ed.) (1973) *Climatic Geomorphology*.

Melbourne University Press for figures 11.6 and 11.7 from A. G. Isachenko (1973) *Principles of Landscape Science and Physical-Geographic Regionalization*.

Methuen, London for figure 9.3 from D. K. C. Jones (1981) *The Geomorphology of the British Isles: Southeast and Southern England*; figure 9.13 from E. G. Bowen (1957) *Wales*; figure 9.13 from C. A. M. King (1976) *The Geomorphology of the British Isles: Northern England*; figure 10.13 from J. Thornes and D. Brunsden (1977) *Geomorphology and Time*.

PWN – Polish Scientific Publishers, Warsaw, for figure 11.5.

Charles Scribners' Sons, New York (an imprint of Macmillan Publishing Co.) for figure 1.2 from R. A. Daly (1926) *Our Mobile Earth*.

SEDES, Paris for figure 12.13 from J. Tricart and A. Cailleux (1965) *Introduction à la Géomorphologie Climatique*.

Tauchnitz, Leipzig for figures 2.1 and 2.2 from A. Penck and E. Brückner (1909) *Die Alpen in Eiszeitalter*.

Whitcombe and Tombs, Christchurch, New Zealand for figure 12.8 from C. A. Cotton (1942) *Climatic Accidents in Landscape Making*.

Wiley, New York for figures 7.3, 7.15 and 7.16 from W. D. Thornbury (1965) *Regional Geomorphology of the United States*.

Yale University Press, New Haven for figure 3.3 from R. A. Daly (1934) *The Changing World of the Ice Age*.

Organizations

Département des Cartes et Plans, Bibliothèque Nationale, Paris, for Plates 5.1, 5.2 and 5.4.

Geological Survey of Great Britain for figure 9.1.

King's College, London, Department of Geography Archives for plate 5.7.

Military Engineering Experimental Establishment, Christchurch, Hampshire, for figure 11.2(A).

NASA for Frontispiece.

National Portrait Gallery, London, for plate 4.1.

US Army Engineer Waterways Experiment Station, Vicksburg, Mississippi, for figure 11.2(B).

University of Chicago Archives for plate 4.4.

University of Chicago, Department of Geology, for plate 4.3.

University of Cincinnati Archives for plate 11.2.

University of Hamburg Archives for plate 11.1.

Individuals

Mme P. Birot for plate 5.5.

Professor A. Cailleux for plate 12.1.

The Executors of Professor A. L. Du Toit for figures 1.4 and 1.5 from (1937) *Our Wandering Continents*.

Professor A. Goudie for plate 5.6.

Professor L. C. King for figures 6.1, 6.2 and 6.3 from (1962) *The Morphology of the Earth*.

Professor M. Klimaszewski for figure 11.5.

Professor E. Juillard for plate 8.1.

Professor G. J. Martin for plates 4.2, 4.5, 4.6, 4.7, 6.1, 6.3, 6.4, 6.5, 10.1 and 10.3. (1974: 1980) quotes.

Helmut Penck for plates 10.2 and 10.5.

Professor P. Pinchemel for plates 5.1, 5.2 and 5.4.

Professor G. Sendler for plate 11.1.

Dr M. Simons for translations of W. Penck (1925).

Professor A. N. Strahler for plate 4.8.

Professor A. Young for figure 10.12 from (1972) *Slopes*.

Mr M. Young, Mr I. Agnew and Miss L. Judge for drawing.

Mr D. A. Blackburn for photography.

Additional

Grateful acknowledgement is given to the following organizations for the use of short, individual quotations, the detailed sources of which are given in the text:

American Geographical Society for Bowman (1926) and Johnson (1929, 1932).

Annals de Géographie for Martonne (1929).

Association of American Geographers for Russell (1949), Bryan (1950) and Strahler (1950).

Blackwell, Oxford for Stoddart (1986).

Geological Society of America for Leverett et al (1922).

Institute of British Geographers for Wooldridge and Linton (1939) and Wooldridge (1958).

La Pensée for Tricart (1956).

Longman, London for Linton (1951) and Tricart (1968).

Macmillan, London for Hettner (1972).

Macmillan, Melbourne for Mabbutt (1968).

Methuen, London for Steers (1945) and Wooldridge (1951).

Thomas Nelson and Sons, Sunbury-on-Thames for Holmes (1944).

Oliver and Boyd, Edinburgh for Du Toit (1937), King (1962), Machatschek (1969) and Young (1972).

Presses Universitaire de France for Birot (1955).

Soviet Geography for Gerasimov (1968).

Whitcombe and Tombs for Cotton (1942).

Wiley, New York for Kuenen (1950) and Thornbury (1954, 1965).

Yale University Press for Daly (1934).

PART I

Global Influences

CHAPTER ONE

Crustal Changes

Introduction

In the previous two volumes in this series little has been written regarding the relationship between the earth's surface relief features and crustal changes. Certain primitive notions on the origin of mountains prior to about 1830 were touched on in volume 1 (part 1); Davis' ideas regarding the tectonic history of the Appalachians appeared in volume 2 (chapter 11); and Walter Penck's tectonic assumptions were set out in volume 2 (chapter 23). It is now necessary, however, to describe some of the more important researches dealing with the tectonic framework of geomorphology between about the middle of the nineteenth century and the general acceptance of plate tectonics in the 1950s and 1960s.

Before studies of Alpine tectonic structures began to dominate views relating to relief-forming processes in the 1870s and 1880s, a number of crustal elevation processes were proposed. Charles Darwin (1838) supported Lyell's uniformitarian principles when he advocated a scheme of slow vertical volcanic pressures taking the form of repeated strokes. Shortly afterwards (1843) an important work on the physical structure of the Appalachians appeared written by two brothers, Henry Darwin Rogers and William Barton Rogers who were born in Philadelphia and later lived in Boston. In 1857 Henry moved to Scotland and became a professor – probably the first native American to attain that status in Europe – in the University of Glasgow. Their analysis shows great skill in unravelling the stratigraphy and structure of the huge asymmetrical chain and of the tightly packed isoclinal folds and thrusts which contribute its Ridge and Valley province. The report was illustrated by clear, true-to-scale cross-sections which were copied in numerous texts, but there was less certainty about the primary dynamic forces, and the wave-like geological structures that operated over such a large region were ascribed to

> an onward, billowy movement proceeding from beneath, and not of a folding due simply to some great horizontal or lateral compression.
>
> (Rogers and Rogers 1843)

A much more ambitious series of works was begun in 1846 by James Dwight Dana who postulated the elevation of mountain ranges and continents by lateral pressures set up by crustal cooling and shrinkage. This reached a climax with a five-part publication in the *American Journal of Science* (1873) on the origin of mountains which he ascribed to the formation of 'geanticlinals or anticlinoria' (elevated by compression) or to 'geosynclinals' (subsidence with sedimentation accompanied by and followed by compression). The shrinkage theory implied that, once formed, surface elevations would tend to persist in the same place giving a continuity to the location of continental and mountain cores, a concept which was only dispelled almost a century later by the widespread recognition of plate movements. This theory of Dana's was later termed that of 'Archaean Protaxis', signifying long-standing organization resulting from the application of pressure. Until the full implications of ideas relating to isostasy were realized, notions such as volcanic uplift, horizontal compression and crustal shrinkage appeared to provide satisfactory explanations for the major crustal movements.

Isostasy

The idea of isostasy, or a state of hydrostatic balance in the crust of the earth, originated in the mid-eighteenth century when great efforts were being made to ascertain the geodetic shape of the globe. During the French expedition to Peru in 1735–44 Pierre Bouguer (1749), who at the tender age of 15 had succeeded his father as royal professor of hydrography at Paris, found that the deflection of a plumb-line was much smaller for the Andes than had been expected from the mass represented by those mountains (1749). In 1750 R. de Boscovitch postulated that this deficiency probably implied a low density mass (or root) at depth which compensated the mass of the mountains. At the end of the eighteenth century, spurred by the contemporary interest in gravity measurements of the spheroid, N. Maskelyne and Charles Hutton compared the density of a Scottish mountain, Schiehallion, and of the earth and found the former to be much less. Their findings supported the concept of a less dense layer below mountains and of a lighter crust of the earth with an uneven distribution of gravity due to an irregular distribution of rock density. By 1837 Charles Babbage could discuss elevation that may be expected to follow 'displacements of isogeotherms' due to a sedimentary blanket being stripped from one area and laid down in another, and his friend, John Frederick Herschel, mentions subsidence caused by the weight of accumulating sediments and considered that such subsidence might be correlated with upheaval of the periphery of the subsident area (see Hall 1859). From about this time it became increasingly common to associate long-continued subsidence and sedimentation with the ultimate formation of mountain ranges.

The mid-nineteenth century witnessed dramatic developments in the isostatic concept. Sir George Airy (1855) and Archdeacon J. H. Pratt (1859) explained the unexpectedly small gravitational effect of the Himalayas by the existence of foundations that buoy up or support the elevated masses with roots of relatively low density, or 'the compensating effect of a deficiency of matter below the mountain mass'. There was a difference between their views, as Airy postulated deep roots of lesser density penetrating a denser layer, whereas Pratt assumed a plane within the crust where compensation was complete or a level below which solid crust gave place to yielding conditions. However, for most geomorphologists their ideas fused into a most convenient concept of a kind of floating crust which consists partly of mountain chains and continents that are buoyed up by unseen foundations (Steers 1929).

In 1865 the idea of isostatic equilibrium was given a new twist by Thomas Francis Jamieson of Edinburgh, a progressive glacialist, who suggested that Scotland had recently sunk under the load of ice-sheets and had subsequently risen when the ice melted. Jamieson later went further by postulating that variations in the thickness of the original ice sheet were reflected in the degree of crustal rebound, as could be observed in variations in the height of the ancient shoreline. So henceforth the crustal balance was firmly linked to variations in density, to weight of sediment accumulation and to surface loading by ice and water. The idea of a 'floating' crust in which mountains and continents were buoyed up isostatically was given some publicity by Joseph LeConte (1872) although he considered the earth's interior was too rigid for the deduction to be more than of theoretical interest.

In Britain Osmond Fisher, a much undervalued geophysicist, in his *Physics of the Earth's Crust* (1881) also supported the notion of a floating crust and went further in considering that mountains with deep-seated foundations would, when degraded by erosion, be resuscitated by uplift. A few years later C. E. Dutton summarized the problems and possibilities of this theme, which he had been contemplating for a long time, and supplied the now popular terminology:

> For this condition of equilibrium of figure, to which gravitation tends to reduce a planetary body, I propose the name isostasy . . . . We may also use the corresponding adjective isostatic.
>
> (Dutton 1889: 53)

Isostasy thus implied the tendency to equal poise or equal hydrostatic pressure and was, he says, in essence the same as the ideas of Babbage and Herschel. Its gist was that great masses of sedimentary deposits displace the earth beneath them and subside while great denudation of the surface will cause uplift. Both Fisher and Dutton disliked the earth-contraction hypothesis as an adequate explanation of earth tectonics and each devised a crude kind of crustal convective movement. For Dutton most of the elevation of the land was considered independent of folding and resulted from the quite unexplained expansion of subcrustal magma (see volume 1, p. 586).

G. K. Gilbert, a close friend of Dutton, was also interested in gravity and isostasy. In his monograph on Lake Bonneville (1890) he included sections on the 'Hypothesis of terrestrial deformation by loading and unloading' and 'The strength of the earth'. The Great Basin, which enclosed the vast Pleistocene ancestor of the Great Salt Lake, was considered to be due to crustal expansion under the influence of large-scale (epeirogenic) displacement, whereas the shorelines of the lake were shown to indicate that its site had subsided under the great load of water and rebounded when most of the load was removed (see volume 1, pp. 567–9).

Seismic measurements soon fuelled speculations regarding possible layers of different density in the earth. In 1897, Emil Wiechert, a famous geophysicist of the University of Göttingen and inventor of an improved seismograph, suggested that a rock mantle about 1,400 km thick surrounded a largely iron core. His estimate was approved by Richard Dixon Oldham, who wrote a detailed report on the great Assam earthquake of 10 June 1897 and from the study of seismic waves provided the first clear evidence that the earth had a central core (Oldham 1906). Later he postulated an Earth with a crust, a shell (or mantle) and a core, each with distinguishable physical properties. The idea was refined by Andrija Mohorovičič (1910) who, from a close examination of waves propagated by an earthquake near Zagreb on 8 October 1909, detected a layer where the speed of transmission suddenly changed – the Mohorovičič discontinuity (Moho). He thought it was about 50–60 km (30–35 miles) thick and marked the transition from continental granitic rocks (sial) to the denser underlying rocks (sima). The introduction of the torsion balance by Baron von Eötvös gave much greater accuracy to gravity measurements over short distances and it began to be recognized that these might indicate departures from isostatic equilibrium and, consequently, of tendencies of the crust to move vertically in compensatory directions (Spencer 1913). Thus the recognition of areal variations in terrestrial gravity enhanced the idea of a vertically mobile crust.

By 1914, Beno Gutenberg, one of Wiechert's students, was writing in virtually modern terms of the density layers in the earth. He determined the existence of a low velocity layer in the upper mantle at a depth between 100 and 200 km, which was destined to prove of great significance in arguments about the nature of movements in the crust. He considered that the crust was thin and merged through the Moho discontinuity into the mantle which was 2,900 km (1,800 miles) thick and surrounded a central core. The outermost 100 km (60 miles) or so comprised the crust, and the uppermost mantle was thought to be relatively cool forming the lithosphere, beneath which was a warmer layer of partial melting.

Also in 1914 Joseph Barrell (1864–1919), professor of structural geology at

Yale and author of several writings on the strength of the earth's crust, postulated two crustal layers: an outer, stronger layer, 50 to 70 miles thick and of varying density, called the lithosphere; and beneath it a zone, 70 to 300 miles thick, of flowage.

> The theory of isostasy shows that below the lithosphere there exists . . . a thick earth shell marked by a capacity to yield readily to long-enduring strains of limited magnitude . . . . Its comparative weakness . . . is its distinctive feature. It may then be called the sphere of weakness – the *asthenosphere*.
>
> (Barrell 1914b: 659)

Barrell tried to explain geological phenomena by the dynamic interaction of lithosphere and asthenosphere. In his opinion isostatic compensation worked only on a large scale, or *in general*, despite the local effects of erosion and sedimentation. A different view was taken by some of his contemporaries, including John F. Hayford (1868–1925) of Northwestern University who was a great admirer of G. K. Gilbert. Hayford (1909) used an areal method rather than one based solely on linear arcs and took into account topographic irregularities and isostasy. He estimated that the most probable depth of isostatic compensation would be about 113 km, whereas some other geologists postulated more than double that depth:

> Hayford's work constituted the first demonstration of the validity of the concept of isostasy . . . . Like Pratt – but unlike Airy – he and his successor in the Coast and Geodetic Survey, William Bowie, believed that isostatic compensation is complete and local;
>
> (Reingold 1972: 188)

The hypothetical mechanism suggested for the compensatory movements involved horizontal flows of rock far below the surface in association with frictional heat and chemical reactions.

The plasticity of some earth layers became rapidly more probable with the discovery of radioactive decay. In 1896 Antoine Henri Becquerel discovered that emissions from uranium salts made an impression on a photographic plate. Within a few years Pierre and Marie Curie had proved the atomic properties and shown the existence of the radioactive elements polonium and radium in uranium ore; and Ernest Rutherford and Frederick Soddy had proposed that the disintegration was responsible for the production of heat. By 1906 Robert John Strutt, later Fourth Baron Rayleigh, had discovered the presence of small units of radioactive elements in a wide variety of rocks. The importance of radioactivity to geology and so to geomorphology was recognized and popularized by John Joly (1857–1933) of Trinity College, Dublin, who from 1903 onward drew attention to it as a probable source of terrestrial heat. Soon he and others had established that the rate of decomposition of radioactive minerals into the final form of inert lead must have remained constant and so provided a means of assessing the age of radioactive rocks. Joly publicized his ideas in two main books, *Radioactivity and Geology* (1909) and *The Surface History of the Earth* (1925), and thereafter the scientific community was assured of the abundance of heat, and therefore of the possibility of fluidity, in the earth's crustal layers.

Isostatic thinking in geomorphology during the first half of the twentieth century was concerned not so much with questioning the operation of some tendency towards isostatic adjustment but with determining to what degree this tendency is opposed by the inherent strength of the crust itself. Hayford (1911), Bowie (1917, 1921, 1927) and Reid (1922) were strong advocates of the efficacy of isostatic forces in bringing about compensation even on a local level in the face of low values of crustal strength, whereas Gilbert (1895), Barrell (1914b, 1919), Born (1923), Putnam (1930) and Gunn (1949: 267, 278) favoured higher estimates for crustal strength and allowed the possibility of isostatic compensation only on a large-scale regional basis. T. C. Chamberlin (1927) even went so far as to deny the existence of hydrostatic conditions in the crust and proposed substituting the term 'elastasy' for isostasy. Estimates of the actual order of crustal strengths began to be made as soon as the compensatory effects of unloading by ice (de Geer 1892) and lake water (Gilbert 1890) were identified. This question is still largely unanswered, but by about 1950 there was a general feeling that crustal strength was lower than was hitherto supposed and Hubbert (1945: 1651) proposed that

> the behaviour of the earth as a whole in geologic time must be very similar to that of the ordinary viscous fluids and extremely soft muds of our everyday experience.

Geosynclines

The first definitive statement of the geosyncline (named by Dana (1873)) as a linear belt of long-continued, shallow water sedimentation subsequently subjected to elevation was given by James Hall (1859). However, his notions of mechanisms of crustal depression (i.e. by weight of sedimentation) and of elevation were distinctly vague, as Dana was quick to point out. Studies by Suess (1875) and Heim (1878) had suggested that mountain structures had been produced by tangential compression directed away from the ocean basins towards the continental shields, involving the deformation of great thicknesses of marginal marine deposits. Marcel Bertrand first formulated the concept of 'facies' (1897), showing that single stratigraphic time units are often composed of different intergrading deposits, the character of which gives rise to useful speculation on the distance, relief and tectonic behaviour of the sedimentary source areas. One result of this was that the tectonic significance of 'flysch' deposits, on which Walter Penck later relied so heavily, became apparent. Of more immediate importance was the association of the detrital sediments in the geosynclinal belts with sources external to the present main continental areas, particularly in North America, and this resulted in the 'borderlands' theory. Beginning with 'Appalachia' (Williams 1897), large landmasses were postulated to have existed on the oceanward sides of all marginal or 'ortho' geosynclines (Stille 1936a, b). During the first quarter of the twentieth century this concept was an accepted part of the tectonic framework of sedimentation and found its most influential exponent in Charles Schuchert (1910, 1923), who added several more similarly large and equally conveniently disappearing borderlands (e.g. Llanoria, Cascadia, Yukonia, etc.) yoked to the other North American orthogeosynclines. Similarly, Barrell (1914a) made a detailed study of the source area of Appalachia which was held to be associated with the deposition of the Upper Devonian rocks of New York State, but it was this same author who about 10 years later began the reduction of the concept of massive borderlands into large 'geanticlines' (Barrell 1925, Thom 1937) before their replacement by island arcs. Up to the early 1920s most American geologists were prepared to believe in the possibility of the existence and disappearance of past landmasses on the site of the present ocean basin margins (Chorley 1963). It is interesting that much of Davis' theory of Pennsylvanian drainage (1889D) (see the complete bibliography of W. M. Davis, volume 2, pp. 793–825) turned on the postulation of initial Appalachian consequent streams flowing north-westward, away from the assumed main tectonic axis lying to the east of the present folded mountains (see volume 2, chapter 11 and volume 3, chapter 7). The problem of reversing this initial drainage was just as difficult for Davis as explanations for the extremely convenient disappearance of the huge borderlands subsequently became for geologists.

The resolution of the problem of borderlands and geosynclinal evolution was effected largely by the gravity measurements which the Dutch geophysicist Vening Meinesz (1934) carried out in the East Indies between 1923 and 1932. In the 1930s the active downbuckling of seismically active geosynclinal belts with volcanic islands arcs and negative gravity anomalies was identified in both the East Indies (Lawson 1932, Umbgrove 1934, 1938, 1947) and the West Indies (Hess 1938). Stille, who had previously (1934) suggested a mechanism whereby continents might grow by the orogenic consolidation of marginal orthogeosynclinal belts, proposed the term 'eugeosyncline' for such volcanically active geosynclinal belts which were regarded as the first indication of real orogenic activity in the geosyncline (Stille 1941; see Kay 1951).

Orogeny

The term orogeny, as Haarmann (1930) pointed out, was persistently misused in the earlier part of the twentieth century to designate the production of geological structures, such as folds and thrusts, associated with lateral compression and crustal shortening. Although such compression might be expected to produce increased crustal elevation, it was clear that many topographic mountains have been the result of the bodily upheaval of previously folded belts, perhaps in a post-compressional phase.

Almost inevitably ideas on crustal shortening took on new proportions in the 1840s when Arnold Escher von der Linth postulated large-scale over-thrusting (*Ueberschiehung*) or overfolding (*Umbiegen*) in the European Glarus Alps (Bailey 1935). Soon overfolds or overthrusts were being proposed in the Appalachians, in Canada, the Lepontine Alps and the Scottish Highlands where horizontal translations of more than 10 miles were demonstrated by Peach and Horne (1884). In 1875 Eduard Suess, who had already demon-strated the thrusting of crystalline rocks over younger sedimentary beds, published his *Die Entstehung der Alpen* in which *Alpen* seemed to mean mountain chains anywhere. He paid especial attention to the global surface pattern of mountain chains which, under the overall influence of a cooling shrinking globe, fashioned their relative movement around or between relatively resistant blocks. Three years later Albert Heim of Zürich produced his two-volume *Untersuchungen über den Mechanismus der Gebirgsbildung* (1878) which he dedicated to his old tutor Escher von der Linth. Heim had acquired from his father a love of mountain scenery and from his mother a taste and talent for drawing. He was probably the first genuine geological artist and could illustrate the most complex structures with brilliant draw-ings, cross-sections and three-dimensional models which influenced the ideas of some of his contemporaries, including Marcel Bertrand (1884). In 1888 Alpine tectonists received spectacular support for their overthrust views when A. E. Törnebohm identified colossal overthrusts of as much as 130 km (80 miles) in Scandinavia. It was therefore more than coincidental that about this time geologists took on a more grandiose view of Alpine lateral tectonics. In 1893 Hans Schardt emphasized the reality of huge composite masses being thrust far northward and of such masses being isolated from their roots by erosion, when he interpreted the Pre-Alps (a mass 120 km long by 40 km broad) as a detached part (outlier) of an overthrust. In other words the Pre-Alps were not, as had been thought, mountains protruding through a local cover of younger rocks but were a mass of translated rocks lying upon a pre-existing local layer. In the first three decades of the present century a number of influential texts (e.g. Lugeon 1901, Staub 1924, Collet 1927, Steers 1929, Heritsch 1929) stressed the importance of the nappe theory in Alpine crustal shortening, while that in the Appalachians (due to disharmonic shallow folding and distributed thrusting) was estimated to have involved a compres-sion of 320 km (Keith 1923).

These researches on crustal movements associated with orogeny naturally had a considerable influence on ideas regarding the origin of mountain ranges, and T. Mellard Reade (1886) ascribed the lateral movements to thermal expansion following regional subsidence and (geosynclinal) sedimentation. A more important work was the first volume of Emile Haug's *Traité de Géologie* which appeared in Paris in November 1907 and was republished four times by 1927. It dealt with 'Les phénomènes géologiques' and its profound erudition, sweeping summaries, and copious illustrations made it, and its companion volumes, an indispensable reference for earth scientists. Its author was born in Alsace and spent most of his life (1881–1927) in Paris at the Sorbonne. He made a notable study of geosynclines (1900) and believed that when subsidence proceeds in a geosyncline, marine regression occurs over the adjacent coastlands, or, conversely, when compression and folding begin in a geosyncline, marine transgression affects the epicontinental areas (see, however, the work of Stille (1924)). For Haug, geosynclines were long, narrow mobile belts of accentuated deep-water sedimentation between stable blocks (some perhaps submerged as parts of oceans) (Haug 1907: 166), very different from the marginal continental model of Schuchert and Stille. In the last chapter of his volume 1 Haug summarized the various theories of orogenesis. Contraction, he says, may be unilateral or bilateral, as was foreseen by Elie de Beaumont (1829, 1852) and in the United States by Dana and LeConte. Albert Heim (1878) provided powerful arguments for contraction when he estimated that the lateral width of the Jura had been diminished by about 5 km and of the Alps by 120 km, which estimate, as Haug pointed out, should be doubled or quadrupled in the light of the subsequent increase in the knowledge of overthrusting. Haug (1907: 515–34) identified the main causes of lateral orogenic thrust as subsidence (i.e. collapse partly due to global shrinkage leading to marginal thrusting), movement of the intergeosynclinal blocks, isostatic uplift and large-scale gravitational sliding (a secondary, not a primary, cause). For Haug, orogenic deformation was most intense at the geosynclinal margins with the axial region remaining 'undamaged or broken by a median geanticline' (Haug 1907: 533).

In the 1920s and 1930s several influential volumes on orogenesis were produced by Leopold Kober, including *Der Bau der Erde* (1921) and *Die Orogentheorie* (1933). Kober was a confirmed global contractionist and contraction provided a constant compressive impulse, although mountain building tended to be cyclic. Isostasy was readily admitted, together with the notion of mountains having 'roots' of lesser density than that of adjoining basal levels. So also was the concept that old rigid continental masses (Kratogens) have increased in area by the addition of newer folded ranges. The growth occurred during a number of major orogenic periods, in each of which there was the following sequence:

1 Contraction leading to broad geosynclinal subsidence and sedimentation.

Figure 1.1 Kober's view of orogenic deformation: A, scheme of Alpine orogenesis –
the orogen; B, cross-section of the eastern Alps showing intense orogenic deformation
accompanying the collision of two forelands, with the African (right) riding over
the European

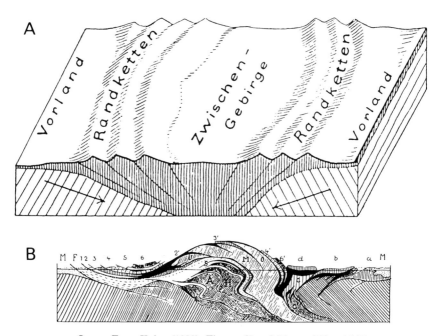

Source: From Kober (1928), Figures 50 and 36, pp. 175 and 142

2 Continued subsidence causes increase of temperature in the geosynclinal
   sediments, a decrease of resistance and the creation of instability.
3 As the two margins (borderlands) of the geosyncline move towards each
   other (usually accompanied with vulcanism and metamorphism) the
   geosynclinal sediments are forced as two border ranges (*Randketten*) over
   the edges (forelands) of the adjacent rigid masses, one on each flank (figure
   1.1). If the geosyncline is wide, or the compression moderate, not all the
   sediments may be deformed and there may remain a median mass
   (*Zwischengebirge*) which may be relatively unaffected, such as the Hungarian
   Plain. If the compression is very severe, the two forelands may collide
   producing complex mountains such as the Alps.
4 The resulting mountains are subjected to erosion and their general uplift
   into a belt of highlands is aided by isostasy.

Kober's views are interesting because they involved the notion of a two-
sided mountain complex (as distinct from Suess' one-sided); they appeared

to explain the location of rift valleys (e.g. the Middle Rhine and East Africa) in stable continental blocks as well as shallow (disharmonic, Jura-type) folding of the thin transgressive sediments on the edges of the stable blocks; they explained the Atlantic Ocean and the Mid-Atlantic Ridge as a foundered orogenic belt; and, finally, they were not opposed to notions of the lateral movement (i.e. drift) of continental masses.

Among the last influential texts on orogenic mechanisms prior to the essentially 'modern' geophysical studies ushered in by Arthur Holmes and Vening Meinesz, were J. A. Steers' *The Unstable Earth* (1929), Walter Bucher's review *The Deformation of the Earth's Crust* (1933) and Reginald Aldworth Daly's *Our Mobile Earth* (1926). Daly, a firm believer in the efficacy of isostasy, and successor to the chair of geology at Harvard after Davis, devoted much space to the nature and genesis of mountain ranges (1926: 221–91). While not ruling out an orogenic mechanism involving 'the horizontal crushing of geosynclinal prisms which lay in front of *slow moving, migrating continents*' (1926: 260), he proposed a large-scale gravitational slipping mechanism as the crust slid off huge domes and dome arches formed by the combination of terrestrial contraction, rotation and erosion (figure 1.2). Present high mountain ranges could therefore have formed at low elevations, even below sea level, and later have been bodily upheaved thousands of feet by the heating of mountain roots and down-slipped crustal blocks, the latter being an interesting forerunner of the idea of plate subduction.

The work of Holmes and of Vening Meinesz will be discussed later in this chapter with respect to continental drift. Holmes (1928–9) employed a complex model of thermally produced convection currents in the substratum below the simatic layer which included convergence under continental margins; compression; the formation of geosynclines, borderlands and foredeeps; mountain building; and the formation of new minor substratum currents due to the development of the thickened mountain root leading to the thinning and destruction of the borderlands. Vening Meinesz's gravity studies promoted the view that modern island arcs lie above a crustal downbuckle (tectogene) (Vening Meinesz 1934) or underthrust (Umbgrove 1947, Gutenberg and Richter 1950). The view that island arcs represent an early orogenic phase and that converging subcrustal convection currents could produce crustal downbuckles and mountains was supported by experimental studies (Kuenen 1935, Griggs 1939) (figure 1.3). Umbgrove (1947) used the concept of the cyclic character of the thermal build-up of convection currents (i.e. 'the pulse of the earth') to explain the supposed episodic character, or periodicity, of orogeny (earlier questioned by Shepard (1923)). About the time of the Second World War many scholars believed that weakened phases of convection current circulation allowed isostatic processes to elevate the folded and thrust orogenic structures

Figure 1.2 Sequence of orogenic development: A, formation of geosyncline; B, sediments crumpled by the sliding of a crustal block; C, low mountain structure after folding; D, later upheaval due to thermal expansion of mountain roots and foundered crustal blocks.

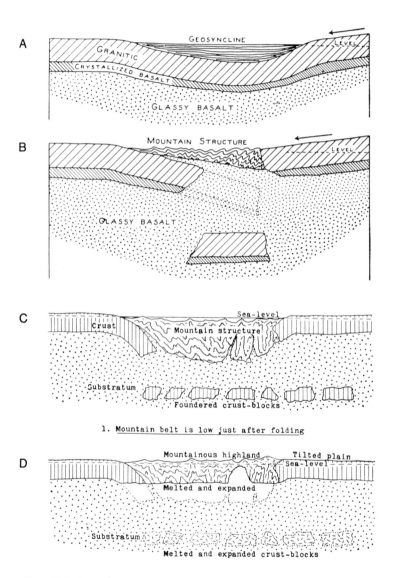

Source: From Daly (1926) Figures 166 and 168, pp. 269 and 276. Reprinted by permission of Charles Scribner's Sons, an imprint of Macmillan Publishing Company from *Our Mobile Earth* by R. A. Daly. Copyright 1926 Charles Scribners' Sons, copyright renewed 1954 Reginald Aldworth Daly).

Figure 1.3 The simulation of the effect of converging convection currents by rotating drums on the deformation of a plastic crust and fluid substratum: A, both drums rotating leading to the development of a tectogene and surface thrust masses similar to those of Kober's (1928) orogen (see figure 1.1); B, one drum rotating, showing the asymmetrical thickening of the root and asymmetrical thrusting.

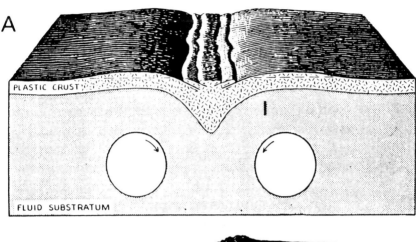

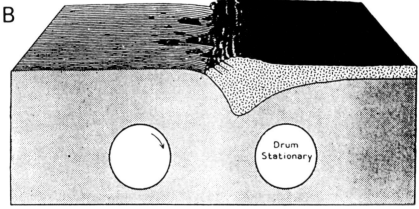

Source: From D. T. Griggs, *American Journal of Science*, Vol. 237(9), Figs. 14 and 15, pp. 642 and 643. Reprinted by permission of *American Journal of Science*).

with their sialic roots to produce greater relief, usually in a post-orogenic phase.

Post-orogeny

Davis (1905J, 1905K, 1920J) believed that orogenic belts were susceptible to considerable intermittent uplift associated with post-orogenic erosion. In Europe this thesis formed the basis for the work of de Martonne in the Transylvanian Alps (1907) and by Johnson (1931) in the American Appalachians. Belief in this post-orogenic history derived from two sources: denudation chronology (see chapters 7, 8 and 9) and the interpretation of crustal movements from associated sedimentation.

Following the development of the facies concept, it was believed that the distribution of sedimentary facies provided the key to an understanding of the pattern of vertical uplift of the source areas. Thus, some workers (e.g. Penck 1919) recognized the possibility of a wide range of such patterns; others preferred to interpret the stratigraphic record as indicating mainly short periods of rapid uplift separating long periods of quiescence (Chamberlin 1909, Stille 1936b); whereas others like Gilluly (1949) believed in widespread and continuous orogenies. There can be no doubt that much of Walther Penck's geomorphic work (see chapter 10) was an attempt to provide physiographic support for a general pattern of uplift which he had previously inferred from stratigraphic evidence. The importance which Penck placed on the ability to identify the movements of the source area from the record of sedimentation is clearly stated in chapter 1 of his *Die morphologische Analyse* (1924) and resulted largely from his studies in the Andes (Penck 1924: 277–9). It is interesting that later few geologists were willing to attempt more than to postulate the occurrence of some rather generalized uplift from the sedimentary record alone, and not to infer the pattern of uplift in any great detail as he did (e.g. Spieker 1946, 1949). In 1917, in a most important paper, Barrell showed that much of the character of the sedimentary record is controlled by the nature of the subsidence of the basin of sedimentation, as distinct from the behaviour of the adjacent source area. Although these behaviours are often so closely linked that it is difficult to distinguish between the two, the work of Barrell began to cast doubt on the simple association between the nature of sedimentation and the pattern of uplift of the source area. This doubt, however, did not really become universal until after 1930 (Bailey 1930, Jones 1938, Krynine 1941) when geomorphologists were forced to adopt much less doctrinaire views regarding the intepretation of patterns of uplift from the sedimentary record (Chorley 1963).

It should not be thought, however, that the traffic in diastrophic ideas has been exclusively in one direction, for denudation chronology played a large part in the development of ideas relating to the post-orogenic history of mobile belts. The success of the cyclic theory of Davis led naturally to the widespread assumption that all erosion surfaces, accordances, straths, terraces, valley-in-valley forms, and nickpoints must be associated with erosive processes working under the control of ancient baselevels. This assumption, together with the observed persistence of quite high relief in belts of even ancient folding, led most geologists to conclude that the orogenic phase, responsible for producing the geological structures, was followed by a prolonged period of quiescence during which varying degrees of peneplanation were achieved (Bucher 1933: 431, 1939: 422). The present high elevation of these surfaces implied that subsequently a bodily movement had taken place in a vertical direction (Penck 1909: 70–1, Stille 1936b: 850–1, de Sitter 1952: 300–1), although Lawson (1932) believed that during compression isotastic equilibrium was constantly being re-established by uplift, and the geometry of the valley-in-valley features implied that the uplift had been intermittent, possibly with decreasing intervals between the uplift phases (Barrell 1917: 747, 763, Campbell 1933: 571–3, Bucher 1939: 422). Some workers (e.g. Daly 1905, Penck 1919), however, proposed alternative methods by which summit accordances and other topographic breaks might be produced, and Walther Penck (1924: 7–8) thought that this discontinuous, jerky interpretation of post-orogenic history was more a product of the means of analysis rather than a physical reality.

The concept of discontinuous uplift of mobile belts following a prolonged quiescence and peneplanation, however, had to face possible rethinking on two counts – those of 'mechanisms' and 'time'. If it is assumed that isostatic compensation is relatively rapidly effected, that the earth's crustal strength is low, and that the orogenic phase produces large negative gravity anomalies in mobile belts, possibly associated with the existence of a low-density 'root', then a long-continued post-orogenic stability seemed to some geologists rather unlikely as tendencies to isostatic uplift operate in the face of some crustal strength (King 1955). Such uplift was associated with the existence of a downbuckled root (Vening Meinesz 1934, Griggs 1939), batholithic intrusion (Umbgrove 1947: 80), the thermal expansion of roots and foundered crustal blocks (Daly 1926: 276) and a variety of other suggested processes.

The second problem confronting the analysis of post-orogenic uplift of orogenic belts was that when it was applied to younger mountain ranges there was a marked lack of time during which widespread erosion surfaces might have been successively developed. Walther Penck's piedmont benchland theory was specifically designed to circumvent both these difficulties, and it is significant that the only important application of this idea in the United States was by Sauer (1929: 212–13) in the California Coast Range. Indeed, it is in such a young tectonic belt that time difficulties become particularly apparent if its history is believed to include successive periods of prolonged stability necessary for erosion surfaces of the Davisian type to develop. Thus,

Willis (1925: 677) postulated evidence of two cycles of erosion in the Gabilan
Range since the end of the Pliocene; Trask (1926: 293) that the Santa Lucia
Coast Range had, during the Pleistocene and subsequently, been almost
completely peneplained, than uplifted about 4,500 feet and eroded to
form the present mountains; and Putnam (1942: 749) in the Ventura region
was forced to compress into the same short timespan the deposition of
4,000–6,000 feet of sediments, their subsequent deformation, their erosion to
late maturity, vertical uplift of about 1,000 feet, and erosion to produce the
present relief. Davis himself began to recognize the problems presented by
the young mountain ranges of the Pacific coast when he moved to California
where he spent his last years, for he wrote at the age of 80:

> the scale on which deposition, deformation and denudation have gone on by
> thousands and thousands of feet in this new-made country is ten- or twenty-fold
> greater than that of corresponding processes in my old tramping ground. On
> shifting residence from one side of the continent to the other, a geologist must learn
> his alphabet over again in an order appropriate to his new surroundings.
>
> (Davis 1930C: 404, Chorley 1963)

Epeirogeny

Associated with the idea of the bodily uplift of eroded mountain ranges was
that of epeirogeny. In his monograph on Lake Bonneville, Gilbert (1890:
340) described broad 'swelling' movements of the land platforms which were
held responsible for the creation of continents. These he named 'epeiro-
genetic', distinguishing them from the mountain-building or 'orogenic'
movements, and collectively terming the two categories of movement 'diastrophic'. The epeirogenic idea grew steadily, and by the 1920s many
geologists were ascribing major transgressions and regressions to differential
continental warping, as did Stephenson (1928) in the Gulf Coastal Plain. The
palaeogeographic maps of Schuchert (1910), Lapparent (1906) and Haug
(1907–11) were influential in this respect, showing graphically the 'coming
and going of epeiric seas' (Bucher 1933: 425). Mechanisms were proposed by
Barrell (1914b) to do with compressional heating, consequent changes in
crustal density and viscosity, and isostatic adjustment; and by Joly (1923)
employing radioactive heat. It was the work of Hans Stille, however, which
brought about by the middle of the 1930s almost a revolution in epeirogenic
thinking.

It had been a basic feature of Suess' (1883–1908) eustatic theory that the
major continental platforms had remained absolutely stable, despite orogenic
movements due to lateral compression in the adjacent mobile belts. Stille
suggested that marginal orogenic compression could be correlated with
continental elevation outside the mobile belts (Stille 1924: 364) and with
consequent marine withdrawals from the continent. This 'synorogenic

strengthening of epeirogenesis' (Stille 1936b: 853–4) implied that the stability of the craton is strikingly linked to that of the geosyncline. This concept became a widely accepted part of geological thought and in more recent years many new and interesting ramifications of the general notion of epeirogeny were proposed. For example, Bucher's alternating 'diastolic' and 'systolic' phases (1939: 428), associated with expansions and contractions of the earth's subcrust, were linked to the vertical movements of continental platforms; and Von Bubnoff (1954) proposed the term 'diktogenesis' for those diastrophic processes involving vertical uplift of areas smaller than those affected by epeirogenesis. Billings (1960: 379, 390–1) was convinced that the uplift of the Colorado Plateaus could not be explained either by isostatic uplift or by compression. More recently L. C. King (1962: 144–6) proposed that much of the surface history of the continents can be explained in terms of the parallel retreat of scarp slopes taking place on huge, broad, arching flexures. This vertical deformation, termed 'cymatogeny' (1962: 191), was believed to produce undulations of uplift or depression up to hundreds of miles across and thousands of feet in vertical displacement. Although King distinguished between this and epeirogeny, making only two references to the latter in his large book, it is difficult to appreciate his distinction which excluded large-scale epeirogenic uplift from cymatogeny, a process which he genetically linked to speculative subcrustal mechanisms. In any event, the idea that continental platforms are susceptible to large-scale vertical movements became so popular by mid-century as to introduce a serious complication to the eustatic theory (see chapter 3) (Chorley 1963).

Continental Drift

Although the concept of floating crustal units contributed to the idea of continental migrations on a large scale, what sparked off and kept the concept of continental displacement alive was the obvious conformity of the outline of the western coast of Africa and the eastern coast of South America (figure 1.4). True, transoceanic similarities were soon shown to be multitudinous but the simple South Atlantic super-fit never lost its appeal. The problem was to devise an acceptable mechanism to explain it (Hallam 1973).

In 1858, Antonio Snider-Pellegrini in his *La Création et Ses Mysteres Dévoilés* postulated that the New World had broken away from the old and supplied a fantastically unsatisfactory catastrophic mechanism. However, by then the idea of lateral migration of continents was a tenet of the uniformitarians as by 1830 Lyell advocated it to help explain climatic change (see chapter 2). The problem became more complicated after 1879 when G. H. (later Sir George) Darwin proposed that the moon had split off from the earth, leaving behind the Pacific Ocean. This notion was used by Osmond Fisher in his *Physics of the Earth's Crust* (1881) as a possible cause of

Figure 1.4 Du Toit's suggested continental restoration for the South Atlantic region under the displacement hypothesis

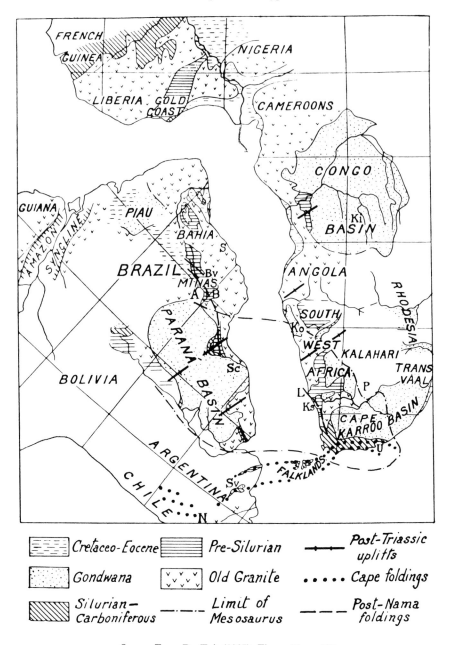

Source: From Du Toit (1937), Figure 13, p. 108.

disintegration and lateral movement of cooled magmatic crust. Like some of his contemporaries Fisher believed in a relatively fluid interior but he went far beyond them in suggesting, as an adequate cause of earth tectonics, convection currents which rose beneath oceans and sank beneath continents. A few years later C. E. Dutton, in discussing isostasy, also favoured a crude kind of internal convection based on continental erosion carrying sediments from the land to the coast, so causing the one to tend to rise and the other to sink, thereby creating an imbalance. The compensatory force was believed to set up a motion from coast to continent which folded the accumulating marginal sediments.

Among geophysicists the idea of subcrustal convective currents remained unpopular for another half a century at least but on the European mainland, especially in German-speaking areas, the possibility of mobile continents floating or slipping on a plastic interior, and hence a polar wandering, gradually became widely accepted. Here Suess (1883–1908) ruled the roost. His wider tectonics were based on the general pattern of mountain belts in the Old World where ranges exhibiting lateral overfolds and thrusts seemed to reflect a subsidence of ocean basins and a tangential thrust towards them from high areas to the north. The proposed cause of the compression was the contraction of the cooling earth which induced land blocks to move isostatically upon a fluid subcrust.

However, the first logical and coherent hypothesis of continental displacement did not arise as would have been expected in Europe. In 1898 Frank Bursley Taylor (1860–1938) of Harvard published a pamphlet containing suggestions on the history of the continents. When the moon (originally a comet) had been captured into the earth's gravitational field, it exerted a tidal force on the earth increasing the latter's speed of rotation. This combined action pulled the continents away from the poles towards the equator. In a longer paper in 1910 and in subsequent writings Taylor showed that he depended heavily on Suess' analytical pattern of Old World mountain belts but rejected isostatic adjustment as a satisfactory cause of crustal mobility on a fluid interior. Instead he favoured a narrow shear zone above a fairly rigid interior. Instead of the wrinkling effects of a contracting globe, he envisaged a broad crustal creep from polar towards lower latitudes (later termed *Pohlflucht*), during which the moving surface was raised and bent into arcuate patterns of mountains where it was obstructed by old immobile blocks (figure 1.5). Details of the argument lie outside our purpose but it is interesting, in retrospect, to notice that even at this time the submarine mid-oceanic ridge was proposed as the site of the rifting between Africa and South America. The lack of recognition of Taylor's concepts must be placed partly on the weakness of his causal explanation and largely on the fact that Wegener, the great protagonist of continental drift, was treading on his heels. In the meanwhile, Howard Bigelow Baker, another American geologist, had

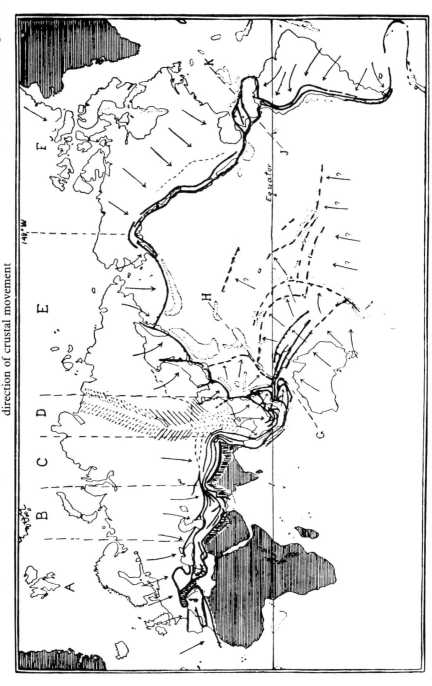

Figure 1.5 Taylor's view of Tertiary continental movements: heavy continuous and dashed lines, mountain ranges and island chains; faint dotted lines, foredeeps and deep ocean basins; vertical shading, foredeeps on land filled with sediment; arrows, the general direction of crustal movement

Source: After Taylor (1910); from Du Toit (1937), Figure 1, p. 13

Plate 1.1 Alfred Wegener (1880–1930)

Source: From Meteorologische Zeitschrift, 1931

suggested in 1908 that all the continents were grouped together around Antarctica at the beginning of the Tertiary era (Hallam 1973).

Alfred Lothar Wegener (1880–1930) warrants special attention because of the details and width of his advocacy of continental displacement (plate 1.1). He was born in Berlin, the son of a doctor of theology who directed an orphanage. After an excellent education at the Universities of Heidelberg and Innsbruck he moved away from astronomy into meteorology and geology. He joined four expeditions to Greenland, twice as leader, and in November 1930, on his fiftieth birthday, left a base on the central ice-cap making for the west coast and was never seen again. As a lecturer in meteorology and geophysics he had shown a dislike of elaborate mathematical detail and a striking ability to disentangle complicated themes by emphasizing the broad issues. He could see both the wood and the trees (Schwarzbach 1986).

Wegener first suspected the drifting aspect of continents in 1910 when he studied the conformity of the coastlines on opposing sides of the South Atlantic but thought the idea was improbable. In the following year, apparently quite unaware of Taylor's work in the United States, he returned to the theme, when other scientists postulated palaeontological similarities, that the opposing coasts had probably once been joined by a land bridge which had foundered. A review of the evidence caused him to revive his earlier suspicions and in 1912 in a lecture and two short articles entitled *Die Enstehung der Kontinente* he envisaged large-scale continental displacement. Early in the First World War he was wounded and used his convalescence to elaborate his theory in *Die Enstehung der Kontinente und Ozeane* (1915) in which he mentioned 'Verschiebung der Kontinente' or continental displace-ment, a term replaced in English usage after 1926 by 'continental drift'. The book was revised and enlarged four times (1920, 1922, 1929 and in 1936 by Kurt Wegener, his brother) and the third edition was translated into at least five main languages.

As new evidence, especially of a palaeoclimatic nature, became available what was regarded as a working hypothesis in 1912 gradually began to accumulate almost into an acceptable theory. The general premise – by no means new – was that at or near the end of the Permian period a supercontinent (Pangaea) had split into separate blocks which drifted slowly, for the most part westward and in some areas equatorward. Deductions from a wide variety of evidence suggested possible interconnections between the continents themselves and world climatic zones or latitude. To give simple examples: evidence of glaciation by a continental ice-sheet in Late-Carboniferous times allowed part of South America, southern Africa, India and Australia to be grouped together near a south pole then in or near South Africa. At that time, tropical features such as laterite and Coal Measures (formed from tropical forests) would represent an equatorial or tropical location (figure 1.6).

Figure 1.6 Climatic changes in the Carboniferous, Eocene and Pliocene/Early Quaternary produced by continental drift: E, ice; G, gypsum; K, coal; L, loess; S, salt; W, desert sandstone; stippled, arid

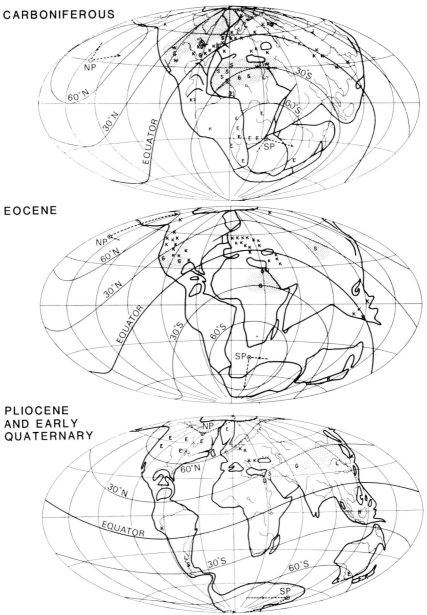

CARBONIFEROUS

EOCENE

PLIOCENE AND EARLY QUATERNARY

Source: From Köppen and Wegener (1924), Figures 3, 14 and 19, pp. 22, 97 and 117

The continental blocks were assumed to be floating in an isostatic state upon a pliable medium and mountains were created where the advancing front of a moving continent encountered the resistance of the ocean floor. But in leaning towards mobility Wegener was faced with devising an acceptable force adequate to cause continental displacement on such a vast scale and in so many varied directions. He thought that a complete solution to the problem was a long way off and suggested two main components and a third subsidiary component of such possible forces:

1 A *Pohlflucht* (polar flight) or differential gravitational force arising from the relative bulge of the earth at the equator which, according to Baron von Eötvös (1848–1919), the great Hungarian geophysicist, would cause a movement of continents equatorward.
2 Tidal forces generated by the pull of the moon and sun causing friction to act like a brake on the rotating earth, and since its effect is greater on protuberances than at lower levels, the continents would tend to lag behind and drift westward.
3 Local and regional bulges from the true spheroid, as shown by gravimetric deviations, should create isostatic forces which tended to move the crust laterally.

However, no scientist could show that these forces were competent to solve the problem and, in the absence of any acceptable mechanism, Wegener's hypothesis foundered. In 1922 Philip Lake expressed the opinion of many reviewers when he wrote:

Whatever Wegener's own attitude may have been originally, in his book he is not seeking truth; he is advocating a cause and is blind to every fact and argument that tells against it. Much of his evidence is superficial. Nevertheless he is a skilful advocate and presents an interesting case.

(Lake 1922: 338)

Time was on Wegener's side, however, and his mobile hypothesis staggered on with the injection of various stimulants in spite of the strong opposition of influential geophysicists such as R. T. Chamberlin in the United States and Sir Harold Jeffreys (1923) of Cambridge, England. The steadily increasing evidence that material under the oceans was denser than that under land-masses was seen to militate against the popular belief in the supposed past existence of transoceanic land bridges.

In 1926 gravimetric instruments of an improved type, installed in a submarine, allowed Felix Andries Vening Meinesz (1887–1966), a Dutch geodesist, to find in the sea floor of the East Indies a long narrow belt with an abnormal gravity deficiency (Vening Meinesz 1934). Such negative gravity belts were generally interpreted as the result of less dense crustal matter being buckled downward into the denser lower layer and held there by some compressive process still operating. Vening Meinesz favoured the idea of a rigid crust and cooling earth in which a warmer sublayer experienced convection currents which dragged on the crust causing local buckling especially in the negative gravity belts of, presumably, descending motion. The advocates of continental displacement also included many biogeographers, who according to Hallam (1973: 26) were so handicapped by 'blithe ignorance' that they became burdens rather than supports.

As we have seen, R. A. Daly accepted a form of drift but thought that the continents slid laterally under gravitational influences from bulges at the equator and at the poles, while between them, in each hemisphere, was an unexplained lower zone. In Europe, tectonists such as E. B. Bailey in Britain and E. Argand and R. Staub on the mainland, were adherents to continental displacement, as would be expected of scientists who associated the lateral movements of continents with mountain building. Staub in 1928 gave the name Laurasia to a great northern landmass which in his scheme interacted with a great southern landmass, Gondwana. He conceded the presence of a general westerly drift but emphasized the supposed alternation in geological time of the effects of centrifugal action or polar flight due to the earth's rotation (which compressed the equatorial geosyncline), and of subcrustal streaming which tended to produce polar drift which drew the two supercontinents apart again. This hypothesis differs from that of Wegener but the estimation of the climatic zones and the relative position of the poles for past epochs and the striking impression of large-scale continental fracturing and dispersal were highly favourable to the concept of continental drift.

There can be no doubt, however, that the most influential supporters of Wegener were Arthur Holmes in Britain and Alexander L. Du Toit in South Africa. Du Toit (1878–1948), who was born in Cape Town, studied and lectured for some time in Glasgow and London before returning to field geology in his homeland. For his own extensive fieldwork he wrote in 1927 a detailed geological comparison of South Africa with South America which was welcomed by Wegener. It foreshadowed Du Toit's greatest contribution to the concept of continental drift in *Our Wandering Continents: An Hypothesis of Continental Drifting* (1937). This volume was a trumpet blast played, for some geologists, with an invigorating virtuosity and assurance and, for others, with a rather disconcerting arrogance, especially when discussing the principal contributions adverse to his hypothesis (Du Toit 1937: 29–35). Indisputably it added a mass of relevant data and eliminated many of Wegener's errors, such as using recent (Quaternary) terminal moraines in North America and Europe as evidence of former transatlantic continuity. It was the first book to make full use of the idea of a Pangaea that had been subdivided – at least since late Palaeozoic times – into Laurasia in the north and Gondwana in the south on either side of a great geosyncline, the 'Tethys Sea'. Du Toit reconstructed Gondwanaland in much greater and accurate

detail than hitherto and altogether made a truly important contribution to
the hypothesis of continental drift. We must be content with one non-
controversial quote:

> An outstanding consequence of the Hypothesis is the orderly and interrelated
> nature of all associated phenomena . . . the critic may, and with justification, ask
> why the Hypothesis has apparently found so few whole-hearted supporters. The
> answer is, first, that it cuts at the basis of customary geological interpretations and
> is hence not particularly welcome, and, second, that no forces have so far been, nor
> according to its opponents can be, invoked competent to move the continents about
> as supposed.
>
> (Du Toit 1937: 4)

Whereas Du Toit was essentially a field geologist, his younger contemporary, Arthur Holmes (1890–1965) was trained as a physicist or geophysicist. He spent some years on research at Imperial College, London, under the
direction of R. J. Strutt and from measurements of the radioactive decay of
igneous rocks drew up a geological chronology. He early favoured the idea of
the lateral displacement of continents but recognized the serious short-
comings of Wegener's advocacy:

> Some of his [Wegener's] evidence was undeniably cogent, but so much of his
> advocacy was based on speculation and special pleading that it raised a storm of
> adverse criticism. Most geologists, moreover, were reluctant to admit the possi-
> bility of continental drift, because no recognized natural process seemed to have the
> remotest chance of bringing it about.
>
> (Holmes 1944: 495)

Holmes set out to devise a scheme of subscrustal convection currents which
might explain continental drift. In 1928–9 he proposed a general model of the
earth's outer layers, in which the uppermost layer consisted mainly of
granites, the intermediate layer mainly of diorites and the lower layer mainly
of peridotite beneath the oceans, with some eclogite beneath the continents.
The earth's crust was believed to comprise the upper and intermediate layers
together with the top of the lower layer, the basal part of the lower layer being
'glassy' or 'thermally fluid', within which a system of convection currents was
assumed to operate. Where currents flow horizontally along the undersurface
of the crust, they would exert a drag on it causing tension where they diverge
and compression where they converge. Orogenic belts would be expected to
form where two converging currents turned downward. Currents, whose
existence was based on a complex set of thermal assumptions associated with
processes of radioactive decay, were proposed to:

1 Rise under major sialic continents, where radioactivity is greatest, and also
   under central ocean basins.
2 Converge and descend under continental margins associated with

Figure 1.7 Crustal development under the action of convection currents: A, currents ascending beneath a continental block and descending at the oceanic margins; B, stretching and breaking of a continental block by convection current drag and the formation along the leading edges of marginal geosynclines, borderlands and foredeeps where the currents descend (crosses indicate dynamic metamorphism); C, the transformation of a geosyncline into a mountain range, the development of a secondary convection circulation beneath the mountain roots and the sinking of the borderland

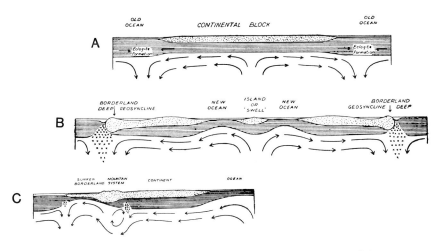

Source: From Holmes (1928–9), Figures 2, 3 and 4, pp. 579 and 582

compression, geosynclines, borderlands, foredeeps, crustal thickening and mountain formation (figure 1.7).

Diverging currents were held to produce sufficient shear stress to stretch, thin and rupture the sialic continents causing parts to drift apart. Crustal thickening associated with mountain building was believed to generate new, subsidiary convection circulations which were responsible for the destruction of borderlands. This hypothetical mechanism received wider publicity when it was incorporated into the popular *Principles of Geology* (1944) in which Holmes stressed that:

> The detailed complexity of convection systems, and the endless variety of their interactions and kaleidoscopic transformations, are so incalculable that many generations of work ... may well be necessary before the (convective) hypothesis can be adequately tested.
>
> (Holmes 1944: 508–9)

Until the revolution in geophysics of the 1950s and 1960s associated with plate tectonics, there was consistent opposition from most American and British scientists to any notion of lateral crustal movements because of the assumed inadequacy of the forces involved (Jeffreys 1923, 1935). In 1928 (Waterschoot van der Gracht 1928), of a gathering of fourteen eminent scientists, only five supported the idea unreservedly and most expressed various degrees of opposition, some (e.g. R. T. Chamberlin) quite heatedly. A generation later Wegener's hypothesis still had not clearly upset the general balance of contemporary geophysical beliefs and a contemporary judgement in 1950 was characteristic:

> If the conclusions of adherents to drift in some form or other, such as Du Toit, Wegmann, Gutenberg and Kirsch, are confronted with the opinion of opponents, for instance Bucher, Umbgrove, Stille, and Cloos, it becomes obvious that neither of the two camps can claim a decisive victory. But the evidence favourable to drift often proves illusive, or at least open to serious doubt, on closer inspection. For the time being most geologists appear to have lost faith in continental drift as a sound working hypothesis.
>
> (Kuenen 1950: 129)

Only a decade later all was to change!

References

Airy, G. B. (1855) 'On the computation of the effect of the attraction of the mountain masses disturbing the apparent astronomical latitude of stations in geodetic surveys', *Philosophical Transactions of the Royal Society of London* 145: 101–4.

Bailey, E. B. (1930) 'A new light on sedimentation and tectonics', *Geological Magazine* 67: 77–92.

—— (1935) *Tectonic Essays: Mainly Alpine*, Oxford: Oxford University Press.

Barrell, J. (1914a) 'The Upper Devonian delta of the Appalachian geosyncline', *American Journal of Science* 4th series, 37: 225–53.

—— (1914b) 'The strength of the earth's crust', *Journal of Geology* 22: 655–83.

—— (1917) 'Rhythms and the measurement of geologic time', *Bulletin of the Geological Society of America* 28: 745–904.

—— (1919) 'The status of the theory of isostasy', *American Journal of Science* 4th series, 48: 291–338.

—— (1925) 'The nature and environment of the Lower Cambrian sediments of the Southern Appalachians', *American Journal of Science* 5th series, 9: 1–20.

Beaumont, E. de (1829–30) 'Recherches sur les quelques-unes de révolutions de la surface du globe', *Annales des Sciences naturelles* 8: 5–25, 284–416; 9: 5–99, 174–240.

—— (1852) *Notice sur les Systèmes de Montagnes*, 3 vols, Paris.

Becquerel, H. (1896) 'Sur les radiations émises par phosphorescence', *Comptes Rendus* 122: 420, 501, 689.

Bertrand, M. (1884) 'Rapports de structure des Alpes de Glaris et du bassin houllier du Nord', *Bulletin de la Société Géologique de la France* 3rd series, 12: 318–30.

—— (1897) 'Structures des Alpes françaises et récurrence de certains facies sédimentaires', *Report of the 16th International Geological Congress, Zürich, 1894*, pp. 163–77.

Billings, M. P. (1960) 'Diastrophism and mountain building', *Bulletin of the Geological Society of America* 71: 363–98.

Born, A. (1923) *Isostasie und Schweremessung: Ihre Bedeutung für geologische Vorgänge*, Berlin: Springer.

Boscovich, R. J. (1750) *De litteraria expeditione per pontificiam ditionem.*

Bouguer, P. (1749) *La Figure de la Terre*, Paris.

Bowie, W. (1917) 'Investigations of gravity and isostasy', *US Coast and Geodetic Survey, Special Publication* no. 40.

—— (1921) 'The relation of isostasy to uplift and subsidence', *American Journal of Science* 5th series, 2: 1–20.

—— (1927) *Isostasy*, New York: Dutton.

Bucher, W. H. (1933) *The Deformation of the Earth's Crust*, Princeton, NJ.

—— (1939) 'Deformation of the earth's crust', *Bulletin of the Geological Society of America* 50: 421–31.

Campbell, M. R. (1933) 'Chambersburg (Harrisburg) peneplain in the piedmont of Maryland and Pennsylvania', *Bulletin of the Geological Society of America* 44: 553–73.

Chamberlin, T. C. (1909) 'Diastrophism as the ultimate basis of correlation', *Journal of Geology* 17: 685–93.

—— (1927) 'Intrageology – Elastasy versus isostasy', *Journal of Geology* 35: 89–94.

Chorley, R. J. (1963) 'The diastrophic background to twentieth century geomorphological thought', *Bulletin of the Geological Society of America* 74: 953–70.

Collet, L. W. (1927) *The Structure of the Alps*, London: Arnold.

Daly, R. A. (1905) 'Accordance of summit levels among Alpine mountains', *Journal of Geology* 13: 105–25.

—— (1926) *Our Mobile Earth*, New York: Scribners.

Dana, J. D. (1873) 'On some results of the earth's contraction from cooling, including a discussion of the origin of mountains and the nature of the earth's interior', *American Journal of Science* 3rd series, 5: (30): 423–43; 6: (31): 6–14, (32): 104–15, (33): 161–72.

Darwin, C. (1838) 'On the connexion of certain volcanic phenomena, and on the formation of mountain-chains and volcanoes, as the effects of continental elevations', *Proceedings of the Geological Society* 2: 654–60.

Du Toit, A. L. (1927) 'A geological comparison of South America with South Africa', *Carnegie Institution of Washington* publication 381, pp. 1–157.

—— (1937) *Our Wandering Continents*, Edinburgh: Oliver & Boyd.

Dutton, C. E. (1889) 'On some of the greater problems of physical geology', *Bulletin of the Philosophical Society of Washington* vol. 11 (reprinted in (1925) *Journal of the Washington Academy of Science* vol. 15).

Fisher, O. (1881) *Physics of the Earth's Crust*, London: Macmillan.

Geer, G. de (1892) 'Quaternary changes of level in Scandinavia', *Bulletin of the Geological Society of America* 3: 65–8.

Gilbert, G. K. (1890) 'Lake Bonneville', *US Geological Survey Monograph* 1.

—— (1895) 'A new light on isostasy', *Journal of Geology* 3: 331–4.

Gilluly, J. (1949) 'Distribution of mountain building in geologic time', *Bulletin of the Geological Society of America* 60: 561–90.

Griggs, D. T. (1939) 'A theory of mountain building', *American Journal of Science* 237: 611–50.

Gunn, R. (1949) 'Isostasy – extended', *Journal of Geology* 57: 263–79.

Gutenberg, B. and Richter, C. F. (1950) *Seismicity of the Earth*, Princeton, NJ.

Haarmann, E. (1930) *Die Oscillationstheorie*, Stuttgart: Ferdinand Enke.

Hall, J. (1859) 'Description and figures of the organic remains of the Lower Helderberg group and the Oriskany sandstone', *New York Geological Survey, Paleontology* vol. 3.

Hallam, A. (1973) *A Revolution in the Earth Sciences: From Continental Drift to Plate Tectonics*, Oxford.

Haug, E. (1900) 'Les geosynclinaux et les aires continentales', *Bulletin de la Société Géologique de la France* 3rd series 28: 617–711.

—— (1907–11) *Traité de Géologie*, 2 vols, Paris: Colin.

Hayford, J. F. (1909) *The Figure of the Earth and Isostasy from Measurements in the United States*, Government Printing Office, Washington, DC.

—— (1911) 'The relations of isostasy to geodesy, geophysics and geology', *Science* 33: 199–208.

Heim, A. (1878) *Untersuchungen über den Mechanismus der Gebirgsbildung*, 2 vols, Basel.

Heritsch, F. (1929) *The Nappe Theory in the Alps*, London: Methuen.

Hess, H. H. (1938) 'Gravity anomalies and island arc structure with particular reference to the West Indies', *Proceedings of the American Philosophical Society* 79: 71–96.

Holmes, A. (1928–9) 'Radioactivity and earth movements', *Transactions of the Geological Society of Glasgow* 18: 559–606, part III.

—— (1944) *Principles of Physical Geology*, Edinburgh: Nelson.

Hubbert, M. K. (1945) 'The strength of the earth', *Bulletin of the American Association of Petroleum Geologists* 29: 1630–53.

Jamieson, T. F. (1865) 'On the history of the last geological changes in Scotland', *Quarterly Journal of the Geological Society of London* 21: 161–203.

Jeffreys, H. (1923) *The Earth: Its Origin, History and Physical Constitution*, Cambridge.

—— (1935) *Earthquakes and Mountains*, London: Methuen.

Johnson, D. W. (1931) *Stream Sculpture on the Atlantic Slope*, New York: Columbia University Press.

Joly, J. (1909) *Radioactivity and Geology*, London.

—— (1923) 'Movements of the earth's surface crust', *Philosophical Magazine* 45: 1167–88.

—— (1925) *The Surface History of the Earth*, Oxford.

Jones, O. T. (1938) 'On the evolution of a geosyncline', *Quarterly Journal of the Geological Society of London* 94: lx–cx.

Kay, M. (1951) 'North American geosynclines', *Geological Society of America, Memoir* 48.

Keith, A. (1923) 'Outlines of Appalachian structure', *Bulletin of the Geological Society of America* 34: 309–80.

King, L. C. (1962) *The Morphology of the Earth*, Edinburgh: Oliver & Boyd.

King, P. B. (1955) 'Orogeny and epeirogeny through time', *Geological Society of America, Special Paper* 62: 723–40.

Kober, L. (1921) *Der Bau der Erde*, Berlin, Gebrüder Borntraeger (2nd edn 1928).

—— (1933) *Die Orogentheorie: Grundlinien eines natürlichen Gestaltungsbildes der Erde*, Berlin: Gebrüder Borntraeger.

Köppen, W. and Wegener, A. (1924) *Die Klimate der geologischen Vorzeit*, Berlin: Gebrüder Borntraeger.

Krynine, P. D. (1941) 'Differentiation of sediments during the life history of a landmass (abstract)', *Bulletin of the Geological Society of America* 52: 1915.

Kuenen, Ph. H. (1935) *The Snellius Experiment: Geological Results (1) 5, Geological Interpretation of Bathemetrical Results*, Utrecht: Keminken Zoon.

—— (1950) *Marine Geology*, New York, Wiley.

Lake, P. (1922) 'Wegener's displacement theory', *Geological Magazine* 59: 338–46.

Lapparent, A. de (1906) *Traité de Géologie*, 5th edn, Paris (1st edn 1881).

Lawson, A. C. (1932) 'Insular arcs, foredeeps, and geosynclinal seas of the Asiatic coast', *Bulletin of the Geological Society of America* 43: 353–81.

LeConte, J. (1872) *Elements of Geology*, New York: Appleton (5th edn 1903).

Lugeon, M. (1901) 'Les grandes nappes de recouvrement des Alps du Chablais et de la Suisse', *Bulletin de la Société Géologique de la France* 4: 723–825.

Martonne, E. de (1907) 'Recherches sur l'évolution morphologique des Alpes de Transylvanie', *Annales de Géographie* 1: 1–247.

Mohorovičič, A. (1910) 'Das Bebenvom 8 Oct. 1909', *Jahrbuch der Meteorologischen Observatoriums in Zagreb* Band 9, Tome 4, Abh. 1.

Oldham, R. D. (1906) 'The constitution of the earth as revealed by earthquakes', *Quarterly Journal of the Geological Society of London* 62: 456–75.

Peach, B. N. and Horne, J. (1884) 'Report on the geology of the north-west of Sutherland', *Nature* 31: 31.

Penck, A. (1909) 'The origin of the Alps', *Bulletin of the Geological Society of America* 41: 65–71.

—— (1919) 'Die Gipfelflur der Alpen', *Sitzungsberichte der Preussischen Akademie der Wissenschaften zu Berlin, Math.-Phys. Klasse* 17: 256–68.

Penck, W. (1924) *Die morphologische Analyse: Ein Kapital der physikahischen Geologie*, Stuttgart: Engelhorns, trans. H. Czech and K. C. Boswell, 1953, London: Macmillan.

Pratt, J. H. (1859) 'On the deflection of the plumb line in India', *Philosophical Transactions of the Royal Society of London* series B, 149: 745–96.

Putnam, G. R. (1930) 'Isostatic compensation in relation to geologic problems', *Journal of Geology* 38: 590–9.

Putnam, W. C. (1942) 'Geomorphology of the Ventura region', *Bulletin of the Geological Society of America* 53: 691–754.

Reade, T. M. (1886) *The Origin of Mountain Ranges*, London.

Reid, H. F. (1922) 'Isostasy and earth movements', *Bulletin of the Geological Society of America* 33: 317–26.

Reingold, N. (1972) 'John Fillimore Hayford 1868–1925', *Dictionary of Scientific Biography*, vol. 6, New York: Scribners, pp. 188–9.

Rogers, H. D. and Rogers, W. B. (1843) *On the Physical Structure of the Appalachian Chain, as exemplifying the laws which have regulated the elevation of great mountain chains generally*, Boston.

Sauer, C. (1929) 'Land forms in the Peninsular Range of California as developed about Warner's Hot Springs and Mesa Grande', *University of California Publications in Geography* vol. 3 (1928–30), pp. 199–248.

Schuchert, C. (1910) 'Palaeogeography of North America', *Bulletin of the Geological Society of America* 20: 427–605.

—— (1923) 'Sites and nature of North American geosynclines', *Bulletin of the Geological Society of America* 34: 151–230.

Schwarzbach, M. (1986) *Alfred Wegener: The father of continental drift*, Madison, WI: Science Tech. Publishers.

Shepard, F. P. (1923) 'To question the theory of periodic diastrophism', *Journal of Geology* 31: 599–613.

Sitter, L. U. de (1952) 'Pliocene uplift of Tertiary mountain chains', *American Journal of Science* 250: 297–307.

Snider-Pellegrini, A. (1858) *La Création et ses Mystères Dévoilés*, Paris.

Spencer, J. W. (1913) 'Relationship between terrestrial gravity and observed earth movements of eastern America', *American Journal of Science* 4th series, 35: 561–73.

Spieker, E. M. (1946) 'Late Mesozoic and Early Cenozoic history of central Utah', *US Geological Survey Professional Paper* 205-D, pp. 117–61.

—— (1949) 'Sedimentary facies and associated diastrophism in the Upper Cretaceous of central and eastern Utah', *Geological Society of America, Memoir* 39: 59–82.

Staub, R. (1924) 'Der Bau der Alpen', *Beiträge zur Geologie der Schweiz*, N.F., vol. 24.

—— (1928) *Der Bewegungsmechanismus der Erde*, Berlin: Gebrüder Borntraeger.

Steers, J. A. (1929) *The Unstable Earth*, London: Methuen (4th edn 1945).

Stephenson, L. W. (1928) 'Major marine transgressions and regressions and structural features of the Gulf Coastal Plain', *American Journal of Science* 5th series, 14: 281–98.

Stille, H. (1924) *Grundgragen der vergleichenden Tektonik*, Berlin: Gebrüder Borntraeger.

—— (1934) 'The growth and decay of continents', *Research and Progress* 1: 9–14.

—— (1936a) 'Die Entwicklung der amerikanischen Kordillerensystems in Zeit und Raum', *Sitzungberichte der Preussischen Akademie der Wissenschaften zu Berlin* 15: 134–55.

—— (1936b) 'The present tectonic state of the earth', *Bulletin of the American Association of Petroleum Geologists* 20: 849–80.

—— (1941) *Einführung in den Bau Amerikas*, Berlin: Gebrüder Borntraeger.

Suess, E. (1875) *Die Enstehung der Alpen*, Wien: Braumüller.

—— (1883–1908) *Das Antlitz der Erde*, 3 vols, Wien: Tempsky.

—— (1897–1918) *La Face de la Terre* trans. E. de Margerie, Paris: Colin, 3 vols (1897, vol. 1; 1900, vol. 2; 1902–18, vol. 3).

—— (1904–24) *The Face of the Earth*, trans. H. B. C. and W. J. Sollas, Oxford, 5 vols (1904, vol. 1; 1906, vol. 2; 1908, vol. 3; 1909, vol. 4; 1924, vol. 5).

Taylor, F. B. (1910) 'Bearing of the Tertiary mountain belt on the origin of the Earth's plan', *Bulletin of the Geological Society of America* 21: 179–226.

Thom, W. T. (1937) 'Position, extent, and structural makeup of Appalachia', *Bulletin of the Geological Society of America* 48: 315–22.

Trask, P. D. (1926) 'Geomorphogeny of the northern part of the Santa Lucia Coast Range, California', *American Journal of Science* 5th series, 12: 293–300.

Umbgrove, J. H. F. (1934) 'The relations between geology and gravity field in the East Indian Archipelago', In 'Gravity Expeditions at Sea, 1923–32', *Netherlands Geodetic Commission, Publication* vol. 2, Delft.

—— (1938) 'Geological history of the East Indies', *Bulletin of the American Association of Petroleum Geologists* 22: 1–70.

—— (1947) *The Pulse of the Earth*, 2nd edn, The Hague: Martinus Nijhoff.

Vening Meinesz, F. A. (1934) 'Gravity expeditions at sea, 1923–32', *Netherlands Geodetic Commission, Publication* vol. 2, Delft.

Von Bubnoff, S. (1954) *Grundprobleme der Geologie*, Berlin: Akademie-Verlag.

Waterschoot van der Gracht, W. A. J. M. van (ed.) (1928) *The Theory of Continental Drift: A symposium on the origin and movement of land masses both intercontinental and intracontinental*, Tulsa, OK, American Association of Petroleum Geologists.

Wegener, A. (1912) 'Die Enstehung der Kontinente', *Petermanns Geographische Mitteilungen* 58: 185–95, 253–6, 305–9.

—— (1915) *Die Enstehung der Kontinente und Ozeane*, Braunschweig (English translation of the 3rd edn as (1924) *The Origin of Continents and Oceans*, London: Methuen: French translation of the 3rd edn (1923) and of the 5th edn (1937).

Wiechert, E. (1897) 'Über die massenverteilung im Innern der Erde', *Nachrichten von der Gesellschaft der Wissenschaften in Göttingen, Math.-Phys. Klasse* N.F., 4: 221–43.

Williams, H. S. (1897) 'On the southern Devonian formations', *American Journal of Science* 4th series, 3: 393–403.

Willis, R. (1925) 'Physiography of the California Coast Ranges', *Bulletin of the Geological Society of America* 36: 641–78.

CHAPTER TWO

Climatic Changes

Along with the gravitational influences of crustal and eustatic changes (see chapters 1 and 3), a combination of climatic changes, dominated by solar energy, was to underlie much geomorphic thinking between the latter part of the nineteenth century and the middle of the twentieth. Chapter 12 of the present volume deals with the broad influences of climate on landforms (i.e. the study of morphogenetic landforms) and, specifically, the effects of former palaeoclimates and climatic changes on previous, now fossil, landforms (i.e. the study of climato-genetic landforms). It is the purpose of the present chapter to take up where volume 1 left off in the middle of the nineteenth century and to sketch the ways in which our understanding of the history of world climates developed during the ensuing century, so as to form a basis for the appreciation of the importance of past and present exogenetic processes in geomorphology. The study of past climates is complicated not only by possible variations in the emission of solar radiation but also by global influences which may have altered the relative receipt of solar radiation upon different parts of the earth, as well as by geographical changes which may have altered the absolute and relative positions of landmasses through geological time, the elevations and relief of continents, the disposition of ocean currents and many other factors having a profound influence on the regional climates of the globe. At the same time it must be remarked that our understanding of the nature and mechanisms of present world climates was revolutionized between the mid-nineteenth and the mid-twentieth centuries, such that accurate descriptions of the climates of the tropics were not available until the latter time.

Palaeoclimatic Studies in the Later Nineteenth Century

In volume 1 (chapters 13, 18 and 24) we traced the rise of the glacial theory until the middle of the nineteenth century. At that time most earth scientists, following Karl Schimper and Charles Lyell, were unwilling to accept that large ice sheets had extended over lowland areas, preferring to believe that glacial 'drift' had been carried by floods or by floating icebergs. In 1832, for example, Bernhardi's suggestions that a Scandinavian ice sheet had previously expanded south into Germany received little support, but around mid-century views were to change rapidly.

Of course, at this time most scholars accepted that there was ample evidence for a recent colder period in the northern temperate latitudes; Desnoyers (1829) and Reboul (1833) introduced the term Quaternary, and in his *Principles* Lyell defined the Pleistocene for marine strata having more than 70 per cent of living mollusc species. However, around 1850 the history of the glacial period came to be recognized as being more complex, with Collomb (1847) identifying two drift layers in the Vosges and Godwin-Austen (1851) interpreting the periglacial deposits in southern England as forming part of a complicated chronology. Evidence of at least two glaciations was soon postulated in a number of European locations, as follows:

Wales: valley glaciation; marine submergence to 2,300 feet above present sea level; marine withdrawal; re-establishment of valley glaciers and moraine formation (Ramsay 1852).

Switzerland: first glacial epoch; diluvian gravels; second glacial epoch; modern epoch (Morlot 1856).

Scotland: glaciation with striations and till; marine submergence to 3,000 feet above present sea level with arctic shells and iceberg deposition; drop of sea level to below present level and valley glaciers expanded; renewed submergence to +40 feet with marine clay deposition and disappearance of glaciers; slow withdrawal of sea level to present position (Jamieson 1860, 1865; see also Chambers 1853).

England: two glacial boulder clays in Norfolk identified separated by interglacial sands (Harmer 1867).

The works of Ramsay (1852) and Jamieson (1860, 1862, 1863, 1865) were particularly important in setting up a Pleistocene chronology (Davies 1969: 298–9) and, with that of Geikie (1863), of dispelling the notion of the marine origin of 'drift'.

On the continent Otto Torell and others had postulated that Scandinavia had been buried beneath a thick ice-cover, and Zittel (1874) had established that much of the Bavarian plain had been under ice. Yet most German geologists held fast to the 'drift' theory. Their diluvial notions were shattered on 3 November 1875 when at a crowded meeting of their geological society, Torell convincingly showed that the central Scandinavian ice-sheet had expanded southward over lowland north Germany and dropped the erratics. Thereafter, accounts of lowland ice-sheets or glacial deposits began to proliferate and to include discussion not only of the probable limits of ice advance but also of the alternating nature of the deposition associated with those margins. Any appreciable amelioration of climate appeared to have caused recession in the thickness and extent of ice-masses. Detailed investigations of glacial and fluvio-glacial deposits in many countries suggested that two main periods of ice advance had been interrupted by a prolonged interval of milder climate which produced a layer rich in organic matter. The findings

Plate 2.1 Eduard Brückner (left) and Albrecht Penck (right) in the summer of 1893 near Films (Vorder Rhein Valley, Graubünden). The photograph was given to Mrs Penck by Brückner entitled 'Albrecht the Squashed'

Source: Photograph courtesy of Penck's daughter-in-law Mrs Anny Penck. From J. Büdel (1982) *Climatic Geomorphology*, Princeton, NJ: Princeton University Press, frontispiece. By permission of Gebrüder Borntraeger, Berlin.

Figure 2.1 The glacial deposits on the north side of the Alps west of Munich. Note the rivers Günz, Mindel and Würm. The Riss lies just to the west of the area shown

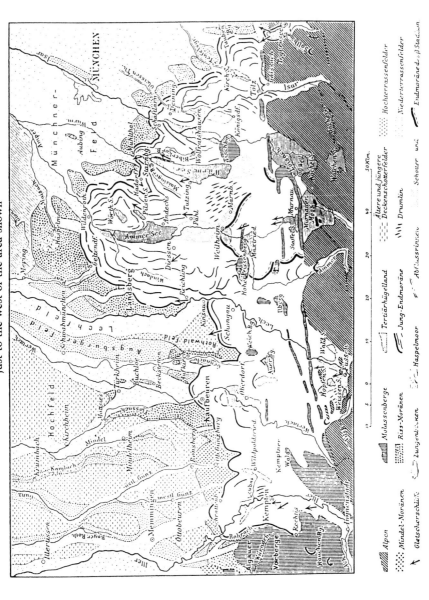

Source: From Penck and Brückner (1909), Vol. 1, Fig. 37, p. 177

Figure 2.2 Cross-section of the valley of the River Inn at Kufstein, Austria

Source: From Penck and Brückner (1909), Vol. 1, Fig. 49, p. 259

of Ramsay in the British Isles, Heer in the Alps and Albrecht Penck and others in Germany and Scandinavia contributed to this opinion. But soon refinements were made and subdivision began to multiply until complicated sequences of major and minor advances and retreats began to be deduced for various countries and hemispherical correlation became well nigh impossible.

Albrecht Penck played an important part in this work. In his *Die Vergletscherung der deutschen Alpen* (1882) he postulated three distinct epochs of glaciation in the German Alps and these were found to be applicable to a much wider area in upper Bavaria. Later investigations with Eduard Brückner, led to their production of the three volumes of *Die Alpen im Eiszeitalter* (1901–9) which was a milestone in the establishment of climatic phases in the Pleistocene period (plate 2.1). They distinguished four ice advances which they named after four Bavarian rivers west of Munich (figure 2.1), in chronological order, Günz, Mindel, Riss and Würm, separated from each other by interglacials. They also suggested chronologies of valley cutting (figure 2.2). These subdivisions were gradually applied worldwide but it was soon recognized that they could not be used rigidly in a universal sense. New modified schemes with a bewildering new terminology sprang up for northern Europe and for North America and numerous local details were added. This may be exemplified from the third edition of *The Great Ice Age* (1894) by James Geikie (see volume 1, Fig. 87, p. 452), in the preface of which he states that:

> Since the last issue of this book, seventeen years ago, the numbers of workers in glacial geology has been greatly increased .... . Nowhere has glacial geology been more actively prosecuted in recent years than in America.
>
> (Geikie 1894: VI–X)

In this connection it should be noted that the Sangamon soil, younger than the last (Wisconsin) glaciation, had been recognized in Illinois in 1873 and the Wisconsin glacial deposits had been separated into early and late divisions on the basis of 'topographical freshness' in 1883 by Thomas Chrowder Chamberlin of the University of Chicago.

Geikie therefore invited Chamberlin, then with the Peary Relief Expedition in Greenland, to outline the state of glacial geology in North America. Chamberlin began by warning the reader that some eminent American geologists only attributed the surficial material of about half the drift-covered area to glaciers and that of the rest to sea-borne ice. He himself accepted the glacial theory and produced two coloured diagrams of the distribution of the ice and of glacial deposits. He explained that:

> The drift deposits of the Great Plain region of North America may be looked upon as a series of sheets overlapping each other in imbricate fashion, the outermost disappearing beneath the next inner, and this, in turn, dipping beneath the succeeding, and so on.
>
> (Chamberlin 1895a: 736)

Chamberlin (1895a: 754–70) went on to describe the northward sequence of progressively younger glacial deposits consisting of the Kansas Formation or outermost drift sheet; the post-Kansas organic interval, estimated at 50,000 years as a minimum and perhaps twice that long; the East-Iowan till-sheet, which like its predecessor is not usually bordered by any definite terminal moraine, and which is connected with the chief deposits of loess; the post-East-Iowan vegetal horizon with oxidation, ferrugination, erosion and other signs of a notable interval; the East-Wisconsin Formation with pronounced glacial features, including massive terminal moraines; Lacustrine Formations of stratified clays, sands and gravels, bordered by beach-lines; the Toronto Fossiliferous Beds on the northern shore of Lake Ontario near Toronto, which indicate an important interglacial episode; and the Champlain deposits of brackish to fresh water. Chamberlin (1895a: 772) concluded:

> There has been a growing disposition among field workers to recognize three glacial epochs, and in many respects this furnishes the most satisfactory interpretation of present data .... There are, however, some evidences in favour of the bipartite view ... the (Lake) Bonneville and Lahontan phenomena indicate two periods of humidity separated by one of aridity.

Elsewhere in the volume James Geikie points out that the occurrence of three glacial advances applied well to both North America and Europe, but in his summary he added:

> Now it has been proved that the maximum glaciation supervened in early Pleistocene times and that three or four separate and distinct cold epochs of diminishing severity succeeded.
>
> (Geikie 1894: 806)

Pleistocene matters moved fast in the United States and by the end of the century Chamberlin had identified and named the Wisconsin (1895a) and Kansan (1895b) glaciations and the Aftonian (pre-Kansan) mild deposits (1895b); while Leverett (1899) had done similarly for the Illinoian glaciation and had named (1898a, b) the Sangamon and Yarmouthian interglacials. Shortly afterwards Shimek (1909) identified and named the Nebraskan glaciation. By the end of the century opposition to the idea of extended Pleistocene continental ice-sheets had effectively ceased in both North America and Europe.

At the same time as Quaternary climatic changes were being investigated in middle and high latitudes, the attention of nineteenth-century scientists was turning to conditions in lower latitudes. The Colorado Plateaus were recognized by Newberry (1861: 47) as having been 'formerly much better watered than they are today'. Major consideration, however, was directed to lake fluctuations as climatic indicators and in 1863 Jamieson proposed that during glaciations in higher latitudes the arid regions would be moister and their lakes would expand. In 1865 Lartet noted the present shrunken size of the Dead Sea and, in the same year as Hull (1885: 182) first used the term 'pluvial' period in respect of the Dead Sea region, Russell (1885) demonstrated the formerly greater extent of what he called 'Lake Lahontan' in Nevada-California. Shortly afterwards Russell (1889: 369) was able to show for the first time the synchronicity of a glacial advance and an arid lake expansion, in respect of Mono Lake in California. This synchronicity was supported by Gilbert (1890: 309–11, 318) in his classic monograph on Lake Bonneville in Utah.

Although, as we have seen, Lyell defined the Pliocene/Pleistocene boundary on the basis of marine fauna, the interpretation of Tertiary and earlier palaeoclimates was based just as firmly on botanical evidence (Dorf 1964). In the 1840s Gilpin (1843) and Corda (1845) showed the tropical climatic significance of Carboniferous fossil plants, and shortly afterwards Heer (1855–9) proposed that warm Tertiary climates had existed in Europe up to the end of the Miocene, with palms growing in the location of Germany, followed by Pliocene cooling. Charles Lyell devoted much space to vicissitudes of climate in his *Principles of Geology*. In the eleventh edition (1872) he found it necessary to rewrite and greatly extend what he had already rewritten and enlarged in the tenth edition (1866). The twelfth edition (1875), the year of his death, remained virtually unchanged. With this customary thoroughness and lucidity he summarizes the geological evidence for climatic change:

> Hooke, about the year 1688, grounded his belief in the reality of the higher temperatures of the waters of the ancient sea on the occurrence of fossil turtles and ammonites in the Portland oolite.
>
> (Lyell, 1875: I: 172)

> Since I first attempted, in the year 1830, to account for vicissitudes of climate by reference to changes in the physical geography of the globe, our knowledge of the subject has greatly increased, and the problem to be solved has assumed a somewhat new aspect . . . in times past the climate of the extra-tropical regions has by no means been always hotter than now, but, on the contrary, there has been at least one period . . . when the temperature of those regions was much lower than at present.
>
> (Lyell 1875: I: 173)

> An examination of the fossils of the Pliocene, Miocene, and Eocene strata, viewed successively in the order of their higher antiquity, afforded us evidence of a temperature continually increasing, in proportion as we receded farther from the Glacial epoch.
>
> (Lyell 1875: I: 211–12)

> In the Secondary or Mezozoic ages, the predominance of reptile life, and the general character of the fossil types of that great class of vertebrata, indicate a warm climate and an absence of frost between the 40th parallel of latitude and the pole . . .

the general character of the molluscs and corals, . . . is in perfect accordance with the inferences deduced from the associated fossil reptiles. If we then carry back our retrospect to the primary or Palaeozoic ages, we find an assemblage of plants which imply that a warm, humid, and equable climate extended in the Carboniferous period uninterruptedly from the 30th parallel of latitude to within a few degrees of the pole. . . .

(Lyell 1875: I: 231–2).

In later years, the geological evidence for the hot climates came to include evaporites, rock salt, gypsum, wind-blown deposits, red-beds and laterites, as well as floral indicators. In 1879 Asa Gray, brushing aside Sir Joseph Hooker's (1848) earlier reticence, believed fossil plants to be 'the thermo-meters of the ages, by which climatic extremes and climate in general through long periods are best measured'. By this time the Norwegian Axel Blytt (1876) had already laid the foundation for climatic interpretation based upon bog stratigraphy. As late as 1892 Seward was reluctant to place reliance on fossil plants as palaeoclimatic indicators, but by this time the nature of Tertiary climatic changes was beginning to be understood (Heer 1868–83), as evidenced by Dutton's (1880–1: 120) view that 'the general tenor of the facts is to the effect that the Miocene was a humid period and the Pliocene a dry one throughout the greater part of the West'.

Nineteenth-century pre-Tertiary palaeoclimatological studies of the nor-thern mid-latitudes concentrated on the tropical conditions of the Carbon-iferous, the aridity of the Triassic and the warmth of the Cretaceous; in connection with the latter Roemer (1852) used strata in Texas to postulate the character of climatic belts and the existence of a Gulf Stream in the Cretaceous. However, the most striking evidence had to do with a proposed Permo-Carboniferous glaciation, particularly in the southern hemisphere, based on fossil tills and striations. In 1855 Ramsay considered that some striated Permian deposits in England were of glacial origin; shortly after-wards Blanford and others (Blanford et al., 1859: 47–50) found moraines of Upper Palaeozoic age in India; later Sutherland (1870) and Woodworth (1912) made glacial discoveries of a similar date in South Africa and South America, respectively, thereby initiating the far-flung discovery of pre-Quaternary southern hemisphere ice ages (David 1896, Coleman 1926) in what later became known as Gondwanaland and providing strong support for the concept of continental drift (see chapter 1). Nevertheless at about 1890 the most popular palaeoclimatic view was of generally warm or mild pre-Tertiary climates becoming colder during the Tertiary and culminating in the Pleistocene ice age.

Attempts at Chronological Estimates

When James Croll (1875: 325) deduced that the last main period of frigid temperatures had persisted from about 240,000 to about 80,000 years B.P.

Figure 2.3 Estimates of the dates of cold and warm phases of the Pleistocene. The Milankovitch curve is for summer radiation at the present latitude 65°N. Temperature changes are expressed in terms of those averages associated with the present latitudes 60–75°N

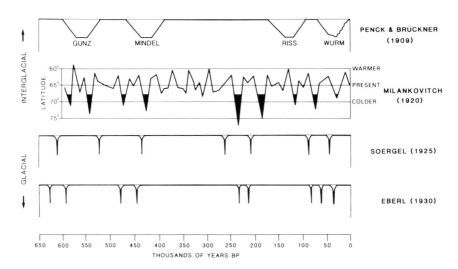

Source: After Zeuner (1959), Figure 48, p. 182 and Figure 55, p. 209

(before present), he was using the then generally accepted rate of denudation (estimated in 1861 by Humphreys and Abbot for the Mississippi basin) of 1 foot of rock in about 6,000 years (see volume 1, pp. 436–9). But he and other geologists considered that the rate of denudation must have been greater during the Glacial Epoch. Joseph Prestwich attempted to use crude estimates of river-valley deepening in Britain and Northern France in glacial and post-glacial times. The rate of deepening compared with the total incision of 150 feet or 180 feet seemed to suggest

> that the Glacial Epoch – that is to say the epoch of extreme cold – may not have lasted longer than from 15,000 to 25,000 years, and ... the so called Post-glacial period, or the melting away of the ice-sheets, to from 8,000 to 10,000 years or less.
> (Prestwich 1888: II: 534)

But a modern critic must admit that Prestwich's excellent results owed more to luck than to judgement or accuracy of measurement. Other crude estimates of post-glacial time were derived from isostatic rebound rates and the rate of recession of wavecut cliffs and waterfalls, such as Niagara and St Antony.

Time estimates were also based upon deposition rates. Steck (1892) estimated a post-glacial time of 20,000 years from the lake delta deposits at

Interlaken and 14,000–15,000 years from a delta in Lake Brienz. Similarly, Heim (1894) gave a post-glacial time estimate of 16,000 years from delta deposits in Lake Lucerne. A more complete chronology was developed by Penck and Brückner (1909: 1169) who estimated relatively Pleistocene mild periods from assumed and comparative rates of weathering and soil development in the Alps and from deduced rates of erosion of Alpine valley deposits. They estimated post-glacial time at one-third the Riss-Würm interglacial and one-twelfth the Mindel-Riss interglacial. As the post-glacial phase, estimated largely on the estimated time to build out a delta into Lake Lucerne, was calculated at about 20,000 years, the interglacials became 60,000 years (Riss-Würm) and 240,000 years (Mindel-Riss) respectively. Subsequently the estimated age of the delta was greatly reduced, as also were Penck's assessments, but his belief that the Ice Age was several hundred thousand years long persisted. Later Soergel (1925) and Eberl (1930) made similar calculations with respect to the terraces of the River Ilm, marginal to the Scandinavian ice, and gravel trains north-east of Lake Constance, respectively (figure 2.3). The latter author also identified pre-Günz cold phases which extended the Pleistocene back to about 1 million years B.P.

In 1931 G. F. Kay estimated the lengths of the interglacials (Leverett 1929) from the depth of surface leaching, assuming post-glacial (i.e. post-Mankato) time to have been 25,000 years, as follows (see also Kay and Leighton 1933):

Till or gravel sheet	Depth of leaching (feet)	Time of exposure of till-sheet (1,000 years)
Mankato	2.5	25 (assumed)
Iowan	5.5	55 (Post-Iowan)
Illinoian	12.0	120 (Sangamon interglacial)
Kansan	30.0	300 (Yarmouthian interglacial)
Nebraskan	20.0	200 (Aftonian interglacial)

These results were supported by the work of Sayles (1931) in Bermuda who estimated 120,000 years and 250,000 years for the Sangamon and Yarmouthian interglacials, respectively. By the middle of the century the following American Pleistocene chronology was popular:

Glacial or interglacial	Substage	Climate	1,000 years B.P.
Recent		cool interglacial	25–0
	Mankato	glacial	
	Cary	glacial	
Wisconsin	Tazewell	glacial	65–25
	Peorian	cool interglacial	
	Iowan	glacial	

Glacial or interglacial	Substage	Climate	1,000 years B.P.
Sangamon		warm interglacial	190–65
Illinoian		glacial	300–190
Yarmouthian		cool interglacial	600–300
Kansan		glacial	700–600
Aftonian		warm interglacial	900–700
Nebraskan		glacial	1,000–900

It should be noted that subsequent ^{14}C dating has led to considerable modifications of the above timescale. Radiocarbon dating was developed by W. F. Libby between 1946 and 1949 (Libby et al. 1949, Libby 1951) enabling dating of selected deposits back to 50,000 B.P. and ushering in the modern era of recent palaeoclimatic research.

Other botanical dating methods included peat bog stratigraphy and tree-ring analysis. The peat bogs in Scandinavia range in age from late glacial down to recent times and contain alternating layers of peat and root-beds ('forest beds') which were investigated in Norway by A. G. Blytt (1876), in Denmark by K. J. V. Steenstrup as early as the 1840s, and in Sweden by R. Sernander. Blytt found a sequence of climatic phases in successive stages from Arctic, through Boreal and Atlantic to the present, which is given in detail in Geikie (1894: 496). These peat bog studies were improved after 1916 when Lennart von Post applied pollen analysis to them. In the first half of the nineteenth century Witham (1833) and Unger (1847) suggested the importance of tree rings in dating. From 1919 onwards dating by means of tree rings (dendrochronology) was developed by Andrew E. Douglass, an American astronomer, who noticed that certain tree rings are relatively wide in wet years and narrow in dry ones. Later Antevs (1928) developed a North American climatic chronology by means of tree rings and this method subsequently actually allowed the refinement or correction of the ^{14}C scale, but it can only apply to the last few thousand years. Yet Geikie's prophecy had almost come to pass:

> The time will come when it will be possible to correlate the various 'forest-beds' of all these separate tracts, and thus trace out in detail the various modifications and changes which have affected the flora of Northern Europe since the disappearance of the last great Baltic glaciers.
>
> (Geikie 1894: 497–8)

The meltwater flow of glacial streams occurs mainly in summer and the resulting sediment deposition is much thicker and coarser in that season than in winter. Counting of this conjunction of thick and thin biannual seasonal layers each year – or the so-called *varves* – allows the final stages of the local

Plate 2.2 Gerard De Geer (1858–1943)

ice-cap to be dated and its duration within the length of the varve deposition determined. In the eighteenth century varves were believed to be annual, but modern semi-annual analysis began as the result of work by Heer (1865) in Switzerland and especially by Gerard De Geer (plate 2.2) in Sweden (1884). De Geer analysed varved sediments at some 1,500 locations in a 30 year period working north for some 600 miles to Lake Ragunda which was accidentally drained in 1796 exposing some 3,000 years of most recent varves, thus anchoring the chronology to the modern date and suggesting a mean rate of ice retreat northward of some 7 miles/100 years. He presented his results at the International Geological Congress at Stockholm in 1910 and in 1912 published a detailed recessional timescale back to more than 12,000 years B.P., later extended to 18,000 years B.P. (De Geer, 1926, 1940). This method was refined by Sauramo (1923, 1929) in Finland and Antevs (1925, 1928) worked out a similar retreat chronology from southern New England to eastern Canada at an average rate of 5 miles/100 years over some 13,500 years with a less continuous varve sequence. In the long run the assessments have proved remarkably precise and have been vindicated by the recent ^{14}C evidence. The only cycles they clearly indicate are of 10 or 11 years and of about 21,000 or 22,000 years. Similar cycles were found in middle Eocene varves beside the Green River in Utah and Colorado by W. H. Bradley in 1929 and in slates in Thuringia, Germany, by H. Korn in 1938, although some palaeoclimatologists are sceptical of the results (Schwarzbach 1963: 93–4).

Some of the assumed astronomical causes of climatic change, to be discussed later, also assisted chronological estimates. The radiation curves constructed for the Swiss Alps by Milankovitch (1930), Soergel (1937) and Beck (1938) enabled former minimum temperature troughs to be tentatively dated. In particular, Milankovitch (1930) gave the following calculations for the past occurrence of summer minima troughs (in 1,000 years B.P.) (figure 2.4):

25, 72 and 115	Last glaciation
187 and 230	Penultimate glaciation
435 and 476	Second glaciation
550 and 590	First glaciation

Causes of Climatic Change

In trying to put into geomorphic perspective the many theories which have been advanced to account for climatic changes it is important to distinguish between those which address themselves to Pleistocene climates, which are of comparatively recent date and generally associated with the present latitudinal distributions, and those climates of earlier times which to a greater or lesser degree seem to violate the present latitudinal organization of climate. The causes which have been proposed fall under the headings of astronomical theories, solar energy theories and terrestrial theories.

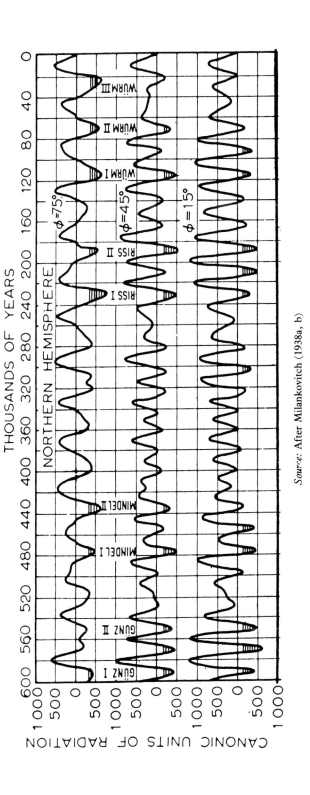

Figure 2.4 Estimated curves of solar radiation in summer for latitudes 15°N, 45°N and 75°N. Cross-lining denotes possible glacial advances

Source: After Milankovitch (1938a, b)

Perturbations of the earth's orbit were recognized by early Greek philosophers, but from a geomorphological point of view the discussion of the astronomical impact on climate begins with John Herschel in 1833 and J. A. Adhémar in 1842. The latter attempted to show that a period with cold winters caused by the precession of the equinoxes and movement of the perihelion would create in one hemisphere an accumulation of ice which on melting formed huge floods capable of transporting erratic boulders. In the 1860s this theme was improved and greatly elaborated by James Croll whose fine summary *Climate and Time in their Geological Relations* became popular. Croll postulated that when the earth was in perihelion at the summer solstice, the northern hemisphere would experience relatively short hot summers and relatively long cold winters which would generate more ice than the summer warmth could melt. This astronomical effect would increase with the eccentricity of the orbit and conditions would be reversed when the summer solstice occurred at aphelion (figure 2.5). The extension of ice in the polar areas of the relevant hemisphere would, by radiation of cold, absorption of heat and condensation of moisture, with concomitant cloudiness, further lower temperatures. The eccentricity of the orbit would also affect prevailing winds and ocean currents. The interglacial periods would be characterized by almost complete withdrawal of glacial ice and by much subaerial denudation of glacial deposits. However, as critics pointed out, it was obvious that extensive glaciations must alternate between the northern and southern hemispheres. Croll immediately found powerful supporters among whom were Albrecht Penck and G. Pilar on the mainland and, particularly, James

Figure 2.5 Diagram to illustrate the combined effect of the eccentricity of the Earth's orbit and the precession of the equinoxes. On the left, the maximum eccentricity coincides with winter in aphelion in the northern hemisphere giving glacial conditions there; on the right, about 10,500 years later, conditions are reversed.

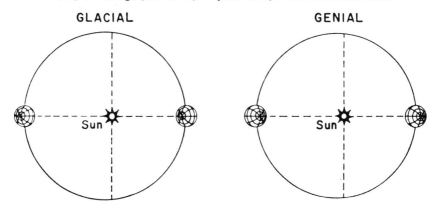

Source: From Ball (1891), pp. 100–1

Figure 2.6 Maps giving a popular explanation of extremes of heat and cold on the continents based on assumed lateral shifts of the landmasses

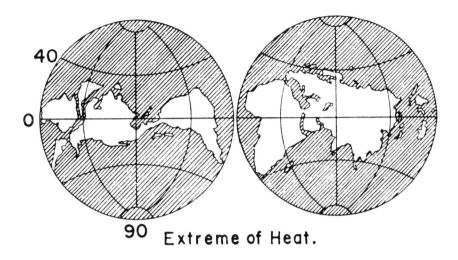

Extreme of Heat.

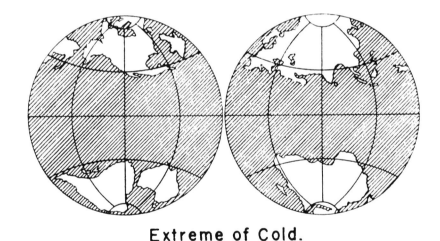

Extreme of Cold.

Source: After Lyell (1875), Vol. 1, Figures 14 and 15, p. 270

Geikie in *The Great Ice Age* (especially in the third edition of 1894). In the interim Croll's calculations were refined by Ball (1891) who added other astronomical variables, although the influential climatologist Hann (1908) did not place importance on astronomical theories of glaciation.

Sir Charles Lyell, too, had considered that the astronomical controls over climate were being exaggerated and in the fully revised eleventh and twelfth editions of his *Principles of Geology* (1872 and 1875) he provided a thorough and lucid account of the problem of accounting for the vicissitudes of climate (Lyell 1872: I: 172–297). He proceeded first to demonstrate with the aid of striking diagrams how rearrangement of land and sea on the globe might produce extremes of heat and cold in global climate (figure 2.6):

> Such extremes may never have occurred, but we may safely conclude that there may sometimes have been an approximation to them in the course of those ages to which our geological records refer.
>
> (Lyell 1872: I: 271)

Today it seems odd that such ideas should come from the greatest uniformitarianist and that Lyell, of all people, should have moved continents with such freedom as to be the first notable geologist to advocate continental drift! He goes on to discuss how far the dominant influence of these changes in the position of land and sea may have been affected by the three recognized astronomical factors: the eccentricity of the earth's orbit, the precession of the equinoxes and the obliquity of the plane of the ecliptic.

The first, the orbital eccentricity, can only make an extreme difference of 14 million miles in the distance the earth is from the sun or a proportion of 1,003 to 1,000. The second, precession, moves round the orbit as a cycle of 21,000 years, so its astronomical or hemispherical condition is reversed in about 10,500 years. Winters when the relevant hemisphere is in perihelion (nearest the sun) ought to be less rigorous than winters in aphelion. Lyell illustrates this with a persuasive diagram (figure 2.7) which has an exaggeration which has haunted textbooks ever since. In it the supposed effect of eccentricity on the position of the sun in the orbital ellipse has been more than doubled; in fact the sun should be almost central. Lyell pointed out that Croll assumed that at maximum eccentricity the total heat received from the sun differs by about one-fifth, assuming that it varies inversely as the squares of the distances. This, as we have already noticed, would exaggerate the cold in the hemisphere in which winter occurred in aphelion and in its high latitudes precipitation would fall as snow. The summer heat, although one-fifth greater, would be insufficient to remove the winter snow because ice-melt would encourage fog and overcast skies. But, Lyell adds, Croll paid little or no attention to the influence of abnormal geographical conditions such as now prevail:

> The simple fact that totally different climates exist now in the same hemisphere and under the same latitude would alone suffice to prove that their occurrence cannot be exclusively due to astronomical influences.
>
> (Lyell 1872: I: 283)

The third astronomical influence, the obliquity of the ecliptic, or inclination of the earth's axis of rotation, was clearly understood by Lyell who knew it

Figure 2.7 Diagram to illustrate the precession of the equinoxes. Here the difference between aphelion (94,500,000 miles) and perihelion (91,500,000 miles) is greatly exaggerated

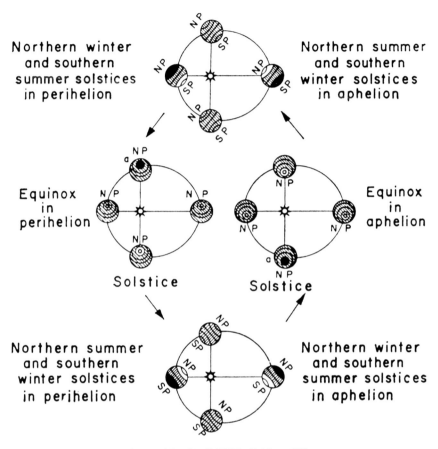

Northern winter
and southern
summer solstices
in perihelion

Northern summer
and southern
winter solstices
in aphelion

Equinox
in
perihelion

Equinox
in
aphelion

Solstice

Solstice

Northern summer
and southern
winter solstices
in perihelion

Northern winter
and southern
summer solstices
in aphelion

Source: After Lyell (1875), Vol I, p. 276

had only a latitudinal influence, strongly polar, and did not alter the total incoming radiation.

Whenever the obliquity is greater than now (23° 28'), more of the arctic and antarctic regions would be exposed to a long night in winter, and consequently the cold of that season would be greater, and under the opposite circumstances the reverse would take place ... . So often as the extreme of possible obliquity happened to combine with the maximum eccentricity and with geographical circumstances of an abnormal character like those prevailing in high altitudes, a greater intensity of cold would be produced than could exist without such a

combination, and so far this would favour a glacial epoch. But when, on the other hand, the obliquity was at its maximum, the cold would be lessened . . . .
(Lyell 1872: I: 293–4)

Despite his detailed consideration of Croll's astronomical theory of climatic change, Lyell believed that terrestrial causes were the prime movers of such change and that these were merely assisted by astronomical variations.

The period between 1890 and the early 1920s was particularly rich in palaeoclimatic speculation and, for example, Schwarzbach (1963) estimated that, of the forty-five different ice-age theories which were proposed during the century prior to 1950, fully two-thirds appeared during these 30 years. Although some astronomical speculations continued in the form of possible cosmic dust effects (Nölke 1908), of gas-filled parts of space (Culverwell 1895) and, of course, from 1913 the early researches of Milutin Milankovitch, theorizing regarding palaeoclimatic mechanisms tended to concentrate on solar and, following Lyell, terrestrial causes. It became clear at quite an early stage that relatively small changes in solar energy could induce glacial periods (Wood 1876, Dubois 1893) and studies of sunspots (Arctowski 1914) were ultimately combined with astronomical processes to produce a composite theory by Huntington and Visher (1922). During this period proposed terrestrial mechanisms for climatic change, chiefly Pleistocene, multiplied, as follows:

1 Continental drift (see chapter 1). In addition to Lyell, other nineteenth-century scholars had proposed a shift of continents with reference to the poles as a cause of climatic change (Evans 1866, Oldham 1886) but it was left to Wegener's (1915) theory and his collaboration with his father-in-law, Wladimir Köppen (1924), to exploit its palaeoclimatic significance. Although continental drift appeared unlikely to account for Pleistocene ice advances and retreats (Brooks 1926, 1949, Coleman 1933, Behrmann 1944), the recognition of both warm and cold climatic anomalies prior to the mid-Tertiary made movement of landmasses a possible cause of palaeoclimatic distributions (Nairn 1961). In particular, the theory of Köppen and Wegener (1924) obviated the absurdity of proposing the existence of a Permo-Carboniferous ice-cap near the equator!

2 Changes in land and sea distribution. The radical possibility suggested by Lyell (1872: I: 271) was also considered by Semper (1896), Schuchert (1921) and Kerner-Marilaun (1930). However, as Arldt (1919) pointed out, the area of land poleward of about 40°N appears to be a critical control over climate and this area does not seem to have changed much during the Tertiary.

3 Changes in land elevation. This mechanism for climatic change achieved considerable vogue in the early years of the present century (Ramsay 1910, 1924, Deeley 1915) and Brooks (1926) used the assumed high elevation

(3,000–5,000 m), plus the incidence of volcanic dust, to explain the Palaeozoic glaciation of Gondwanaland. In the 1890s Warren Upham had devised an orographic–isostatic theory in which mountain building caused the glacials and the isostatic depression by the weight of the ice-sheets brought about interglacials. The strongest supporter of this thesis was T. C. Chamberlin (1922) who believed the present mean elevation of the continents (2,500 feet) to be greater than the average during the geological past and that the average elevation of uplifted major peneplains is probably not much in excess of one-third this elevation. On the other hand, Dacqué (1915: 449) believed the present continental elevation to be less now than during warmer Trias/Jurassic, Late Devonian and Ordovician times; and Simpson (1930) pointed to the repeated general global temperature similarities between the two hemispheres irrespective of elevation.

4 Changes in ocean circulation (Hull 1897, Chamberlin 1899, 1906). Willis (1932) proposed the diversion of warm currents as a mechanism and Wundt (1944) the blocking of the Gulf Stream, whereas Behrmann (1944) postulated its glacial strengthening.

5 A long-term decrease in atmospheric carbon dioxide throughout geological time due to the growth of limestone and carbonaceous deposits was believed to have had a depressing effect on world temperatures (Arrhenius 1896, Chamberlin 1899).

6 An increase in volcanic dust might well lower temperatures (Sarasin and Sarasin 1902, Huntington and Visher 1922). This mechanism formed part of the most influential composite theory by Brooks (1926) who assumed that glaciation could be explained without continental drift by a combination of continentality, orogenesis, elevation, diversion of ocean currents, land barriers and incidence of volcanic dust to the extent that each orogenic phase was believed to have been closely followed by a glacial epoch.

After a spate of palaeoclimatic works in the 1920s employing terrestrial causes (Brooks 1926, Coleman 1921, 1925, 1926, Huntington and Visher 1922, Schuchert, 1921) the interwar period was increasingly dominated by recourse to astronomical and solar mechanisms of climatic change (Nairn 1961). In 1904 Ludwig Pilgrim had tabulated the variations in solar radiation for about 1 million years prior to 1850. Milutin Milankovitch (1920) improved on these calculations to supply his friend, Wladimir Köppen, with curves showing the changes of solar radiation at the earth's surface in the summer half-year at latitudes 55°N, 60°N and 65°N for the last 650,000 years. These simulated fluctuations (figures 2.3 and 2.4) produced four glacial ages, were non-periodic and without hemispheric alternation. They were immediately adopted by Köppen and Wegener (1924) in their influential *Die Klimate der geologischen Vorzeit*. Milankovitch, working on slightly improved astronomical statistics, produced a new set of tables and curves in

1930, 1938(a, b) and 1941, which differed from the older curves mainly in the greater intensity of certain maxima and minima (Zeuner 1959: 180–90, Berger *et al.* 1984). Thereafter scientists could study separately the astronomical effect on the summer half-year and winter half-year and the radiation curves for every tenth degree of latitude between 5° and 75°. However, the general acceptance of this astronomical theory was not to come for another quarter of a century. In 1940 G. S. Simpson, a meteorologist, expressed doubts as to the efficacy of Milankovitch's mechanism alone for producing glacial temperatures, a view supported by Woerkom 13 years later. Previously Simpson (1930, 1934) had attempted to explain multiple glaciations by differences between the curves of solar radiation, temperature and precipitation, on the one hand, and of snowfall on the other. The Cambridge physicists Hoyle and Littleton (1939) had even invoked the possible effect of interstellar matter on the earth's climate. Despite these differences of opinion, by mid-century astronomical and solar mechanisms of climatic change had achieved ascendancy over terrestrial ones.

Descriptive Palaeoclimatology

The first half of the twentieth century witnessed great advances in the understanding of palaeoclimates (e.g. Eckardt 1921, Arldt 1919–22, Kerner-Marilaun 1930), particularly of Tertiary and Quaternary times. Descriptive palaeoclimatology was to assume increasing geomorphic significance as it became clear that distinctive climates commonly support distinctive assemblages of landforms (morphoclimatology) and that past climates have left their imprint on many present landforms (climato-genetic geomorphology) (see chapter 12).

The rich literature of the period relating to Tertiary palaeoclimates clearly pointed to warmer conditions than at present during the Early and Middle Tertiary in the northern mid-latitudes. Berry (1916) interpreted the Early Eocene flora of the American Gulf Coast as being similar to that of present subtropical rain forests; Kryshtofovich (1929) gave a classic description of warm Tertiary flora in Asia; Reid and Chandler (1933) described the warm flora of the London Clay; Hollick (1936) demonstrated that Alaskan Tertiary flora indicated warm conditions; and Chaney (1940) also used floral evidence to indicate high Tertiary temperatures generally in North America. More precisely, Ihering (1927) gave the following estimates for mean palaeotemperatures in Central Europe: Early Eocene 68°F, Late Eocene and Oligocene 74°F, Miocene 72°F and Pliocene 60–55°F. By mid-century it was generally recognized that the Tertiary was warmer than at present with summers tending to be dry and springs wet in Europe (Schwarzbach 1963: 178–80). The tropical zone extended 10–15° in latitude farther north and 10° farther south than at present and the polar tree limit was shifted 20–30°

northward in the northern hemisphere and 10° southward in the southern hemisphere. Polar ice-caps did not exist during the Early Tertiary and the general world climate was warmer and more uniform than at present. Cooling began during the Miocene in Europe, where fossil leaves show evidence of frost action, and by the Pliocene the climate was somewhat as at present, after which it got much colder during the Quaternary.

Quaternary palaeoclimatic studies were naturally more detailed (e.g. Kessler 1925). In 1916 Post reconstructed a climatic history using bog pollen; 2 years later Enquist (1918) dated the beginning of Fennoscandian post-glacial times from the deglaciation of the end moraines; correlations were suggested between glaciations and lower-latitude pluvial periods on the basis of African lake fluctuations (Nilsson 1931, Wayland 1934); stone orientations were used to determine the direction of ice movement (Richter 1932); the study of Quaternary climates from deep-sea cores began in the 1930s; and Wissman (1938) investigated the detailed climatic significance of the Chinese loess.

Work on the glaciation of the North German Plain continued apace during the first half of the twentieth century, with Gagel (1913) using weathered zones to study the three major interglacials and Wervecke (1927) proposed the existence of evidence for six glaciations. By mid-century the following Quaternary sequence had emerged for north Germany, with percentages suggesting the relative southerly extensions of the Scandinavian ice (Zeuner 1959) (figure 2.8):

Würm (last) glaciation
- Post-glacial
- Pomeranian phase – glacial (87.5%)
- Masurian phase – subarctic
- Weichsel phase – glacial (92.5%)
- Rixdorf phase – cold to temperate
- Warthe phase – glacial (95%)
- Mild phase
- Cold phase
- Eem phase – mild

Saale (third) glaciation (97.5–102%)
Great interglacial – temperate
Elster (second) glaciation (100%)
Interglacial
Donau-Günz (first) glaciations

In addition, the simple Penck–Brückner four Alpine glaciations had multiplied to pre-Günz (five glacial stages), Günz (two stages), Mindel (two stages), penultimate 'interglacial' (two stages), Riss (two stages), last 'interglacial' (? stages), Würm (three stages), post-Würm (one stage) – giving at least seventeen stages in all!

Figure 2.8 One hundred years of progress in the subdivision of the Pleistocene in Europe

MID 19th CENTURY	1874	1901-9 PENCK and BRUCKNER	1943 GENERAL CHRONOLOGY		1943 ALPS	1943 SCANDINAVIA	1943 EAST ANGLIA	1943 SEA LEVEL
ICE-AGE	SECOND GLACIATION	WÜRM GLACIATION	FOURTH GLACIATION	3rd GLACIAL PHASE	WÜRM 3	POMERANIAN		−30m?
				INTERSTADIAL OSCILLATION		MESURIAN INT.		−12m?
				2nd GLACIAL PHASE	WÜRM 2	WEICHSEL	HUNSTANTON	−70m?
				INTERSTADIAL OSCILLATION		RIXDORF INT.		+1·3m
				1st GLACIAL PHASE	WÜRM 1	WARTHE	LITTLE EASTERN?	−100m
		THIRD INTERGLACIAL	LAST INTER-GLACIAL	SECOND PART OF INTERGLACIAL	LAST INTERGLACIAL	LAST INTERGLACIAL		LATE MONASTIRIAN +7·5m
				MINOR COOL PHASE				MAIN MONASTIRIAN +18m
				FIRST PART OF INTERGLACIAL				
		RISS GLACIATION	THIRD GLACIATION	2nd GLACIAL PHASE	RISS 2	SAALE	GREAT CHALKY	VERY LOW
				INTERSTADIAL OSCILLATION				
				1st GLACIAL PHASE	RISS 1			
	INTERGLACIAL	GREAT INTERGLACIAL		GREAT INTERGLACIAL, WITH COOLER PHASES	GREAT INTERGLACIAL	GREAT INTERGLACIAL	INTERGLACIAL	TYRRHENIAN +32m
	FIRST GLACIATION	MINDEL GLACIATION	SECOND GLACIATION	2nd GLACIAL PHASE	MINDEL 2	ELSTER	NORTH SEA DRIFT	VERY LOW
				INTERSTADIAL OSCILLATION				
				1st GLACIAL PHASE	MINDEL 1			
		FIRST INTERGLACIAL	FIRST INTERGLACIAL	FIRST INTERGLACIAL	FIRST INTERGLACIAL	FIRST INTERGLACIAL	CROMER FOREST BED	MILAZZIAN +60m
		GÜNZ GLACIATION	FIRST GLACIATION	2nd GLACIAL PHASE	GÜNZ 2		WEYBOURNE CRAG	LOW
				INTERSTADIAL OSCILLATION				
				1st GLACIAL PHASE	GÜNZ 1		NEWER RED CRAG	
PLIOCENE	PLIOCENE	PLIOCENE		VILLAFRANCHIAN (PLIOCENE)	DONAU	PLIOCENE	PLIOCENE	SICILIAN +100m

Source: After Zeuner (1959)

In another classic area, eastern England, work identifying four glaciations (Solomon 1932) and on the earlier climatic deterioration (Boswell 1931) led to the following sequence emerging by the Second World War:

Hunstanton glaciation (youngest)
Interval
Little Eastern glaciation
Interval
Great Chalky Boulder Clay glaciation (British ice source)
Interglacial with higher sea level than at present
North Sea Drift glaciation (Scandinavian ice source)
Crag Series (increasing cold phases) (oldest)

Towards the middle of the twentieth century the general view of the Quaternary period was that it lasted about 1 million years; that during the glacial stages the temperatures in Western Europe were about 4°C lower than at present and in northern temperate latitudes 8–12°C lower; that during these stages continental glaciers expanded to about three times their present area; that there were at least three major Central European glaciations (Mindel, Riss and Würm) preceded by two or more cold phases (Donau and Günz); that the European interglacial stages had climates similar to, or slightly warmer than, today's but that each interglacial was colder than its predecessor; that higher-latitude glacial advances were associated with pluvial conditions in tropical deserts; and that, since the end of the last glaciation (more than 10,000 years B.P.), the temperature rose in Europe to be 2–3°C greater than at present during the 'climatic optimum' (7,000–5,000 years B.P.) and then tended to fall irregularly (Schwarzbach 1963: 208–9).

After this time advances in palaeoclimatology were to be revolutionized by the development of geochemical methods such as the use of ^{14}C isotope dating (Libby et al. 1949, Libby 1951) and the use of $^{18}O/^{16}O$ ratios in marine calcium carbonates to estimate oceanic palaeotemperatures (Urey et al. 1951). The geomorphic significance of palaeoclimates was similarly revolutionized by the work of Julius Büdel (1951) (see chapter 12) who proposed a pattern of displacement of climatic zones during the last (Würm) glacial stage compared with that of more recent times. This explains why, in many areas, geomorphic evidence of present climates is superimposed on that of palaeoclimates (Nairn 1961, Stoddart 1969). It was Büdel's view that during the Würm glaciation the decrease of temperature, relative to that of present times, was greatest in higher latitudes and least in lower latitudes, allowing the following generalized shifts of climatic belts at the longitude of Western Europe (figure 2.9):

1 North of about 30°N the belts were displaced equatorward.
2 Around the core of the subtropical deserts (23.5°N) the belts contracted from north and south.

Figure 2.9 The suggested displacement of climatic and vegetational zones in the longitudinal belt 0° to 15°E between the equator and the north pole during the last (Würm) glaciation compared with the present

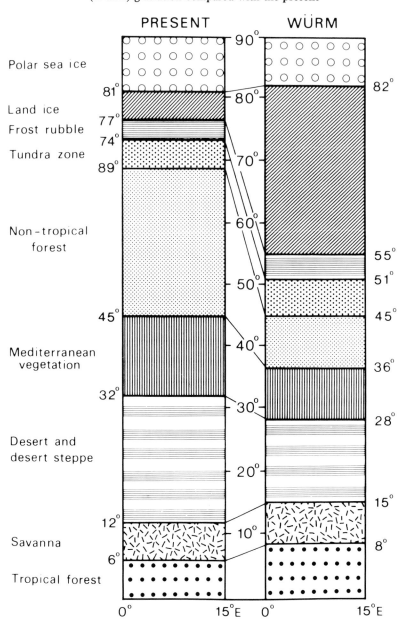

Source: After Büdel (1951), Fig. 1

3 In the lowest latitudes the humid savanna and tropical rainforest belts expanded poleward some 2–3° of latitude, although being somewhat cooler and more humid than today. The result of this is that in the higher latitudes present geomorphic tendencies relating to climate (i.e. morphoclimatic processes) are working on landforms showing evidence of previous Pleistocene evolution under colder and drier conditions; on the northern and southern margins of the subtropical deserts more arid present conditions are superimposed on more humid palaeolandforms; and about 6–8°N present savanna conditions have invaded a belt which was tropical forest during the Würm. Büdel pointed out that, although the short postglacial timespan has not been long enough to obliterate the geomorphic evidence of colder Pleistocene conditions, most present landform assemblages exhibit relict features characteristic of earlier climates. We shall return to this important theme of climatic geomorphology in chapter 12.

References

Adhémar, J. A. (1842) *Les Révolutions de la Mer, Deluges Périodiques*, Paris (2nd edn 1860).

Antevs, E. (1925) 'Retreat of the last ice-sheet in eastern Canada', *Memoir of the Geological Survey of Canada* no. 146.

—— (1928) 'The last glaciation', *American Geographical Society, Research Series* no. 17.

Arctowski, H. (1914) 'About climatical variations', *American Journal of Science* 4th series, 37: 305–15.

Arldt, T. (1919–22) *Handbuch der Paläeogeographie*, 2 vols, Leipzig: Borntraeger.

Arrhenius, S. (1896) 'On the influence of carbonic acid in the air upon the temperature of the ground', *Philosophical Magazine* 5th series, 41: 237–76.

Ball, R. (1891) *The Cause of an Ice Age*, London.

Beck, P. (1938) 'Studien über das Quartärklima im Lichte astronomischer Berechnungen', *Eclogae geologicae Helvetiae*, Basle, 31: 137–72.

Behrmann, W. (1944) 'Das Klima der Präglazialzeit auf der Erde', *Geologische Rundschau* 34: 763–76.

Berger, A. *et al.*(eds)(1984) *Milankovitch and Climate*, 2 vols, Dordrecht: Reidel.

Berry, E. W. (1916) 'The Lower Eocene floras of southeastern North America', *US Geological Survey Professional Paper* 91.

Blanford, W. T., Blanford, H. F. and Theobald, W., Jr. (1859) 'On the geological structure and relations of the Talcheer coal field in the district of Cuttack', *Memoir of the Indian Geological Survey* 1: 33–89.

Blytt, A. G. (1876) *Essays on the Immigration of the Norwegian Flora during Alternating Rainy and Dry Periods*, Oslo: Cammermeyer.

Boswell, P. G. H. (1931) 'The stratigraphy of the glacial deposits of East Anglia in relation to early man', *Proceedings of the Geologists' Association of London* 42: 87–111.

Bradley, W. H. (1929) 'The varves and climate of the Green River epoch', *US Geological Survey Professional Paper* 158-E, pp. 87–110.

Brooks, C. E. P. (1926) *Climate Through the Ages*, London: Benn (2nd edn 1949).

Büdel, J. (1951) 'Die Klimazonen des Eiszeitalters', *Das Eiszeitalter* (Stuttgart), 1: 16–26.

Chamberlin, T. C. (1883) 'Preliminary paper on the terminal moraine of the second glacial epoch', *US Geological Survey, 3rd Annual Report* (1881–2), pp. 291–402.

—— (1895a) 'Glacial phenomena in North America', in J. Geikie *The Great Ice Age*, New York: Appleton, pp. 724–75.

—— (1895b) 'The classification of American glacial deposits', *Journal of Geology* 3: 270–7.

—— (1896) 'Editorial', *Journal of Geology* 4: 872–6.

—— (1899) 'An attempt to frame a working hypothesis of the cause of glacial periods on an atmospheric basis', *Journal of Geology* 7: 545–84, 667–85, 751–87.

—— (1906) 'On a possible reversal of deep-sea circulation and its influence on geological climates', *Journal of Geology* 14: 363–73.

—— (1922) 'The age of the earth from the geological viewpoint', *Proceedings of the American Philosophical Society* 61: 247–71.

Chambers, R. (1853) 'On glacial phenomena of Scotland and parts of England', *Edinburgh New Philosophical Journal* 54: 229–81.

Chaney, R. W. (1940) 'Tertiary forests and continental history', *Bulletin of the Geological Society of America* 51: 469–88.

Coleman, A. P. (1921) 'Paleobotany and the earth's early history', *American Journal of Science* 5th series, 1: 315–99.

—— (1925) 'Late Paleozoic climates', *American Journal of Science* 5th series, 9: 195–203.

—— (1926) *Ice Ages Recent and Ancient*, London: Macmillan.

—— (1933) 'Ice ages and the drift of continents', *Journal of Geology* 41: 409–17.

Collomb, E. (1847) *Preuves de l'Existence d'Anciens Glaciers dans les Vallées des Vosges*, Paris: Masson.

Corda, A. J. (1845) *Beitrage zur Flora der Vorwelt*, Prague: J. G. Calve' sche Buchhandlung.

Croll, J. (1875) *Climate and Time in their Geological Relations: A Theory of Secular Changes in the Earth's Climate*, London: Daldy Isbister (4th edn 1890).

Culverwell, E. P. (1895) 'A criticism of the astronomical theory of the ice age and of Lord Kelvin's suggestions in connection with a genial age at the pole', *Geological Magazine* N.S., decade IV, 2: 55–65.

Cunningham, F. F. (1977) 'Lyell and uniformitarianism', *Canadian Geographer* 21: 164–74.

Dacqué, E. (1915) *Grundlagen und Methoden der Paläogeographie*, Leipzig, Teubner.

David, T. W. E. (1896) 'Evidences of glacial action in Australia in Permo-Carboniferous time', *Quarterly Journal of the Geological Society of London* 52: 289–301.

Davies, G. L. (1969) *The Earth in Decay: A History of British Geomorphology 1578–1878*, London: Macdonald.

Deeley, R. M. (1915) 'Polar climates', *Geological Magazine* N.S., decade VI, 2: 450–5.

Desnoyers, J. (1829) 'Observations sur un ensemble de depôts marins plus récens que les terrains tertiares du bassin de la Seine', *Annales des Sciences Naturelles (Paris)* 16: 171–214.

Dorf, E. (1964) 'The use of fossil plants in palaeo-climatic interpretations', in A. E. M. Nairn (ed.) *Problems in Palaeoclimatology*, London: Interscience, pp. 13–31.

Dubois, E. (1893) *Die Klimate der geologischen Vergangenheit und ihre Beziehungen zur Entwickslungsgeschichte der Sonne*, Leipzig (English edition as (1895) *The Climates of the Geological Past*, London).

Dutton, C. E. (1880–1) 'The physical geology of the Grand Canyon district', *US Geological Survey, 2nd Annual Report* pp. 47–166.

Eberl, B. (1930) *Die Eiszeitenfolge im nördliche Alpenvorlande*, Augsburg.

Eckardt, W. R. (1921) 'Die Paläoklimatologie, ihre Methoden und ihre Anwendungen auf die Paläobiologie', *Handbuch der biologischen Arbeitsmethoden*, vol. 10.

Enquist, F. (1918) 'Die glaziale entwickslungs – geschichte nordwest-skandinavian, *Sveriges Geologiska Undersökning* series C, no. 285, rsbok, vol. 12, no. 2.

Evans, J. (1866) 'On a possible cause of climatic changes', *Geological Magazine* 3: 171–4.

Gagel, K. (1913) 'Die Beweise für eine mehrfache vereisung Norddeutschlands in diluvialer zeit', *Geologische Rundschau* 4: 319–62, 444–502, 588–91.

Geer, G. De (1884) 'Kronologi för istiden', *Geologiska Föreningens i Stockholm Förhandlingar* 7: 512–3.

—— (1912) 'A geochronology of the last 12,000 years', *Comptes Rendus, 11th International Geological Congress, Stockholm, 1910*, 1: 241–58.

—— (1926) 'On the solar curve as dating the ice age, the New York moraine and Niagara Falls through the Swedish timescale', *Geografiska Annaler* 4: 253–84.

—— (1940) 'Geochronologia Suecia principles', *Kungliga Svenska Vetenskapsakademiens Handlingar* 18:(6).

Geikie, A. (1863) 'On the phenomena of the glacial drift in Scotland', *Transactions of the Geological Society of Glasgow* vol. 1.

Geikie, J. (1874) *The Great Ice Age*, London: Stanford (3rd edn 1894).

Gilbert, G. K. (1890) 'Lake Bonneville', *US Geological Survey Monograph* 1.

Gilpin, T. (1843) *An Essay on Organic Remains, as Connected with an Ancient Tropical Region of the Earth*, Philadelphia: Butler.

Godwin-Austen, R. A. C. (1851) 'On the superficial accumulation of the coasts of the English Channel', *Quarterly Journal of the Geological Society of London* 7: 118–36.

Gray, A. (1879) 'Plant archaeology' *The Nation* nos 742 and 743, 18 and 25 September.

Hann, J. (1908) 'Allgemeine Klimalehre', *Handbuch der Klimatologie*, vol. 1, 3rd edn, Stuttgart.

Harmer, F. W. (1867) 'On a third boulder clay in Norfolk', *Quarterly Journal of the Geological Society of London* 23: 87–90.

Heer, O. (1855–9) *Flora tertiaria Helvetiae*, vols I–III, Zürich, Winterthur.

—— (1865) *Die Urwelt der Schweiz*, Zürich: Schulthess.

—— (1868–83) *Flora fossilis arctica*, vols I–VII, Zürich: Druck und Verlag Von F. Schulthess.

Heim, A. (1894) 'Geologische Nachlese. No. 2. Über das absolute Alter der Eiszeit', *Vierteljahrsschrift der Naturforschenden Gesellschaft in Zürich* 39: 180–6.

Herschel, J. F. W. (1833) *A Treatise on Astronomy*, London: Longman.

Hollick, A. (1936) 'The Tertiary floras of Alaska', *US Geological Survey Professional Paper* 182.

Hooker, J. D. (1848) 'On the vegetation of the Carboniferous Period as compared with that of the present', *Geological Survey of Great Britain, Memoir* 2, part 2.

Hoyle, F. and Littleton, R. A. (1939) 'The effect of interstellar matter on climatic variation', *Proceedings of the Cambridge Philosophical Society* 35: 405–15.

Hull, E. (1885) *Mount Seir, Sinai and Western Palestine*, London: Bentley.

—— (1897) 'Another possible cause of the geological epoch', *Quarterly Journal of the Geological Society of London* 53: 107–8.

Huntington, E. and Visher, S. S. (1922) *Climatic Changes, Their Nature and Causes*, New Haven, CT: Yale University Press.

Ihering, H. von (1927) 'Das Klima der Tertiarzeit', *Zeitschrift für Geophysik* 3: 365.

Jamieson, T. F. (1860) 'On the drift and rolled gravel of the North of Scotland', *Quarterly Journal of the Geological Society of London* 16: 347–71.

—— (1862) 'On the ice-worn rocks of Scotland', *Quarterly Journal of the Geological Society of London* 18: 164–84.

—— (1863) 'On the parallel roads of Glen Roy, and their place in the history of the glacial period', *Quarterly Journal of the Geological Society of London* 19: 235–59.

—— (1865) 'On the history of the last geological changes in Scotland', *Quarterly Journal of the Geological Society of London* 21: 161–203.

Kay, G. F. (1931) 'Classification and duration of the Pleistocene period', *Bulletin of the Geological Society of America* 42: 425–66.

Kay, G. F. and Leighton, M. M. (1933) 'Eldoran epoch of the Pleistocene period', *Bulletin of the Geological Society of America* 44: 669–74.

Kerner-Marilaun, F. (1930) *Paläoklimatologie*, Berlin: Gebrüder Borntraeger.

Kessler, P. (1925) *Das eiszeitliche Klima und seine geologischen Wirkungen im nicht vereisten Gebiet.* Stuttgart.

Köppen, W. and Wegener, A. (1924) *Die Klimate der geologischen Vorzeit*, Berlin: Gebrüder Borntraeger.

Korn, H. (1938) 'Schichtung und absolute Zeit', *Neues Jahrbuch für Mineralogie, Geologie und Paläontologie* B, Bd 74A, pp. 50–186.

Kryshtofovich, A. N. (1929) 'Evolution of the Tertiary flora in Asia', *New Phytologist* 28: 303–12.

Lartet, L. (1865) 'Sur la formation du bassin de la mer morte ou lac asphaltite', *Comptes Rendus, Academie des Sciences, Paris* 60: 796–800.

Leverett, F. (1898a) 'The weathered zone (Sangamon) between the Iowan loess and the Illinoian till sheet', *Journal of Geology* 6: 171–81.

—— (1898b) 'The weathered zone (Yarmouth) between the Illinoian and Kansas till sheets', *Journal of Geology* 6: 238–43.

—— (1899) 'The Illinois glacial lobe', *US Geological Survey Monograph* 38.

—— (1929) 'The Pleistocene in northern Kentucky', *Kentucky Geological Survey* series 6, 31: 1–80.

Libby, W. F. (1951) 'Radiocarbon dates, II', *Science*, NS 144: 291–6.

Libby, W. F., Anderson, E. C. and Arnold J. R. (1949) 'Age determination by radiocarbon content: World-wide assay of natural radiocarbon', *Science* 109: 227–8.

Lyell, C. (1875) *Principles of Geology*, 2 vols, 12th edn, London: Murray (11th edn 1872).

Milankovitch, M. (1920) 'Théorie mathématique des phénomènes thermiques produits par la radiation solaire', *Académie Yougoslave des Sciences et des Arts (Zagreb)*.

—— (1930) 'Mathematische Klimalehre und astronomische Theorie der Klimaschwankungen', in W. Köppen and R. Geiger (eds) *Handbuch der Klimatologie*, vol. 1(A).

—— (1938a) 'Neue Ergebnisse der astronomischen Theorie der Klimaschwankungen', *Bulletin de l'Académie des Sciences Mathématiques et Naturelles (Belgrade)*, vol. 4(A).

—— (1938b) 'Astronomische mittel zur Erforschung der erdgeschichtlichen Klimate', *Handbuch der Geophysik* 9: 593–698.

—— (1941) *Kanon der Erdbestrahlung und seine Anwendung auf das Eiszeitproblem*, Belgrade: Royal Serbian Academy.

Morlot, A. de (1856) 'Notice sur le Quaternaire en Suisse', *Société Vaudoise des Sciences Naturelles, Bulletin des Séances (Lausanne)* 4: 41–5.

Nairn, A. E. M. (1961) 'The scope of palaeoclimatology', in A. E. M. Nairn (ed.) *Descriptive Palaeoclimatology*, New York: Interscience, pp. 1–7.

Newberry, J. S. (1861) *Report upon the Colorado River of the West* part III, Geological Report, Washington, DC.

Nilsson, E. (1931) 'Quaternary glaciations and pluvial lakes in British East Africa', *Geografiska Annaler* 13: 249–348.

Nölke, F. (1908) *Das Problem der Entwicklung unseres Planetensystems*, Berlin: Springer (2nd Edn 1919).

Oldham, R. D. (1886) 'Probable changes of latitude', *Geological Magazine* N.S., decade III, 3: 300–8.

Penck, A. (1882) *Die Vergletscherung der deutschen Alpen*, Leipzig: Barth.

Penck, A. and Brückner, E. (1909) *Die Alpen im Eiszeitalter*, 3 vols, Leipzig: Tauchnitz.

Post, L. von (1916) 'Forest tree pollen in south Swedish peat bog deposits', *Lecture to the 16th Convention of Scandinavian Naturalists, Kristiana* (trans. in (1967) *Pollen and Spores* 9: 375–401).

Prestwich, J. (1886–8) *Geology: Chemical, Physical and Stratigraphical*, 2 vols, Oxford.

Ramsay, A. C. (1852) 'On the superficial accumulations and surface markings of North Wales', *Quarterly Journal of the Geological Society of London* 8: 371–6.

—— (1855) 'On the occurrence of angular, subangular, polished, and striated fragments and boulders in the Permian breccia of Shropshire, Worcestershire, etc.', *Quarterly Journal of the Geological Society of London* 11: 185–205.

Ramsay, W. (1910) 'Orogenesis und klima', *Oversight Finska vetenskaps – societetet Förhandlinger* vol. 52A.

—— (1924) 'The probable solution of the climate problem in geology', *Geological Magazine* 61: 152–63.

Reboul, H. (1833) *Géologie de la période Quaternaire, et introduction à l'histoire ancienne*, Paris: Levrault.

Reid, E. M. and Chandler, M. E. J. (1933) *The London Clay Flora*, London: British Museum of Natural History.

Richter, K. (1932) 'Der Bewegungsrichtung des Inlandeises, rekonstruiert aus den Kritzen und Längsachsen der Gescheibe', *Zeitschrift für Geschiebeforschung und Flachlandgeologie* vol. 8.

Roemer, F. (1852) *Die Kreidebildungen von Texas und ihre organischen Einschlüsse*, Bonn.

Russell, I. C. (1885) 'Geological history of Lake Lahontan, a Quaternary lake of northwestern Nevada', *US Geological Survey Monograph* 11.

—— (1889) 'Quaternary history of Mono Valley, California', *US Geological Survey, 8th Annual Report 1886–87* part 1, pp. 261–394.

Sarasin, P. and Sarasin, F. (1902) 'Über die mutmassliche Ursache der Eiszeit', *Verhandlungen der Naturförschenden Gesellschaft in Basel* 13: 603–15.

Sauramo, M. (1923) 'Studies on the Quaternary varve sediments in southern Finland', *Bulletin de la Commission Géologique de la Finlande* no. 60.

—— (1929) 'The Quaternary geology of Finland', *Bulletin de la Commission Géologique de la Finlande* no. 86.

Sayles, R. W. (1931) 'Bermuda during the Ice Age', *Proceedings of the American Academy of Arts and Sciences* 66: 382–467.

Schuchert, C. (1921) 'Evolution of geologic climates', *American Journal of Science* 5th series, 1: 320–4.

Schwarzbach, M. (1950) *Das Klima der Vorzeit*, Stuttgart: Enke.

—— (1963) *Climates of the Past: An introduction to paleoclimatology*, London: Van Nostrand (English version of the 2nd edn of *Das Klima der Vorzeit*).

Semper, M. (1896) 'Das paläothermale Problem, spezill die klimatischen Verhaltnisse des Eozans in Europa und im Polargebiet', *Zeitschrift der Deutschen geologischen Gesellschaft*, Hanover, 48: 261–349.

Seward, A. C. (1892) *Fossil Plants as Tests of Climate*, London: Clay.

Shimek, B. (1909) 'Aftonian sands and gravels in Western Iowa', *Bulletin of the Geological Society of America* 20: 399–408.

Simpson, G. C. (1930) 'The climate during the Pleistocene Period', *Proceedings of the Royal Society of Edinburgh* 50: 262–96.

—— (1934) 'World climate during the Quaternary Period', *Quarterly Journal of the Royal Meteorological Society* 60: 425–78.

—— (1940) 'Possible causes of change of climate and their limitations', *Proceedings of the Linnean Society* 152: 190–219.

Soergel, W. (1925) 'Die Gliederung und absolute Zeitrechnung des Eiszeitalters', *Fortschritte der Geologie und Palaeontologie*, Berlin, 13: 125–251.

—— (1937) *Die Vereisungskurve*, Berlin.

Solomon, J. D. (1932) 'The glacial succession on the north Norfolk coast', *Proceedings of the Geologists' Association of London* 43: 241–71.

Steck, T. (1892) 'Die Denudation im Kandergebiet', *Jahresbericht der Geographischen Gesellschaft in Bern* 11: 181–8.

Stoddart, D. R. (1969) 'Climatic geomorphology: Review and reassessment', *Progress in Geography* 1: 159–222.

Sutherland, P. C. (1870) 'Notes on the ancient boulder-clay of Natal', *Quarterly Journal of the Geological Society of London* 26: 514–6.

Unger, F. (1847) *Chloris protogaea: Beiträge zur Flora der Vorwelt*, Leipzig.

Urey, H. C., Lowenstam, H. A., Epstein, S. and McKinney, C. R. (1951) 'Measurement of paleotemperatures and temperatures of the upper Cretaceous of England, Denmark, and the south-eastern United States', *Bulletin of the Geological Society of America* 62: 399–416.

Wayland, E. J. (1934) 'Rifts, rivers, rains, and early man in Uganda', *Journal of the Royal Anthropological Institute*, London, 64: 333–52.

Wegener, A. (1915) *Die Enstehung der Kontinente und Ozeane*, Braunschweig (*La Genèse des Continents et des Oceans*, French translation of the 3rd edn 1923 and of the 5th edn 1937; *The Origin of Continents and Oceans*, English translation of the 3rd edn 1924, London: Methuen).

Werveke, L. van (1927) 'Norddeutschland war wenigstens viermal vom Inlandeise bedeckt', *Zeitschrift der Deutschen geologischen Gesellschaft*, Berlin, 79: 135–55.

Willis, B. (1932) 'Isthmian links', *Bulletin of the Geological Society of America* 43: 917–52.

Wissmann, H. von (1938) Uber Lössbildung und Wurmeiszeit in China, *Geographische Zeitschrift* 44: 201–20.

Witham, H. T. M. (1833) *The Internal Structure of Fossil Vegetables*, Edinburgh: Adam & Charles Black.

Woerkom, A. J. J. van (1953) 'The astronomical theory of climate changes', in H. Shapley (ed.) *Climatic Change*, Cambridge, MA: Harvard University Press, pp. 147–57.

Wood, S. V. (1876) 'The climate controversy', *Geological Magazine* N.S., decade II, 3: 442–51.

Woodworth, J. B. (1912) 'Geological expedition to Brazil and Chile', *Museum of Comparative Zoology, Harvard University, Cambridge, Massachusetts, Bulletin* 56.

Wundt, W. (1944) 'Die Mitwirkung der Erdbahnelemente bei der Entstehung der Eiszeiten, *Geologische Rundschau* 34: 713–47.

Zeuner, F. E. (1945) *The Pleistocene Period*, London: Ray Society (2nd edn 1959 London: Hutchinson).

Zittel, K. von (1874) *Aus der Urzeit: Bilder aus der Schöpfungsgeschichte*, Munich: Rudolph Didenbourg.

CHAPTER THREE

Eustatic Changes

The Eustatic Theory

As the nineteenth century drew to a close it fell to the Austrian scholar Eduard Suess to take stock of the momentous accumulation of geological information and speculation which the previous decades had yielded. The result of more than 20 years of study specifically directed to this end was the three-volume *Das Antlitz der Erde* (1883–1908). Although this huge composition is remembered now as a synthesis of nineteenth-century geology and as the greatest geological work since Lyell's *Principles,* one of its most important and lasting effects was the statement of what has come to be known as the eustatic theory. Forming a groundswell to the whole discussion and being the explicit subject of volume 2 (1888) was the assumption that, apart from the orogenic belts, evidences of continental transgressions and regressions point to a remarkable synchronicity of swings of sea level in widely spaced areas of the globe. The apparently extensive and uniform character of these events prompted Suess to suggest (1906: 537–8) that, apart from certain local movements, the continental areas of the world had been absolutely stable throughout geological time, and that the main onlaps and offlaps of the ocean had resulted from worldwide shifts of sea level due mainly to changes in the capacity of the ocean basins. On examination, these geological evidences of sea-level changes seemed to indicate that the positive eustatic movements of the marine transgressions were generally of long duration and that they were interrupted from time to time by much shorter offlaps associated with negative movements. The mechanism which Suess proposed to explain these rhythmical movements, apart from references to changes in the bodily disposition of ocean waters, perhaps due to the gravitational attraction of alternating polar ice-caps (Adhémar 1842, Suess 1904–24; see 1906: 19, 553), involved the slow filling of the ocean basins by sedimentation, punctuated by increases in the capacity of the ocean basins by rapid subsidences of the ocean bottom (Suess 1904–24; see 1906: 538–44). This basic theme of geological history suffered only slight elaborations due to vertical absolute movements of the crust which took place mostly in localized belts due to lateral compression (i.e. upfolding and downfaulting, etc.). Suess' impressive scholarship, the massive mould into which he cast his beliefs, the timing of his work, its symmetry, and not least the imaginative, at times almost mystical (Suess 1904–24; see 1906: 1, 556), manner of its presentation which seemed so in keeping with the nineteenth-century tradition of German romanticism, all conspired to give added importance to *Das Antlitz der Erde* (Chorley 1963).

Much of Suess' volume 2 (1888) was devoted to an analysis of previous work relating to sea-level changes and, in particular, to the conflict which had existed between those (Neptunists) who believed in the dominance of absolute sea-level movements in earth history and those (tectonists) who placed much more emphasis on localized vertical displacements of the earth's surface. Much of this work has been summarized in our volume 1, particularly in connection with speculations regarding the existence of marine fossils at high elevations; oceanic sedimentation; the transportation and deposition of 'diluvial' (i.e. glacial) material; marine planation; the sculpting of terrain by marine action; raised beaches and marine benches; river terraces; and drowned coastal features. For example, Suess referred to the eighteenth-century glacial diluvial ideas of Emmanuel Swedenborg (1719), the sedimentary filling hypothesis of Eustachio Manfredi (1746) and the dislocation mechanism proposed by Benoît de Maillet and Anders Celsius (see volume 1, pp. 162, 197), the latter calculating that the waters of the Baltic Sea and 'Northern Ocean' were subsiding at a rate of about 40 Swedish inches per century. However, in the early nineteenth century both John Playfair and Leopold von Buch decided that Sweden was in the process of being elevated, and the former expressed a clear preference for imputing changes of level to vertical motions of the land rather than to oscillations of sea level which would be worldwide (Playfair 1802: 446–7). Nevertheless these diastrophists were greatly outnumbered by the many distinguished contemporary geologists, such as Cuvier (volume 1, pp. 75–7), Brongniart and Prévost (volume 1, p. 144), who believed in the sinking of the sea rather than in the elevation of the land. Neptunists continued to hold sway throughout the early nineteenth century but their dominance was being steadily undermined. In 1822, Karl von Hoff, a German stratigrapher, expressed his belief that the sinking of the sea was universal and uniform but in 1834 he discredited Buckland's hypothesis of a universal flood and conformed to the new theory of elevation (von Zittel 1901: 188) (volume 1, pp. 138–9). Robert Chambers (1848: 320) believed that sea-level changes were caused by 'some distant ocean bed sinking, then rising, then sinking again' (volume 1, pp. 315–18), Nathaniel Southgate Shaler (1875) was convinced that sea level rather than that of the land is subject to general movement, and Desborough Cooley (1876: 428) was also sceptical of large-scale terrestrial movements (volume 1, pp. 593–4).

By the middle of the nineteenth century the reality and extent of past sea-level movements were so much a part of popular thinking that Tennyson could write:

There rolls the deep where grew the tree.
O earth, what changes hast thou seen!
There where the long street roars, hath been
The stillness of the central sea.

(*In Memoriam*, CXXIII)

It was puzzling that Suess considered that both Darwin and Lyell were opposed to the eustatic theory, in that the former's theory of coral atoll formation involved subsidence of the floors of oceans, thereby presumably increasing their capacity, and that Lyell's support for the transport of drift and erratics by icebergs involved notions of widespread recent marine inundation (volume 1, chapter 18). Lyell also adhered to Ramsay's (1846) marine planation theory and thought that a tremendous amount of denudation, such as the cutting of the main valleys in the English Weald, would be best explained by marine dissection of rising landmasses (volume 1, chapters 12, 16 and 17). Even in the final editions of his *Principles* Lyell (1872, 1875: I: 192) maintained that the distribution of erratics was attributable to the transport of ice, either on land or by floating bergs at a period when those parts of the present continents were submerged beneath the sea. His distrust of Hutton's dominant fluvialism and his insistence on the great denudational power of the sea and of the existence of former high sea levels – as indicated by 'drift' deposits – tended to foster eustatic concepts. Cunningham (1977: 172) has pointed out that only when Lyell's idea of an elevated sea level during the Pleistocene gave way to that of a diminished sea level could it become clear how marine drift at considerable altitudes could have been dredged up from former sea floors.

It would not be correct, however, to identify in Lyell a wholehearted supporter of the embryonic eustatic theory. In his *Principles* (1872, 1875: II: 164–78) Lyell provides a copiously illustrated account, in a chapter on lands experiencing earthquakes and volcanic eruptions, of elevation and subsidence in the Bay of Baiae near Naples as shown by horizontal bands of borings by marine bivalves on three tall marble columns of the Temple of Serapis (Figure 3.1). He considered that the apparent changes in sea level were due to variations in subterranean heat and to crustal magma accumulation or drainage. In this connection he made the following comment on general sea-level changes:

> the interminable controversies to which the phenomena of the Bay of Baiae gave rise, have sprung from an extreme reluctance to admit that the land, rather than the sea, is subject alternately to rise and fall.

> (Lyell 1875: II: 178–9)

On Scandinavia Lyell included a whole chapter concerning a country where the elevation 'has naturally been regarded as very singular and scarcely creditable' (Lyell 1875: II: 180–97) because of its freedom from violent

Figure 3.1 Columns of the Temple of Serapis at the Bay of Baiae near Naples showing the former high sea level marked by the borings of marine bivalves

Source: From Lyell (1875), I, Frontispiece

earthquakes. He visited Sweden in 1834 and from his own observations and those of others suggested that the rate of upheaval was greatest in the north, very small near Stockholm and replaced in the south by subsidence. In Norway it was not clear whether any land was then rising but, in the 1840s, the finding of two superimposed non-parallel planes of upraised ancient sea-coast implied some elevation and tilt. The different heights at which horizontal raised beaches containing recent shells had been observed along the western and northern coasts show that the assumed 'movement has not been always uniform or continuous but they do not establish the fact of any sudden alterations of level' (Lyell 1875: II: 196).

The preparation and reception of volume 2 of *Das Antlitz der Erde* was greatly facilitated by the publication of the scientific results of the *Challenger* expedition which, under the scientific direction of C. Wyville Thomson, examined the topography, geology, physics, chemistry and biology of the major ocean basins of the world between 1872 and 1876 (Thomson and Murray 1880–95). Von Zittel termed this expedition 'the grandest scientific event of the nineteenth century' (1901: 217), and the publication of the many volumes of results attest to the massive nature of the investigation. Two particular aspects of the *Challenger* results were especially important in the eustatic connection: the publication in 1891 of the volume on deep-sea deposits drew attention to the importance and widespread character of land-derived deposits in the ocean basins; and the mapping of the ocean deeps marginal to the land areas prompted speculation regarding the deformation and changes in capacity of the ocean basins. By the outbreak of the Second World War more than eighty deep-sea expeditions had been mounted, but none surpassing that of the *Challenger* in geomorphic importance, not even the Dutch submarine investigations in the East Indies during the 1920s and 1930s (see chapter 1). Only after 1945 was a new technological era of oceanographic research inaugurated.

Eduard Suess

Throughout the later nineteenth century Eduard Suess (plate 3.1) acted as a magnet for ideas connected with changes of sea level or eustasy. He was the prime advocate and tireless advertiser of this theme, for which he played a role similar to that played by W. M. Davis for the erosional cycle. He was born on 20 August 1831 in London where his father had a wool business. Three years later the family moved to Prague where eventually he went to university and soon became a skilful linguist, speaking English, French and German, to which he added, later in life, Russian. This faculty enabled him to take charge of the foreign correspondence of the Society of the Friends of Science and of part of that of the Museum of Natural History, which not only brought him stimulating international contacts, but also gave him the key to

Plate 3.1 Eduard Suess (1831–1914)

Source: From *Journal of Geology*, 1914, p. 813

the rich international geological literature of the nineteenth century. Through the museums at Prague and geological excursions he devoted himself increasingly to geology and especially to the eastern Alps. In 1861 he became a full professor of geology at the University of Vienna and not long afterwards was elected a member of the national parliament (Hobbs 1914). When the bastions and other fortifications of medieval Vienna were dismantled he wrote a book on the soils exposed by the excavations (Suess 1862). Within the next two decades he earned fame and the great gratitude of the Viennese by bringing them a new freshwater supply from springs in the mountains along a 70 mile gravity aqueduct and by freeing the city of major flŏods by organizing the digging of a Danube canal. His scientific investigations went hand in hand with his academic, cultural and international interests. Although lacking formal training in geology, he acquired considerable skill from contacts with his friends and students in the field and from his omnivorous reading. In 1875 he produced his *Die Entstehung der Alpen* which was in fact a general review of the origin and structure of mountain chains and contained the germs of many ideas which he revised or elaborated later.

In 1883 Suess signed a publication contract to write *Das Antlitz der Erde* which was to appear in three volumes. The first part duly appeared in 1883 but the entire work was not completed until 1909 and in the intervening 26 years many of its syntheses and hypotheses were revised. It was particularly fortunate for the 'Face of the Earth', as it was entitled in the English translation by H. B. C. and W. J. Sollas (Suess 1904–24), that the most influential translation into French (1897–1918) was undertaken by Emmanuel de Margerie. Through the personal influence of de Margerie the three great threads were first woven together which later became the fabric of twentieth-century French regional geomorphology, particularly that of Baulig (see chapter 8). These threads were, first, an appreciation of the work of the late-nineteenth-century western geologists and of W. M. Davis; second, an acceptance of the regional concept of Paul Vidal de la Blache; and, third, an understanding of the eustatic theory. De Margerie, himself a former collaborator of Albert Heim (1888) and a student of Marcel Bertrand, edited the French translation of Suess' work with the aid of seventeen assistants, and added thousands of new footnotes to illuminate the text and more than 500 figures. After about 1900 *La Face de la Terre* became the definitive edition of *Das Antlitz der Erde*. Volume 2, entitled *The Oceans* in the English translation (1906), displays an immense knowledge of the foregoing geological literature and, among other terms, introduced that of *eustatisch* (Suess 1906: 690), translated as *eustasy*, signifying a general and uniform change of world sea level. As part of a clever piece of advocacy or special pleading, Suess was utterly ruthless in incorporating any evidence, however trivial, which might appear to favour changes of sea level and in incidentalizing or belittling what is assumed to be the ineffectual impact of the 'theory of elevation'. Suess' great work must be read several times, at long intervals, before its bias can be fully appreciated.

Suess was always a confirmed and militant eustatist. During the 1860s when studying the ancient shorelines of the Tertiary beds on the edge of the Bohemian massif north of Vienna, he decided that their regular height was attributable to the lowering of sea level rather than to the uplift of the mountain block. In 1869, when he was a member of the national delegation at the opening of the Suez Canal, he believed that the Suez isthmus was a result of a recent fall of sea level rather than uplift of the land. In 1885, after visiting the interior of Norway near Tromso, he thought that the stepped terraces upon the sides of the fiords and other valleys were horizontal and indicated a fall of sea level. He came to the conclusion that the ancient massifs were stable and that sea level had varied, so creating eustatic levels which could be found on the peripheries of all oceans and seas except those, such as the Baltic, which had an interior location. In typical phraseology, he sums up the idea in his farewell address delivered at the Geology Department in Vienna University on 13 July 1901, more than 40 years after his first professional lecture there:

> in Bohemia particularly there is an extraordinary interruption of marine deposits extending upward into the Middle Cretaceous, whereas in the Alps all these great epochs are represented by marine strata. This same transgression of the Middle and Upper Cretaceous shows again in Galicia, then far into Russia, on the other side of the French Central Plateau, on the Spanish Meseta, in large parts of the Sahara, in the valley of the Mississippi, and northward over this region to the vicinity of the Arctic Sea, in Brazil, finally on the shores of central and southern Africa, in east India; and, in fact, over such extraordinarily vast regions that it became impossible longer to explain such transgressions of the sea, according to the older views of Lyell, by means of the elevation and depression of continents.
>
> Through this and similar observations the newer idea has recently come into prominence that such general change must have occurred either in the shape of the hydrosphere or in its entire volume. It was seen that by the forming of a new oceanic depth, due to sinking, a certain amount of hydrosphere was drawn off into the new depression, and that at the same time there appeared to be a general land elevation, or, more correctly, there must have resulted a general sinking of the beach lines. The older view of the numerous oscillations of the continents has also given way more and more to the teachings of marine transgressions, and through the denudation of continents, a more exact examination into the actual mountain movements has become possible.

(Suess 1904: 267–8)

Before his death in Vienna in 1914 he had attracted a large multitude of followers and students, of whom Albrecht Penck was the most distinguished.

In volume 2 of *Das Antlitz der Erde* Suess set out to establish his eustatic theory in preference to the 'theory of elevation'. The former, he believed, involved a new school of thought which enquired seriously into the question of the invariability of the conditions which determine the equilibrium of the oceans:

> Very different causes have been suggested from time to time as capable of producing a universal alteration in the form of the oceanic surface: in all of these the force of gravity plays a leading part.
>
> (Suess 1906: 17)

Suess (1906: 24) refers approvingly to Chambers' (1848) use of the phrase 'changes of relative level' instead of elevation and subsidence, and states that he himself will talk of *displacements of the strand-line* and of *negative or positive displacement of the strand-line*. He goes on to recommend three methods of studying changes in sea level:

> The first consists of tracing the *extension of the ancient seas* . . . . The second is found in a comparative study of the *nature of the sedimentary formations* . . . . The third is the *examination of the existing coast*.
>
> (Suess 1906: 24)

He continues with a brief comparison of the structure of the shores of the Atlantic and Pacific Oceans and then launches into a detailed attack upon ideas held by diastrophists concerning changes of relative land elevation. These test subjects include Lyell's Temple of Serapis at Puzzuoli, Scandinavia, the western Mediterranean, and coral reefs. In Suess' view these are all examples of strictly local application and that the negative eustatic phases were sudden and spasmodic, affording no proof of the tectonic 'theory of elevation'. Suess considered the raised strandlines of the Norwegian fiords of the west coast to be horizontal but did not make any levelling measurements. They were caused, he thought, by retreating ice and not by oscillations of sea level and 'still less of the solid land' (1906: 361). In a long chapter (1906: 393–430) on the Baltic and North Sea during the historic period, he invoked any physical factors which would belittle the 'theory of elevation'. The Baltic was viewed as an incompletely closed arm of the sea in which density, mainly salinity, alters its surface height, such that its level undergoes short or longer periods of oscillation which depend upon climatology and hydrostatics and not on tectonic mechanisms:

> Thus as regards *the general secular elevation of the Scandinavia peninsula, the source and origin of the theory of elevation, definite evidence is entirely wanting.*
>
> (Suess 1906: 415)

After this stand, Suess continues in successive chapters to discuss the western Mediterranean, where he does not detect any proofs of secular continental elevation or subsidence within the historic period; and the northern seas, where it is thought that the strand-line stood higher towards the close of the glacial period than at present. The evidence seemed to suggest to Suess a recent accumulation of water towards the equator and a diminution towards the poles.

In a final survey the causes of sea-level changes are dealt with in detail, and Suess summarizes his eustatic theory with striking clarity:

> as soon as we recognize the Ocean basins as sunken areas, the continents assume the character of horsts, *and the wedge-like outlines of Africa, India, and Greenland, all pointing towards the south, find their explanation in the conjunction of fields of subsidence which reach their greatest development in the same direction* .... *The crust of the earth gives way and falls in; the sea follows it.*
>
> (Suess 1906: 537)

> *The formation of the sea basins produces spasmodic eustatic negative movements.*
>
> (Suess 1906: 538)

> *the theory of secular oscillations of the continents is not competent to explain the repeated inundation and emergence of the land.* The changes are much too extensive and too uniform to have been caused by movements of the earth's crust.
>
> (Suess 1906: 540)

> The formation of sediments causes a continuous, eustatic positive movement of the strand-line.
>
> (Suess 1906: 543)

> Movements like these, which present themselves as oscillations, and extend around all coasts and under every latitude in complete independence of the structure of the continents, cannot possibly be explained by elevation or subsidence of the land. Even as the transgressions of the ancient periods are much too extensive and uniform to have been produced by movements of the lithosphere, so too are the displacements of the strand-line in the immediate past.
>
> (Suess 1906: 500)

> The persistent continuance of a continental surface is in the main the result of local subsidences of the earth's crust, which time after time open up fresh abysses for occupation by the sea, and lower the general level of the strand.
>
> Thus there is an alternate play of forces. The effect of eustatic subsidences and the deposition of sediments is cumulative, and in the course of geologic periods the eustatic negative movements obtain the predominance. In this matter the foldings of the mountain chains plays only a secondary part.
>
> (Suess 1906: 553)

The following remark which occurs more than 500 pages earlier in the text, confirms the tenets of this conclusion:

> great and general negative movements are from time to time produced by the formation of fresh oceanic abysses, or by the addition of new areas of subsidence to abysses already in existence, and it is important to bear in mind that movements of this kind surpass all others in importance.
>
> (Suess 1906: 27)

Yet we get a hint that Suess was not entirely happy with his findings when near the end of his summary he writes:

> The essential problem of this volume has been to determine in what manner and to what extent movements of the Ocean are dependent on those of the earth. In all probability the Ocean is subject to an independent movement which in the course of long periods causes an alternation of positive and negative phases at the equator.
> (Suess 1906: 553)

However in the preface to de Margerie's French translation of Suess' work, Marcel Bertrand, the famous geologist, wrote:

> The creation of a science, like that of a world, demands more than a single day; but when our successors write the history of our science, I am convinced that they will say that the work of Suess marks the end of the first day, *when there was light.*

Von Zittel (1901: 320) was more cautious:

> Many of the hypotheses suggested by Suess will probably not endure criticism of the future. Yet there can be no doubt that even the expression of an hypothesis having due respect to all known data marks an important step . . . rouses each one from the particular to the general, and brings him back with renewed vigour and mental insight to the particular.

Eustatism and Twentieth-century Diastrophic Thought

The publication in 1894 of Albrecht Penck's *Morphologie der Erdoberfläche* gave considerable assistance to the eustatic theory. Although assuming more liberal crustal movements than did Suess, Penck held that the continental blocks are generally stable, that oceanic sedimentation could account for a sea-level rise of as much as 1 metre in 32,000 years, that glacial eustatism was a reality, and that the marginal parts of ocean basins are liable to deformation on a scale which changed their capacity. The existence of the deep ocean trenches marginal to the continents, identified by the *Challenger* expedition, provided a direct explanation for the dramatic and episodic drops of sea level which Suess had postulated. In 1900 Emile Haug suggested (p. 628, see also 1907: 165) that these 'foredeeps' were downbuckles of geosynclinal character occurring at an early stage of orogenic deformation in the particularly mobile belts immediately marginal to the stable continental shields. It thus became natural to associate relatively sudden increases in the capacity of the ocean basins, together with negative eustatic movements, with the downbuckling of these foredeeps during the orogenic periods of earth history.

By the beginning of the twentieth century the eustatic theory had become a basic article of geological faith and, as such, an important constraint on geomorphological thought. This faith found expression in the important paper by T. C. Chamberlin entitled 'Diastrophism as the ultimate basis of correlation' (1909) in which eustatism was associated with other aspects of diastrophism:

> in general, the relatively upward movements of diastrophism have been located continuously in the continents, and the broad downward movements continuously in the ocean basins, and that, setting aside incidental features, the dominant effect of the successively diastrophic movements has been to restore the capacity of the ocean basins and to rejuvenate the continents. This conclusion seems to me to be strongly supported by the general course of geologic history, wherein sea-transgressions and sea-withdrawals have constituted master features.
>
> (Chamberlin 1909: 688)

Chamberlin's paper marked the coming of age of the eustatic theory, however, and thereafter any geomorphologist who believed that the effects of changes in baselevel are reflected in the associated landforms had to be prepared to take account of the probability of considerable fluctuations of sea level in the geological past (Chorley 1963).

The overriding importance of the eustatic theory also dominated work in stratigraphy for at least the first three decades of the twentieth century, and the construction of palaeogeographic maps increased the general impression of the synchronicity of transgressions and regressions. In Europe, de Lapparent (1906) and Haug (1908–11) were mainly responsible for emphasizing the large and apparently synchronous swings of sea level which, as Suess had suggested, seemed to dominate geological history. In the United States, Charles Schuchert (1910, 1923) followed the eustatic ideas of Suess in a virtually unmodified form (1910: 479), and the palaeogeographic maps of Willis (1909) did much to foster the eustatic notion. The most spectacular, if not the most influential, extension of Suess' theory was the 'pulsation theory' proposed by Grabau (1913, 1920, 1936), who believed that the stratigraphic record prompted the recognition not only of synchronous worldwide eustatic changes but also that these changes followed a rhythmic pulsation (1936: 540):

> I believe, however, that the conviction is growing among stratigraphers that the great oceanic transgressions and retreats, which involve not only one section but all continents, are due primarily to rise and fall of the sea level, the continental [epeirogenic] movements being secondary or contributory factors, which guide the transgressions and locally accelerate or retard both transgressions and regressions. ... If it can be shown that such transgressions and regressions are essentially simultaneous in all parts of the earth, then they can be explained only by a periodic rise and fall of the sea level. It is this rhythmic pulsation, which can be shown to have taken place, that I would express by the terms 'pulsation theory'. ...

The multiplicity of continental onlaps and offlaps, evidence for which exists in the geological record, became especially apparent during the rapid development of oil drilling in the Gulf Coastal Plain during the 1920s. With the establishment of the first subsurface laboratory in 1919 and the increasing use of micropalaeontology after 1920 (Landes 1951), stratigraphic correlations were possible over large areas and these too appeared to show the historical dominance of transgressions and regressions (Levorsen 1931). Following the recognition of suites of cyclical sedimentation which Willis (1893) related to cycles of erosion, the work on cyclothems further emphasized swings of sea level (Chorley 1963).

In the 1930s many influential geologists subscribed to the view that '. . . the major movements of the strand line . . . have affected all continents in the same sense at the same time' (Bucher 1933: 428) and that these changes had been caused primarily by changes in the capacity of the ocean basins (Moore 1936: 1802–4). This eustatic view provided a tempting key to a chronological correlation of landforms on a continental, or even an intercontinental scale, because of the general belief in the important control exercised by grand baselevel. In this regard, however, American opposition to the corollaries of the eustatic theory was particularly interesting. In 1922 F. Leverett, W. C. Alden, D. W. Johnson and W. M. Davis joined to attack the simple eustatic correlations which had been made by Europeans. This opposition was summed up by Johnson (1932: 298) a decade later:

> At present all that can be said with assurance is that the studies thus far made by different observers do not appear to us conclusive as to the validity of any theory respecting correlation and attitude of ancient marine levels in America.

After the mid 1930s the diastrophic tide began to turn against the eustatic theory. Localized continental movements of an epeirogenic character were increasingly invoked to explain marine transgressions and regressions; doubts began to be expressed as to the ability of geologists to date accurately these extensive and long-continued transgressions, and this was to lead to increasing scepticism as to the correlation of transgressions on a worldwide scale which the eustatic theory demanded; and notions of crustal instability multiplied. In the early part of the century there was much support for the concept of 'periodic diastrophism', developed by Suess and Chamberlin (1909: 689), which postulated that the main orogenies were synchronous and worldwide, as evidenced by the major unconformities which allowed the geological periods to be defined. From this concept Davis adopted the idea of short periods of uplift being separated by long periods of quiescence. However, criticism of worldwide orogenic periodicity by Shepard (1923) and support for the concept of more continuous crustal deformation by Gilluly (1949) were symptomatic of a growing disbelief in the reality of long periods of absolute quiescence and in the fundamental stability of the earth's crust. Recent uplift of the Alps had been inferred by Albrecht Penck (and Brückner 1909: 68) on the basis of the gradients of valleys, and that, associated with the isostatic recovery of the Baltic region, had long been apparent. The most spectacular evidence for recent and contemporaneous land movements, in many instances involving areas long considered as the acme of stable continental platforms, came, however, from the differences observed between successive precise geodetic levellings. Such levelling was begun in Finland before the end of the nineteenth century, but even in less obviously unstable areas quite large vertical changes have been inferred. Schmidt (1922) inferred subsidences of up to 90 cm in France during the period 1860–90 (Gilluly 1949: 566, since questioned by the latter); Stille (1936: 855–6) showed a recent uplift of the Rhine massif of about 1 mm per year, a subsidence of the sub-Alpine area of Bavaria of 1–6 cm in 20 years, and suggested that Munich had moved about 12 cm nearer the Alps in the previous 50 years; Gilluly (1949) referred to an uplift amounting to 8 inches during the period 1906–44 for an arch in southern California; and Meshchéryakov and Sinyaguina (1956: 77) proposed that contemporary tectonic movements are responsible for current rates of uplift of parts of the western Russian platform of up to 10 mm per year. It is interesting that, in general, Russian workers had long been impressed by the importance of recent and contemporary tectonic movements of the crust (Karpinskiy 1894, Lichkov 1934, Nikolayev 1949, Dumitrasko *et al.* 1951: 76) (Chorley 1963). When Gutenberg concluded as early as 1941 (p. 730) that contemporary tide gauge readings did not show regular or simultaneous sea-level changes at all stations, few earth scientists were really surprised.

Pliocene and Pleistocene Eustatism

Nineteenth-century attitudes to glacial eustatism were distinctly ambivalent. Although Charles Maclaren (1842) suggested that sea level was lower than at present during ice advances, Lyell's notion of iceberg transport of till and erratics required a higher sea level. The submergence theory was dominant in Britain during the 1860s with Jamieson (1862) and Archibald Geikie (1865) believing in a late-glacial submergence to at least the present 2,000 foot contour level (Davies 1969: 299–301), although both supported the land-ice origin of most glacial deposits. Glacial submergence was believed to have been greatly assisted by isostatic depression of the crust under the weight of ice (Jamieson 1865, Ricketts 1871–2, Jamieson 1882), but the belief in high glacial sea levels was supported by Ramsay (1881) to a height of more than 2,000 feet in North Wales, by Prestwich (1893) to a height of more than 1,000 feet in the late glacial period, by Hull (1910) and, as late as 1920, by Gregory who thought the Irish eskers had been formed in submarine conditions. Jamieson (1865) suggested that the Scottish raised beaches were due to post-glacial rebound of the crust, indicating a glacial submergence of at least 100 feet with reference to present sea level, while marine clays at over 500 feet indicated warmer interglacial inundations. The picture was confused by many land tills containing shelly marine deposits up to quite high elevations, as in North Wales (Ramsay 1881), suggesting considerable sea depths during deposition. However, this began to be resolved in 1870 by James Croll's suggestion that shelly tills in the north of Scotland had been dredged up from the North Sea basin by westward-flowing Scandinavian ice (see also Peach and Horne 1879, 1880). In the interval between the first edition of the *Great Ice Age* in 1874 and the second edition in 1877 James Geikie discarded the sea-ice theory of glacial deposition and after the 1880s glacial submergence and iceberg deposition were not important elements in British geomorphological thought (see Davies 1969).

In 1914 W. B. Wright dedicated his *The Quaternary Ice Age* to 'the memory of T. F. Jamieson of Ellen, originator of the isostatic theory of the Quaternary oscillation of sea-level'. Wright (1914: 406–26) includes a separate chapter on this theme which, he states, was eagerly taken up and applied with remarkable success by geologists in Scandinavia and America to the problems of their own countries. He adds that if it is assumed that the gravitational pull of an ice-sheet upon sea level along its margins is negligible and that there is a considerable timelag in total isostatic recovery, the systems of warped shorelines in Scandinavia, Scotland and Canada, which rise higher and higher on the land as the centres of ice dispersion are approached, favour strongly the isostatic theory. Similarly Joseph LeConte (1891, 1897) thought that isostasy had played a role in the oscillations of the Canadian shield during the Glacial Epoch.

Naturally, the magnitude and speed of glacial eustatic effects attracted more attention than those of the relatively slow isostatic uplift, particularly as the former were strongly advertised by Albrecht Penck (1894) and by Penck and Brückner (1909). In a special account of changes of sea level Penck calculated the approximate water content of Quaternary ice-sheets in order to gauge its possible effect on general sea level. He estimated the total area of these ice-sheets in the northern hemisphere was equivalent to 7.5 per cent of the total area of the ocean. The average thickness of the ice was estimated – very moderately – at 1,000 m (3,280 feet) and mean depth of the sea at 3,500 m. If the specific gravity of the ice is 9/10 that of the sea, the Quaternary ice-sheets in the northern hemisphere were equivalent to $7.5 \times 1000/3500 \times 9/10$, or 1.9 per cent of the ocean mass. The removal of this amount would presumably lower the ocean surface by $1.9/100 \times 3500/1$ or about 66.5 m or, roughly, 210 feet. Wright discussed these suggestions and pointed out that the estimates will vary widely with the presumed thickness of the ice and that no account has been taken of ice-masses in the southern hemisphere, which posed a special problem as to whether the glaciation of the two poles was contemporaneous. However, assuming that the present area of the Antarctic ice-sheet is about 4.5 million square miles and that its average thickness is about 1,000 m:

Calculating as before, we get a volume of ice equivalent to a lowering of the sea-level
of about 40 metres or 130 feet ... if the great Quaternary ice-sheets of the world
were contemporaneous, a lowering of sea-level of about 400 feet is to be expected.

(Wright 1914: 415–16)

The problem continued to worry geologists and oceanographers, and estimates of the possible maximum drop of sea level during the Pleistocene
proliferated. They included: 83 m (Antevs 1929); 75–85 m (Daly 1934: 47);
103 m (Penck 1934); 90–110 m (Kuenen 1950: 536); 85–120 m (Flint 1957:
260). These were, of course, little better than educated guesses.

Geomorphic evidence for sea-level changes included raised beaches,
marine benches and notches; submerged forests; drowned valleys and other
subaerial features; valley fills and river terraces. It is natural that during the
eustatic heyday measurement and correlation of these coastal and near-coastal
features attracted enormous attention, particularly in regard to Pleistocene
eustatism not only outside the areas subjected to glacial isostasy but
also within them, encouraged by Suess' belief in the horizontality of the
Norwegian raised beaches. Nowhere was interest in sea-level changes more
intense than in France, where the French edition of Suess' volume 2 in 1900
became the custodian of the eustatic flame.

The association of river terraces and adjacent coastal benches was strongly
developed in French geological literature after 1862 when Scipion Gras
published his *Description géologique du département du Vaucluse*, in which he
included (pp. 262–71) a theory of Quaternary phenomena. He supposed that
at the close of the Tertiary, a rise of sea level had caused the existing valleys to
be filled up. Then, when the sea began to retreat it deposited sheets of
alluvium upon the plateaus. When the lowering of sea level was quickened,
the valleys began to be deepened, often on the site of the old unfilled valleys.
The general retreat of sea level included some periods of standstill during
which the rivers achieved a state of equilibrium which brought about the
formation of terraces arranged one beneath another (Bourdier 1959: 22).

This theme was developed and publicized by Léon de Lamothe who was
born in Metz in 1849 and died, a general of artillery, at Grenoble in 1936
(Gignoux 1937). After leaving the Ecole Polytechnique, among other activities he began to investigate the alluvial terraces of the Upper Moselle basin in
the Vosges. As became an artillery expert, his topographic measurements
were meticulous and naturally he was in no hurry to publish his findings
which did not appear until 1897 in a scientific 'note' of sixty pages. In this he
imagined a huge lake in the Upper Moselle basin which formed alluvial beds
on its periphery which he tried to delineate. Soon after this publication,
Lamothe was sent to the garrison in Algeria where his attention turned to the
ancient terraces and 'raised' beaches in the basins of the Isser and some other
rivers draining to the Algerian coast of the Mediterranean Sea (1899). Having
carefully measured the altitude of these features he considered that they

occur at similar heights over so long a stretch of coastline that they were best explained by Suess' idea of eustasy and were in keeping with the new fluvial concepts of baselevel of erosion and the profile of equilibrium as was then being taught by the leading French geologists such as Margerie, Noë and Lapparent.

Lamothe's long article on terraces in the Isser basin was well received by geologists, most of whom liked his attempts at mathematical exactitude. He distinguished some higher levels up to 325 m (presumably likely to be pre-Pleistocene) and several predominant lower levels, the chief of which (in metres above the River Isser and the Mediterranean, respectively) were as follows:

River terraces	Coastal platforms
93–95	98–100
55–57	55
28–30	30
15–16	15

In 1901 (a, b) Lamothe presented to leading French scientists a new idea on the role of eustatic oscillations of erosional baselevel upon the creation of systems of terraces in some river valleys. The old beach levels or shorelines at the mouth of the Isser were, he considered, continued upstream by alluvial terraces which maintained the same relative altitude as the coastal beaches. But the altitude of these river-valley terraces was related not to the height of the old raised coastal beaches but to the height of the alluvial floor of the present river. By means of true-to-scale diagrams Lamothe demonstrated the mathematical necessity for the agreement of the altitudinal range of the shorelines and the river terraces but he felt that the altitude of the coastal features could only be measured in relation to mean sea level, a base which could not be adequately fixed. However, if the river-valley floor was substituted for mean sea level the relative elevation of river terraces could be measured to the nearest metre. Such accuracy would verify the correctness of Suess' eustatic theory and the method would be especially desirable on coasts like the French Atlantic where raised marine beaches are fragmentary or almost absent but fluvial terraces are strongly developed. Unfortunately Lamothe went on to suggest that variations of sea level were the only factors involved in the formation of river terraces. Up until 1930 Lamothe continued to publish his eustatic researches on the lower Somme (1916, 1918), the Vosges, the Algerian coast (1904, 1911, 1912) and the Rhône valley downstream of Lyons (1915). During this time his findings were complemented by those of Maurice Gignoux and subsumed into the writings of Charles Depéret. In 1913 Gignoux provided details of the late Pliocene and Quaternary deposits of southern Italy and Sicily. He found that levelling (flattening) occurs commonly at about 80–100, 60, 35 and 15 m and, among other conclusions, suggested that a new maximum level would tend to mask former lower levels and therefore existing benches would, of necessity, suggest regularly decreasing baselevels.

Depéret, who was Gignoux's tutor, was born at Perpignan, in Rousillon, in 1858 and rose to prominence, being for several decades a professor at Lyons and regarded throughout France and the Mediterranean countries as one of the great masters of science (Gignoux 1930). At the opening of the twentieth century he took command of a civilian group which aimed at measuring the height of terraces and shorelines. In 1903 (see also 1906), in collaboration with M. Caziot, he published a study of Pliocene and Quaternary shorelines near Nice. This stressed the relative altitudes of the benches and was obviously inspired by the ideas expressed by Lamothe who, strange to say, it did not mention (Depéret 1903). In the following year Depéret (1904) does refer to his learned friend General de Lamothe, but it was not until 1918 that he admitted publicly that the General's theory had preceded his own. Indeed, many geologists continued to attribute priority to Depéret until Bourdier's (1959) percipient discussion on 'the origins and success of an illusory geological theory: eustatism applied to alluvial terraces' put the priority right. From 1918 onward Depéret (1918–22, 1921, 1922, 1923) attempted to provide a general coordinated chronology or classification of Quaternary time based on marine bench levels in the western Mediterranean. Similarly Lamothe (1916, 1918) tried to correlate these old shorelines with four river terraces near the lower Somme in northern France. By 1923 Depéret's idea of Quaternary levels in Europe consisted of four alluvial terraces which corresponded with the four marine stages postulated by Albrecht Penck and four glaciations. His final scheme, which became widely accepted (Chaput 1928), is given in a slightly modified form below:

above 100 m	pre-Sicilian
80–100 m	Sicilian (cold fauna)
55–60 m	Milazzian (slightly warmer than today)
35 m	Tyrrhenian (warm fauna)
15 m	Monastirian (warm fauna)
3–4 m	(called Flandrian transgression by Dubois (1924))

These levels and names were meant to indicate not shorelines but stages corresponding to cycles of sedimentation.

French savants for the most part seemed unwilling to dare to express openly criticisms of the scheme but scholars abroad were less inhibited. In a symposium (Leverett et al. 1922) the American authors F. Leverett, W. C. Alden, D. W. Johnson and W. M. Davis attacked the simplicity of eustatic correlations made by the Europeans, supported by Emmanuel de Martonne.

Davis (1920O: 483–5) stated his disbelief in the general application of the eustatic theory by Lamothe (see also Davis 1905K). Alden considered it impossible to substitute eustatic for diastrophic movement, as Depéret had done. Similarly Johnson thought:

> Observations in America and Europe do not indicate the existence of a fairly definite series of marine terraces and river terraces, at more or less uniform elevations above the present sea level and present river channels ... he accordingly believes, with the majority of geologists, that differential movements of the land, together with the normal degrading of alluvium-filled valleys, better account for the observed facts.
>
> (in Leverett *et al.* 1922: 481)

Indeed, Davis was over 80 years old when during the 'watches of the night' on 27 April 1931 he slept badly and began to accept the view that many marine platforms might be associated with climatic oscillations of sea level during the Glacial period (see volume 2, p. 686).

In France Maurice Gignoux in his *Géologie Stratigraphique* (1926) expressed discrete reservations on the whole validity of the Lamothe–Depéret concepts. After Depéret's death in 1929 the critics surfaced in strength and many no longer hid their scepticism; however, the theory of eustatically controlled river terraces persisted for a long time in educational texts largely because of its simplicity and the illusion of mathematical certainty given by precise altimetric measurements. As Gignoux said drolly, you only had to fix an altimeter on the handlebar of your bicycle to become a specialist in one of the most difficult branches of geology! Lamothe had shown the power of statistics but before long most scientists had shown that his theory was largely illusory. The type localities were far removed from each other and it was hard to correlate their respective levels, and, more damningly, eventually at least two of them (Milazzian and Monastirian) were thought to be almost certainly tectonically disturbed. In place of eustatic control, the idea of climatically influenced terraces resumed popularity and was successfully incorporated by Bourdier (1938, 1959) into his concept of 'climato-sédimentaires' cycles.

In the British Isles fundamental conditions differed markedly from those in the Mediterranean. The greater part of these islands had been covered by ice on several occasions during the Quaternary and had experienced complicated glacial isostatic depression and recovery as well as the effects of glacially induced eustatic changes of sea level. Wright (1914: 363–86), who discusses at length the raised beaches and submerged forests, stresses the greater ignorance of what happened here to sea level during the Ice Age. However, there is abundant evidence for a late glacial submergence of Scotland which attained at most about 100 feet and affected a relatively limited area. It created a '100-foot' raised beach and some locally developed clays, with an Arctic fauna which had been studied carefully by T. F. Jamieson (1865). This shoreline had been traced widely, especially in the Western Isles where immense shingle beaches, and in places a pronounced bedrock notch, have been created. Locally a lower beach at between 65 and 45 feet occurs and may also be late glacial. More widely on the foreshore between tide-marks submerged forests have been found:

> These prove in the clearest possible manner the occurrence of a post-glacial period of emergence, in which forests flourished throughout the country.
>
> (Wright 1914: 377)

In the northern parts of the British Isles there is, at heights of 0 feet to 35 feet, a so-called '25-foot' raised bench which is apparently the result of a quite distinct and much later submergence than the 100 foot bench:

> It is by far the most clearly marked post-glacial shoreline in Scotland. Its rock-notches, terraces, and storm beaches are everywhere well preserved and easily identifiable throughout ... an oval area embracing the greater part of Scotland, north-eastern Ireland and north of England.
>
> (Wright 1914: 382–3)

During the late 1920s and 1930s a great effort was made to correlate marine benches and fluvial terraces on a global scale. Among the numerous contributions were those of F. Leverett (1928) on Pennsylvania and New Jersey, Wright (1928) on the British Isles and of E. Chaput (1928) on Atlantic France. The last-named appeared in the highly praised *First Report* of a special Commission of the International Geographical Union which had been set up in 1925 expressly to decipher the possible worldwide correlation of Pliocene and Pleistocene marine benches and fluvial terraces (Beckinsale 1972: 122–7). By 1952, when it was replaced by a more restricted goal relating to erosional levels around the Atlantic, this Commission had published seven reports totalling over 920 pages of text. In the interval much progress had been made with details of, for example, the terraces of the eastern seaboard of the United States south of the ice-invaded northern areas (Dubois 1925, Leverett 1928, Antevs 1929, Cooke 1930). However, this literary explosion did not hide doubts which began to appear on the wisdom or possibility of attempting universal terrace correlation solely on the basis of altitude above present sea level, nor did it dispel the reality of differential land movements in recent geological history (Barrell 1915). The frustration was expressed clearly by Henri Baulig in his introduction to the *Sixth Report* of the International Commission:

> It does not seem that these meritorious efforts, any more than the innumerable notes and memoirs published throughout the world, have caused notable progress to be made in this matter ... there is rather the impression of a persistent disagreement, not to say a growing confusion. This is unquestionably due to the intrinsic difficulty of the problem, which is greater than it appears at first sight, and to the diversity of the regions studied. But one cannot help noticing that, when

confronted with the same problems, research workers belonging to different countries or to different 'schools' show a marked, sometimes exclusive preference for one or another method of collecting and interpreting data. It seems that each group is shutting itself up within a circle of conceptions barely penetrable by outside influence.

(Baulig 1948: 3) (translated)

At the same time the reality of recent lower sea levels was maintained by studies of submerged coastal forests (Reid 1913, Wright 1914, 1937), by the increasing recognition of submarine features such as a bench at −200 m. in the Mediterranean and elsewhere (Boule *et al.* 1906–11) and the probability of the drowning of a complete river system in the present Sunda Sea (Molengraaff 1930a, b). Submarine canyons were imputed by some geologists to fluvial erosion during a large Pleistocene fall of sea level (Shepard 1933, 1934), while other scholars like Joly (1925) proposed falls of many hundreds of metres.

During the 1930s, however, the straightforward application of glacial eustatism began to be critically examined, even by the most ardent of eustatists. Baulig (1927, 1935), in opposition to the Davisian approach, suggested that a lowering of sea level might be accompanied not only by valley cutting downstream but also by aggradation upstream (see also Sandford (1929) on the Nile Valley) and concluded:

Absolute shifts of sea-level appear inexplicable except by deformation of the oceanic basins, which in turn implies changes in the absolute position of some at least of the lands. So that epeirogenic and isostatic movements on the one hand and eustatic movements on the other, far from excluding each other, appear correlative.

(Baulig 1935: 31–2)

R. A. Daly, a confirmed glacial eustatist (1925), in his influential work *The Changing World of the Ice Age* stated:

there can be no simple criterion of the eustatic emergence, based upon the naïve idea that this process gives strandmarks at constant height all over the world.

(Daly 1934: 180)

Steers, who in 1929 greatly publicised French eustatism, wrote:

it is difficult to separate clearly the effects of local and general movements. Eustatic movements are supported by the well-known occurrence of transgressions in geological history, but after all they are qualitative rather than quantitative. There certainly does seem to be a rhythm of the beaches from the Sicilian onwards, and a distinct correlation between them and river terraces. In such places the eustatic view seems best to explain the facts. However it is probably unsafe to extend the reasoning too far.

(Steers 1945: 255)

Figure 3.2 Darwin's cross-sections of Pacific coral islands showing the formation of a barrier reef (No. 4) and of an atoll (No. 5)

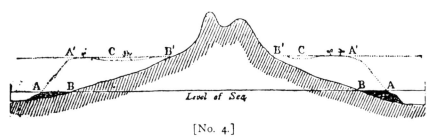

[No. 4.]

AA—Outer edge of the reef at the level of the sea.
BB—Shores of the island.
A'A'—Outer edge of the reef, after its upward growth during a period of subsidence.
CC—The lagoon-channel between the reef and the shores of the now encircled land.
B'B'—The shores of the encircled island.
N.B.—In this, and the following woodcut, the subsidence of the land could only be represented by an apparent rise in the level of the sea.

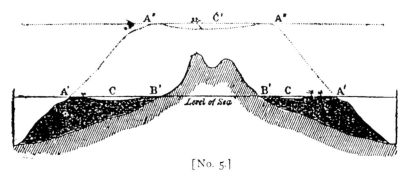

[No. 5.]

A'A'—Outer edges of the barrier-reef at the level of the sea. The cocoa-nut trees represent coral-islets formed on the reef.
CC—The lagoon-channel.
B'B'—The shores of the island, generally formed of low alluvial land and of coral detritus from the lagoon-channel.
A"A".—The outer edges of the reef now forming an atoll.
C'.—The lagoon of the newly formed atoll. According to the scale, the depth of the lagoon and of the lagoon-channel is exaggerated.

Source: From Darwin (1842)

The Coral Reef Problem

On 17 November 1835 Charles Darwin looked from the slopes of Tahiti towards the nearby island of Morea consisting of a central volcanic peak surrounded by a lagoon bordered by a barrier reef. In his mind's eye he saw that it would only be necessary to allow slow subsidence and drowning of the peak along with the vertical growth of the reef at about sea level to produce an atoll (Stoddart 1973: 313). This correlation between amount of subsidence and form of coral reefs was formalized by Charles Darwin (1838, 1842) who distinguished three different forms of coral reefs: fringing reefs, barrier reefs and annular reefs or atolls. He assumed that these forms had evolved with time by growing upward upon a gradually subsiding seafloor (figure 3.2). The rate and continuity of subsidence mattered little as long as the subsidence did not outstrip the rate of upward growth of the coral reef. There would be variants on the main theme; for example, a bank at the right depth (under 120 feet) could give rise to a reef, and a submerged crater at a suitable depth – a comparative rarity – could encourage an atoll-shaped reef in that subsidence.

Although Charles Lyell (1831: II: 296) had a glimmering of the idea of ocean floor subsidence, the subsidence theory of atoll formation was Darwin's own idea which he began to deduce before his visit to Tahiti as the result of studying the effects on the shores of South America of the intermittent elevation of the land, together with the marginal deposition of sediment. The simple substitution of the upward growth of coral for continued sediment deposition gave rise to his subsidence theory of barrier reefs and atolls which he developed on the homeward voyage of the *Beagle* by way of Tahiti, Australia and the Indian Ocean. Professor John W. Judd in his critical introduction to a re-issue of Darwin's *Coral Reefs, Volcanic Islands, South American Geology* (1890) recalled a remarkable conversation he had had with Darwin soon after the death of Lyell in 1875. With characteristic modesty Darwin said that he never fully realized the importance of his subsidence theory till he had discussed it with Lyell shortly after the return of the *Beagle*. On hearing an outline of the new theory, Lyell was so overcome with delight that he danced about and threw himself into the wildest contortions, as was his manner when excessively pleased. At Lyell's instigation on 31 May 1837, Darwin read to the Geological Society, London, a paper entitled 'On certain areas of elevation and subsidence in the Pacific and Indian Oceans, as deduced from the study of coral formations', a published report of which appeared later (1838).

Darwin's theory was strongly supported by James Dwight Dana who was geologist and marine zoologist on the Wilkes' oceanic expedition of 1838 to 1842. When climbing on Tahiti he saw that the deep gorges had obviously been cut by stream erosion. Such ravines, he noticed, would on subsidence at their seaward ends create an irregular coastline with great embayments, such as usually occurred around islands lying within barrier reefs. Thus Dana, whose early opinions later appeared as the popular *Corals and Coral Islands* (1872), added a significant proof of subsidence to Darwin's theory. Not surprisingly Darwin on reading it is said to have found it 'wonderfully satisfactory' and to have been 'astonished at my own accuracy!' (Gilman

1899: 306). Dana also tried to refine Darwin's (1842) estimate of the extent of the subsident area of the Pacific seafloor and came to the striking conclusion that the subsidence had increased from south to north-east and had occurred over so great an area that it constituted one of the great secular movements of the earth's crust and must have affected the entire sphere (Stanton 1971: 552).

During the late nineteenth and early twentieth centuries the subsidence coral theory tended to slip in popularity because the great amount of ocean floor subsidence it required was hard to justify on geophysical grounds and also because there was considerable evidence of recent elevation of land in some coral seas. Seafloor subsidence was strangely opposed by Suess, despite the fact that it provided a striking example of possible changes in capacity of the ocean basin. In this respect *The Face of the Earth* tells us more about Suess' advocacy than it does about coral reefs. With regard to the concepts of Darwin and Dana it states that:

> It needs no proof to show that the rising of the sea in the torrid zone would meet the facts in precisely the same way as a subsidence of the sea floor.
>
> (Suess 1906: 308)

Suess referred to many authorities who had criticized adversely the subsidence theory because coral growth is practically confined to the outer edge of the reefs while the interior part dies. He also pointed to evidence of limestone cliffs, caves and plateaus situated well above sea level. What is more, the presence of some islands with limestone platforms at 90 to 100 m above the sea level might be attributed to the former existence of a continuous limestone plateau which was dissected when sea level was at least 90 to 100 m higher than at present. He is not enamoured of Murray's (1879–80) sedimentation hypothesis (see below):

> The mantling over of submarine mountains with limestone is ill suited to explain the isolated atolls in the middle of the Ocean, and volcanic cones would by this means very seldom reach the zone of coral growth.
>
> (Suess 1906: 319)

Suess, however favoured the idea of wave erosion of the summits of young volcanoes, so forming a platform suitable for coral growth. The reefs generally, he suggested, have grown up under a predominantly positive movement or rising sea level:

> in spite of the extremely valuable information accumulated by recent observers, I believe, in agreement with F. von Richthofen ..., that we must still accept as a true explanation the fundamental idea of the subsidence theory, according to which the larger coral reefs have been built up under the influence of a widely distributed oscillation of the strand, with a positive excess.
>
> (Suess 1906: 321)

If we leave Suess and turn to other main opponents of Darwin's theory it becomes apparent that most of them assume that sea level is virtually unchanging and that coral reefs are thin veneers built upon suitable submarine platforms. Murray (1879–80) and Semper tried to explain the numerous submarine platforms required mainly by postulating the truncation of exposed peaks by wave action. Pelagic oozes would, they thought, accumulate on these platforms up to a depth at which reef-building corals could take over. Murray considered that barrier reefs had started as fringing reefs and had grown outward from the land, partly on the initial submarine slopes of the landmass and partly on their own detritus. The lagoons inside the interior edges of reefs were attributed mainly to solution of the inner, less active or decaying parts of the reefs. However, other investigators were not slow to point out that seawater is not an excellent solvent and that if considerable deposition of detritus is assumed it would probably soon fill up the lagoons, which did not seem to have happened. Other authors suggested that barrier reefs could grow up off any shelving shore, that wave action would cut platforms around islands and on continental coasts on which coral reefs might develop. Yet it was soon admitted that the cliffs which this idea demanded were by no means always present and that the presence of a reef offshore would greatly lessen or inhibit the continued erosion of a strand-line it protected. The concept of coral growth on erosion-formed submarine platforms and decapitated or submerged peaks was carried a step farther by J. S. Gardiner (1903–6) who in the Maldives and Laccadives had found atolls based on what appeared to be a marine planation level at depths of 140 to 170 fathoms. Coral growth at this level was apparently made possible by the presence of special sedentary species which were capable of gradually heightening the submarine platform until it reached about 50 fathoms where ordinary reef-building corals could take over.

In 1915 eustatism returned to the coral reef problem in a big way with the publication of R. A. Daly's (Plate 3.2) glacial control theory (see volume 2, pp. 596–7). Six years previously he had visited Hawaii and had been impressed with the narrowness of its coral reefs. He also noticed traces of recent ice action on the high volcanic peak of Mauna Kea. The proximity of glaciation and reef growth in a sea where the winter warmth today is only just above that needed for reef building, led him to conclude that during the Ice Age the surrounding sea would be too cool for coral reefs to survive. In other words, the existing reefs here were probably younger than the last glaciation and so had been subjected to oscillation of sea level. During a glacial epoch, the ocean would have been chilled and its level lowered by, Daly assumed, as much as 33 or 38 fathoms. At maximum low level the exposed dead corals would be eroded into benches or platforms; when the ice-sheets melted and sea level rose, any surviving corals would again begin to form reefs. Because corals facing the ocean would have the best food supply, isolated islands

Plate 3.2 Reginald Aldworth Daly (1871–1957)

Source: From *Proceedings of the Geological Society of America*, 1958, p. 114

Figure 3.3 Sections illustrating aspects of the glacial control theory of coral reefs. Diagonal shading indicates the stable submarine bank: 1, eustatic rise of sea level during the last interglacial period giving new coral growths (black), mainly at the outer edge; 2, sea level falls and temperature decreases during the onset of the last glacial period, during which corals largely die and reefs are eroded; 3, at the last glacial maximum reefs are destroyed and the bank is eroded to below the lower sea level; 4, the post-glacial rise of sea level, together with a temperature increase allows renewed coral growth particularly at the outer reef margins; 5, the present features of an atoll encircling a flat-floored lagoon

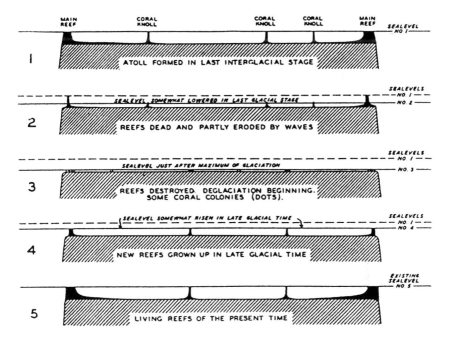

Source: From R. A. Daly (1934) *The Changing World of the Ice Age*, Fig. 130, p. 227. Copyright © Yale University Press. Reproduced by permission.

would generate atolls while larger land masses, like Queensland, would favour the development of linear offshore reefs. All forms of new reef would, according to this theory, consist partly or largely of veneers of new coral and lagoons would be of about the same depth (figure 3.3). It seemed difficult, Daly assumed, not to postulate general crustal stability in the coral seas during the time when the present reefs, and the surfaces on which they grew up, were formed. Daly persisted with glacial themes and in *The Changing World of the Ice Age* (1934) added the isostatic effect associated with the return of a great mass of water to the ocean from ice-melt resulting in depression of the ocean floor. He listed the following points in favour of the glacial control theory:

1 The uniform level and shallowness of lagoon floors.
2 Little sign of material which might have built up the subsiding lagoon floor having been carried over the top of the barrier reef.
3 The similarity in lagoon depth with that of the bank outside the reef.
4 The unlikelihood that reef growth could have kept pace with sea-level changes throughout glacial times.
5 The unlikelihood that isostasy would have allowed long-continued subsidence.
6 The narrowness of the reefs and their inner break of slope at the junction with the lagoon floor.
7 The ambiguous character of the Funafuti bore (1896–8) and the relatively thin coral limestone in the Michaelmas Cay bore (1926) on the Great Barrier Reef.

Had Suess lived one more year – he died aged 83 in 1914 – he would undoubtedly have been elated by Daly's eustatically based theory.

Like Daly's, W. M. Davis' interest in coral reef evolution dated from before the First World War (see volume 2, chapter 25) and it is typical of his methods that he had given theoretical support to Darwin's subsidence theory before he himself had even seen an atoll (volume 2, pp. 590–6, 605–13). He arrived at this view by inductive reasoning, but it must be admitted that his opposition to the glacial control theory (volume 2, pp. 597–8, 600–3, 613–16) may have been partly fuelled by a personal dislike of his Harvard colleague. Davis' work on coral reefs culminated in his 'grand old book' *The Coral Reef Problem* in 1928. Davis was indefatigable in travel and universal in vision. He distinguished major zones with respect to coral reefs: the coral seas proper where, in spite of changes of sea level, the absence of cliffs seems to indicate continuous coral growth throughout the Pleistocene; the cooler and cold seas where reef-forming communities cannot survive; and in between these two zones, between about latitudes 25° and 30° in each hemisphere, a transitional area wherein coral reefs ceased to grow in the chilly water during the Ice Age. In these transitional belts, as for example in the north-west of the Hawaian group, many islands are strongly embayed, and cliffs are common as well as submarine platforms and shallow banks which may be pre-glacial coral reefs planated by wave action during a lower sea level.

Davis demonstrated that the depths of the drowned coral features and lagoons vary considerably by up to more than 100 fathoms and that in fact there is no great difference in depth between what were regarded as the geologically stable and unstable areas of the Pacific. The soundings given by Daly in embayments failed to represent the true depths of those features because many of them have been shallowed by detrital infilling. If, as should be assumed, the embayments do represent great submergence, cliffs should have developed on the promontories between them, and their relative rarity on islands in the chief coral seas seems to point to protection by reefs from the strong attack of waves from the open ocean. Davis obviously was much more in favour of land movement than of eustasy. Subsidence, he thought, dealt more easily with the true depth of embayments and with the disposal of sediments, especially in a subsiding lagoon. His ideas enlarged the subsidence theory and incorporated the possibility of eustatic fluctuations and of crustal isostatic adjustments. His physiographical approach largely re-established the ocean floor subsidence theory, which was 'not only rejuvenated, but, to a large extent, justified' (Steers 1945: 282).

At the time of Davis' overall survey two deep bores had been drilled into coral reefs, one on Funafuti atoll in 1896–8 and another through much thinner coral limestone on Michaelmas Cay, near Cairns, in the Australian Great Barier Reef in 1926. The Funafuti boring reached a depth of about 1,114 feet and passed through only limestone of organic origin. Its upper 748 feet pierced largely broken and unconsolidated limestone and its lower part mainly dolomitic limestone. This inconclusive bore was drilled near the outer rim of the atoll and apparently had passed through a great deal of talus. After the Second World War the deep cores of undisturbed shallow reef limestone obtained at Bikini Atoll in 1947 (779 m), Eniwetok in 1951–2 (1,411 m) and Muraroa in 1964–5 (500m) demonstrated considerable, long-continued and differential subsidence of these islands. Together with the discovery in the Pacific of numerous drowned and bevelled volcanic islands (guyots), these findings gave great support to Darwin's theory of island subsidence, despite the occurrence of a few reefs which have been tectonically uplifted. Post-war recognition of this fact and of the relative amounts of volcanic island subsidence came to accord well with plate-tectonic notions of seafloor spreading away from a mid-oceanic ridge (Stoddart 1973). It remains ironic, however, that Suess, the prime opponent of seafloor subsidence, should have generated his major sea-level changes by means of the deformation of the ocean basins.

Conclusion

The peak of the eustatic theory was marked in geomorphology by the publication of Baulig's *Le Plateau Central de la France* (1928) (see chapter 8). Dedicated to Eduard Suess and to Paul Vidal de la Blache, it was one of the highly influential series of doctoral regional monographs inspired by the latter. Baulig recognized evidence of erosion surfaces at more or less uniform levels, chiefly at 380, 280 and 180 m (Baulig 1928: 432–43), to which extrapolated river profiles seemed also to be graded. Although postulating a mainly subaerial origin for these surfaces, Baulig believed that they must have been cut with reference to higher sea levels (mainly Pliocene), such that:

As Suess remarked it is upon the widespread character of certain events, notably of marine transgressions and regressions, that geological chronology rests, and it could well be that upon a suitably verified and confirmed theory of eustasy that one day universal correlations in morphology may be established.

(Baulig 1928: 544)

In retrospect Baulig (1957) came to believe that his morphological application of the eustatic theory was badly received. However, with the exception of most American geomorphologists, whose epeirogenic ideas had been earlier applied to the Massif Central by Demangeon (1910) and Briquet (1911), and of the tectonically minded Germans, the idea that the eustatic theory might provide a key to the large-scale correlation of landforms received quite widespread support for a time (Chorley 1963).

As has been shown in chapter 1, by the 1930s many geological advances were making inroads into the eustatic theory. In particular, submarine canyons began to be viewed as essentially submarine features produced by turbidity currents (Daly 1936, Kuenen 1950: 509–25) and extensive oceanic borderlands were being notionally replaced first by geanticlines and then by island arcs. At the same time criticisms of the mechanisms by which great eustatic changes might be produced (e.g. de Martonne 1929: 125–32) remained unanswered and, in general, the notion of eustatic movements being superimposed on a largely rigid crust began to break down and, with it, the possibility of large-scale correlation of landforms on a simple height/time basis. British geomorphologists stood their ground longest under the rising tide of criticism and as late as mid-century S. W. Wooldridge could write (see chapter 9):

It is not too much to claim that, in its contribution to earth-history, geomorphology seems within grasping distance of a great unifying generalization. Just at the point where the stratigraphic record fails or becomes incomplete, an alternative principle of inter-regional correlation is offered in the fact that old sea-levels have engraved their mark on the margins of the lands and that the same levels are recognizable inland as 'terraces' or 'platforms'. The latter-day record of the continents seems to have been one of 'uplift' starting long before Pleistocene times – or, more probably, in view of the uniformity of the record, of successive negative eustatic shifts of sea-level. Here, in one of its major fields of advance, geomorphology converges on geophysics.

(Wooldridge 1951: 171–2)

When Baulig made his similar statement almost a quarter of a century previously there was still enough faith in the absolute recent stability of the major continental areas to sustain the hope that landform correlations could be made purely from evidence of elevation. By the middle of the century the decline of the eustatic theory had cast grave doubt on this 'great unifying generalization'.

References

Adhémar, J. A. (1842) *Les Révolutions de la Mer, Deluges Périodiques*, Paris (2nd edn 1860).

Antevs, E.(1925) 'Retreat of the last ice sheet in eastern Canada', *Geological Survey of Canada, Memoir* 146.

—— (1929) 'Quaternary marine terraces in non-glaciated regions and changes of level of land and sea', *American Journal of Science* 5th series, 17: 35–49.

Barrell, J. (1915) 'Factors in movements of the strand line and their results in the Pleistocene and Post-Pleistocene', *American Journal of Science* 4th series, 40: 1–22.

Baulig, H. (1927) 'La Crau et la glaciation Würmienne', *Annales de Géographie* 36: 33–9.

—— (1928) *Le Plateau Central de la France et sa Bordure Méditerranéenne*, Paris: Colin.

—— (1935) 'The changing sea level', *Institute of British Geographers, Publication* no. 3.

—— (1948) *Problèmes des terrasses*, Paris: Colin.

—— (1957) 'La géomorphologie en France jusqu'en 1940', in G. Chabot (ed.) *La Géographie Française au Milieu du XX Siècle*, Paris: Baillière, pp. 27–35.

Beckinsale, R. P. (1972) 'The International Geographical Union and the development of geomorphology' in *Geography through a Century of International Congresses*, International Geographical Union, pp. 114–30.

Boule, M., Cartailhac, E., Verneau, R. and Villeneuve, L. de (1906–11) *Les Grottes de Grimaldi*, vol. 1, Imprime de Monaco.

Bourdier, F. (1938) 'Essai de synthèse sur le Quaternáire du Sud-Ouest', *Bulletin des Etudes locales de la Charente* no. 182.

—— (1959) 'Origines et succès d'une théorie géologique illusoire: L'eustatisme appliqué aux terrasses alluviales', *Revue de Géomorphologie Dynamique* 10: 16–29.

Briquet, A. (1911) 'Sur la morphologie de la partie mediane et orientale du Massif Central', *Annales de Géographie* 20: 30–43, 122–42.

Bucher, W. H. (1933) *The Deformation of the Earth's Crust*, Princeton, N. J.: Princeton University Press.

Chamberlin, T. C. (1909) 'Diastrophism as the ultimate basis of correlation', *Journal of Geology* 17: 685–93.

Chambers, R. (1848) *Ancient Sea Margins, as Memorials of Changes in the Relative Level of Land and Sea*, Edinburgh.

Chaput, E. (1928) 'Les terraces des régions atlantiques françaises', *International Geographical Union, Committee on Pliocene and Pleistocene Terraces* report 1, pp. 69–94.

Chorley, R. J. (1963) 'The diastrophic background of twentieth century geomorphological thought', *Bulletin of the Geological Society of America* 74: 953–70.

Cooke, C. W. (1930) 'Correlation of coastal terraces', *Journal of Geology* 38: 557–89.

Cooley, W. D. (1876) *Physical Geography: Or the terraqueous globe and its phenomena*, London: Dulau.

Croll, J. (1870) 'The boulder-clay of Caithness a product of land ice', *Geological Magazine* 7: 209–14, 271–8.

Cunningham, F. F. (1977) 'Lyell and uniformitarianism', *Canadian Geographer* 21: 164–74.

Daly, R. A. (1915) 'The glacial-control theory of coral reefs', *Proceedings of the American Academy of Arts and Sciences* 51: 155–251.

—— (1925) 'Pleistocene changes of level', *American Journal of Science* 5th series, 10: 281–313.

—— (1934) *The Changing World of the Ice Age*, New Haven, CT: Yale University Press.

—— (1936) 'Origin of submarine canyons', *American Journal of Science* 5th series, 31: 408–20.

Dana, J. D. (1872) *Corals and Coral Islands*, London: Sampson Low.

Darwin, C. (1838) 'On certain areas of elevation and subsidence in the Pacific and Indian Oceans, as deduced from the study of coral reefs', *Proceedings of the Geological Society of London* 2: 552–4.

—— (1842) *The Structure and Distribution of Coral Reefs*, London (2nd edn 1874).

—— (1890) *Coral Reefs, Volcanic Islands, South American Geology*, ed. J. W. Judd, London: Ward Lock.

Davies, G. L. (1969) *The Earth in Decay: A history of British geomorphology 1578–1878*, London: Macdonald.

Demangeon, A. (1910) 'Le relief du Limousin', *Annales de Géographie* 19: 120–49.

Depéret, C. (1903) 'Sur les anciennes lignes de rivages pliocenes et quaternaires des côtes française de la Mediterranée, *Comptes Rendus de l'Academie des Sciences* 136: 1639.

—— (1904) 'Observations à la note du Général de Lamothe sur les anciennes lignes de rivage de la côte algérienne et de la côte niçoise', *Bulletin de la Société Géologique de la France* series 4, 4: 38–9.

—— (1906) 'Les anciennes lignes de rivage de la côte française de la Mediterranée', *Bulletin de la Société Géologique de la France* series 4, 6: 207–30.

—— (1918–22) 'Essai de coordination chronologique générale des temps quaternaires', *Comptes Rendus de l'Academie des Sciences* 166: 480, 636, 884; 167: 418, 979; 168: 868, 873; 170: 159; 171: 212; 174: 1502, 1594.

—— (1921) 'La classification du quaternaire et sa correlations avec les niveaux préhistoriques', *Comptes Rendus de l'Academie des Sciences* pp. 125–7.

—— (1922) 'Essai de classification générale des temps quaternaires', *Congr. Géol. Intern.*, 13th session, Belgium, fasc. 3, pp. 1409–28.

—— (1923) 'La classification des temps quaternaires et ses rapports avec l'antiquité de l'homme en Europe', *Revue Générale des Sciences* 15 March.

Depéret, C. and Caziot, M. (1903) 'Note sur les gisements pliocènes et quaternaires marins des environs de Nice', *Bulletin de la Société Géologique de la France* series 4, 3: 321.

Dubois, G. (1924) 'Recherches sur les terrains quaternaires du nord de la France', *Memoir de la Société Géologique du Nord*, Lille, 8: 1: 1–356.

—— (1925) 'Sur la nature des oscillations du type Atlantique des lignes des riviages quaternaires', *Bulletin de la Société Géologique de la France* 25: 857–78.

Dumitrashko, N. V., Kaminin, L. G. and Meshchéryakov, Y. A. (1951) 'De l'état actuel de la géomorphologie et de ses problèmes', *Izvestiya Akademii Nauk SSSR, Ser. Geog.* pp. 71–82 (in Russian).

Flint, R. F. (1957) *Glacial and Pleistocene Geology*, New York: Wiley.

Gardiner, J. S. (1903–6) *The Fauna and Geography of the Maldive and Laccadive Archipelagoes*, 2 vols, Cambridge.

Geikie, A. (1865) *The Scenery of Scotland Viewed in Connexion with its Physical Geography*, London: Macmillan.

Geikie, J. (1874) *The Great Ice Age and its Relation to the Antiquity of Man*, London: Stanford (2nd edn 1877).

Gignoux, M. (1913) 'Les formations marine pliocènes et quaternaires de l'Italie du sud et de la Sicile', *Annales de la Université de Lyon* N.S., vol. 1(36).

—— (1926) *Géologie Stratigraphique*, Paris: Masson (4th edn 1950).

—— (1930) 'Charles Depéret (1854–1929), *Bulletin de la Sociéte Géologique de la France* 30: 1043–73.

—— (1937) 'Notice à la mémoire du Général L. de Lamothe', *Bulletin de la Société Géologique de la France* series 5, 7: 203–8.

Gilluly, J. (1949) 'Distribution of mountain building in geologic time', *Bulletin of the Geological Society of America* 60: 561–90.

Gilman, D. C., (1899) *The Life of James Dwight Dana*, New York: Harper.

Grabau, A. W., (1913) *Principles of Stratigraphy*, New York.

—— (1920) *Textbook of Geology*, 2 vols, London: Harrap.

—— (1936) 'Oscillation or pulsation', *Rept. XVI International Geological Congress, Washington 1933*, vol. 1, pp. 539–53.

Gras, S. (1862) *Description géologique du département de Vaucluse*, Paris.

Gregory, J. W. (1920) 'The Irish eskers', *Philosophical Transactions of the Royal Society* Series B, 210: 115–51.

Gutenberg, B. (1941) 'Changes in sea level, post-glacial uplift and mobility of the earth's interior', *Bulletin of the Geological Society of America* 52: 721–72.

Haug, E. (1900) 'Les geosynclinaux et les aires continentales', *Bulletin de la Société Géologique de la France* 3rd series, 28: 617–711.

—— (1907) *Traité de Géologie: Les phénomènes géologiques*, vol. 1, Paris: Colin.

—— (1908–11) *Traité de Géologie: Les périodes géologiques*, vol. 2, Paris: Colin.

Heim, A. and de Margerie, E. (1888) *Les Dislocations de l'Ecorce Terrestre*, Zürich: Wurster.

Hobbs, W. H. (1914) 'Edward Suess', *Journal of Geology* 22: 811–17.

Hoff, K. E. A. von (1822–34) *Changes . . . in the Earth's Surface Configuration since Historic Times, . . . and Earth Revolutions beyond the Domain of History*, 3 vols, Göttingen (vol. 1, 1822, Changes of land and sea; vol. 2, 1824, Volcanoes, earthquakes and geysers; vol. 3, 1834, Degradation of the land).

Hull, E. (1910) *Reminiscences of a Strenuous Life*, London: Hugh Rees.

International Geographical Union (1928–48) *Reports of the Commission on Pliocene and Pleistocene Terraces*, no. 1, 1928; no. 2, 1930; no. 3, 1931; no. 4, 1934; no. 5, 1938; no. 6, 1948.

—— (1956) *Premier Rapport de la Commission pour l'Etude et la Corrélation des niveaux d'Erosion et des Surfaces d'Aplanissement Autour de l'Atlantique*, 4 vols, 9th General Assembly, Rio de Janeiro.

Jamieson, T. F. (1862) 'On the ice-worn rocks of Scotland', *Quarterly Journal of the Geological Society of London* 18: 164–84.

—— (1865) 'On the history of the last geological changes in Scotland', *Quarterly Journal of the Geological Society of London* 21: 161–203.

—— (1882) 'On the cause of the depression and re-elevation of the land during the glacial period', *Geological Magazine* N.S., decade 2, 9: 400–7, 457–66.

Johnson, D. W. (1932) 'Principles of marine correlation', *Geographical Review* 22: 294–8.

Joly, J. (1925) *The Surface History of the Earth*, Oxford.

Karpinskiy, A. P. (1894) 'General character of the oscillation of the earth's crust in the region of European Russia', *Izvestiya Akademii Nauk* no. 1.

Kuenen, Ph. H. (1950) *Marine Geology*, New York: Wiley.

Lamothe, L. de (1897) 'Note sur les terrains de transport du bassin de la Haute-Moselle et de quelques vallées adjacentes' *Bulletin de la Société Géologique de la France* series 3, 25: 378–438.

—— (1899) 'Note sur les anciennes plages et terrasses du bassin de l'Isser (Département d'Alger) et de quelques autres bassins de la côte algérienne', *Bulletin de la Société Géologique de la France* series 3, 27: 257–303.

—— (1901a) 'Sur le rôle des oscillations eustatiques du niveau de base dans la formation des systèmes de terrasses de quelques vallées', *Comptes Rendus de l'Academie des Sciences* 132: 1428–30.

—— (1901b) 'Etude comparée des systèmes de terrasses des vallées de l'Isser, de la Moselle, du Rhin et du Rhône: Preuve que leur formation est due à des oscillations eustatiques du niveau de base', *Bulletin de la Société Géologique de la France* series 4, 1: 297–384.

—— (1904) 'Note sur les relations stratigraphiques qui paraissent exister entre les anciennes lignes de rivage de la côte algérienne et celles signalées sur la côte niçoise', *Bulletin de la Société Géologique de la France* series 4, tome 4, 14–38.

—— (1911) 'Les anciennes lignes du rivage du Sahal d'Alger et d'une partie de la côte algérienne', *Memoires de la Société Géologique de la France* 4th series, vol. 1, memoir 6.

—— (1912) 'Aux sujets du deplacement de la ligne de rivage le long des côtes algériennes pendant le Post-Pliocène', *Bulletin de la Société Géologique de la France* series 4, tome 12, 343–8.

—— (1915) 'Les anciennes nappes alluviales et les terrasses du Rhône et de l'Isère dans la region de Valence', *Bulletin de la Société Géologique de la France, Notes et Mémoires* pp. 3–89.

—— (1916) 'Les anciennes lignes de rivage du bassin de la Somme et leur concordance avec celles de la Mediterranée occidentale', *Comptes Rendus de l'Academie des Sciences* 162: 948–51.

—— (1918) 'Les anciennes nappes alluviales et lignes de rivages du bassin de la Somme et leurs rapports avec celles de la Mediterranée occidentale', *Bulletin de la Société Géologique de la France, Notes et Mémoires* 4th series 18: 3–58.

Landes, K. L. (1951) *Petroleum Geology*, New York: Wiley.

Lapparent, A. de (1906) *Traité de Géologie*, 5th edn, Paris (1st edn 1881).

LeConte, J. (1891) 'Tertiary and Post-Tertiary changes of the Atlantic and Pacific Oceans. With a note on the mutual relations of land elevation and ice accumulation during the Quaternary period', *Bulletin of the Geological Society of America* 2: 323–30.

—— (1897) 'Earth-crust movements and their causes', *Bulletin of the Geological Society of America* 8: 113–26.

Leverett, F. (1928) 'Results of glacial investigations in Pennsylvania and New Jersey in 1926 and 1927 (abstract)', *Bulletin of the Geological Society of America* 39: 151.

Leverett, F., Alden, W. C., Johnson, D. W. and Davis, W. M. (1922) 'Review and critique of the Penck-Brückner, de Geer, Déperet systems (Symposium)', *Bulletin of the Geological Society of America* 33: 472–85.

Levorsen, A. I. (1931) 'The Pennsylvanian overlap in the United States', *Bulletin of the American Association of Petroleum Geologists* 15: 113–48.

Lichkov, B. L. (1934) 'On the epeirogenic movements of the earth's crust in the Russian Plain', *Akademii Nauk SSSR, Geomorf. Inst.* trudy vypusk 10.

Lyell, C. (1831–3) *Principles of Geology*, 1st edn, 3 vols, London: Murray (11th edn, 2 vols, 1872; 12th edn, 2 vols, 1875).

Maclaren, C. (1842) 'The glacial theory of Professor Agassiz', *American Journal of Science* series 1, 42: 346–65.

Manfredi, E. (1746) *De Aucta Maris Altitudine*.

Martonne, E. de (1929) 'La morphologie du Plateau Central de la France et l'hypothèse eustatique', *Annales de Géographie* 38: 113–32.

Meshchéryakov, Y. A. and Sinyaguina, M. I. (1956) 'Les mouvements contemporains de l'écorce terrestre et les méthodes de leur étude', in *Essais de Géographie, recueil des articles pour le XVIIIe Congrès International Géographique*, Acad. des Sci. de URSS, Moscow, pp. 71–81.

Molengraaff, G. A. F. (1930a) 'The coral reefs of the East Indian Archipelago, their distribution and mode of development, *Proceedings of the 4th Pacific Science Congress (Java)*, vol. 2, pp. 55–89.

—— (1930b) 'The recent sediments of the seas of the East Indian Archipelago, with a short discussion on the condition of those seas in former geological periods', *Proceedings of the 4th Pacific Science Congress (Java)*, vol. 2B, pp. 989–1021.

Moore, R. C. (1936) 'Stratigraphic evidence bearing on problems of continental tectonics', *Bulletin of the Geological Society of America* 47: 1785–808.

Murray, J. (1879–80) 'On the structure and origin of coral reefs and islands', *Proceedings of the Royal Society of Edinburgh*, 10: 505–18.

Nikolayev, N. I. (1949) 'The most recent tectonics in the USSR', *Komitet Izucheniyu Chetvertich, Perioda*, trudy 1.8.

Peach, B. N. and Horne, J. (1879) 'The glaciation of the Shetland Islands', *Quarterly Journal of the Geological Society of London*, 35: 788–812.

—— (1880) 'The glaciation of the Orkney Islands', *Quarterly Journal of the Geological Society of London* 36: 648–63.

Penck, A. (1894) *Morphologie der Erdoberfläche*, 2 vols, Stuttgart: Engelhorn.

—— (1934) 'Theorie der Bewegung der Strandlinie', *Sitzungsberichte Preussischen Akademie der Wissenschaften, Math.-Phys. Klasse* 19: 321–48.

Penck, A. and Brückner, E. (1909) *Die Alpen im Eiszeitalter*, 3 vols, Leipzig: Tauchnitz.

Playfair, J. (1802) *Illustrations of the Huttonian Theory of the Earth*, Edinburgh: Creech (facsimile by University of Illinois Press with an introduction by George W. White, 1956).

Prestwich, J. (1893) 'On the evidences of a submergence of Western Europe and the Mediterranean coasts', *Philosophical Transactions of the Royal Society* series A, 184: 903–84.

Ramsay, A. C. (1846) 'The denudation of South Wales', *Memoir of the Geological Survey of Great Britain*, vol. 1, London: HMSO, pp. 297–335.

—— (1881) 'The geology of North Wales', *Memoir of the Geological Survey of the United Kingdom*, 2nd edn, London.

Ramsay, W. (1924) 'On relations between crustal movements and variations of sea level during the late Quaternary time', *Bulletin de la Commission Géologique de Finlande* no. 66.

—— (1930) 'Changes of sea level resulting from the increase and decrease of glaciation', *Fennia* 52: (5): 1–62.

Reid, C. (1913) *Submerged Forests*, Cambridge.

Ricketts, C. (1871–2) 'Presidential address', *Proceedings of the Liverpool Geological Society*, vol. 2, session 13, pp. 4–35.

Sandford, K. S. (1929) 'The Pliocene and Pleistocene deposits of Wadi Quena and of the Nile Valley between Luxor and Assiut (Qau)', *Quarterly Journal of the Geological Society of London* 85: 493–548.

Schmidt, M. (1922) 'Neuzeitliche Erdkrustenbewegungen in Frankreich', *Sitzungsberichte der Bayerische Akademie der Wissenschaften, Math.-Phys. Klasse* pp. 1–6.

Schuchert, C. (1910) 'Palaeogeography of North America', *Bulletin of the Geological Society of America* 20: 427–605.

—— (1923) 'Sites and nature of North American geosynclines', *Bulletin of the Geological Society of America* 34: 151–230.

Shaler, N. S. (1875) 'Notes on some of the phenomena of elevation and subsidence of the continents', *Proceedings of the Boston Society of Natural History* 17: 288–92.

Shepard, F. P. (1923) 'To question the theory of periodic diastrophism', *Journal of Geology* 31: 599–613.

—— (1933) 'Submarine valleys', *Geographical Review* 23: 77–89.

—— (1934) 'Canyons off the New England coast', *American Journal of Science* 5th series, 27: 24–36.

Stanton, W. (1971) 'James Dwight Dana', *Dictionary of Scientific Biography*, vol. 3, New York: Scribners, pp. 549–54.

Steers, J. A. (1929) *The Unstable Earth*, London: Methuen (4th edn 1945).

Stille, H. (1936) 'The present tectonic state of the earth', *Bulletin of the American Association of Petroleum Geologists* 20: 849–80.

Stoddart, D. R. (1973) 'Coral reefs: The last two million years', *Geography* 58: 313–23.

Suess, E. (1862) *Der Boden der Stadt Wien*; Wien: Braumüller.

—— (1875) *Die Enstehung der Alpen*, Wien: Braumüller.

—— (1883–1908) *Das Antlitz der Erde*, 3 vols, Wien: Tempsky.

—— (1897–1918) *La Face de la Terre*, trans. E. de Margarie, Paris: Colin, 3 vols (vol. 1, 1897; vol. 2, 1900; vol. 3, 1902–18).

—— (1904) 'Farewell lecture on resigning his Professorship', *Journal of Geology* 12: 264–75.

—— (1904–24) *The Face of the Earth*, trans. H. B. C. and W. J. Sollas, Oxford: Oxford University Press, 5 vols (vol. 1, 1904; vol. 2, 1906; vol. 3, 1908; vol. 4, 1909; vol. 5, 1924).

Swedenborg, E. (1719) *Om wattrens höjd veh förra werldens starka ebb och flod. Bevis utur Sverige*, Stockholm.

Thomson, C. W. and Murray, J. (1880–95) *Report on the Scientific Results of the Voyage of H.M.S. Challenger during the Years 1873–1876*, 41 vols, London: HMSO (Narrative, 1882–5, 2 vols; Zoology, 1880–9, 32 vols; Botany, 1885–6, 2 vols; Physics and Chemistry, 1884–9, 2 vols; Deep Sea Deposits, 1891, 1 vol; Summary of Scientific Results, 1895, 2 vols).

Willis, B. (1893) 'Conditions of sedimentary deposition', *Journal of Geology* 1: 480.

—— (1909) 'Paleogeographic maps of North America', *Journal of Geology* 17: 203–8.

Wooldridge, S. W. (1951) 'The progress of geomorphology', in G. Taylor (ed.) *Geography in the Twentieth Century*, London: Methuen, ch. 7.

Wright, W. B. (1914) *The Quaternary Ice Age*, London: Macmillan (2nd edn 1937).

—— (1928) 'The raised beaches of the British Isles', *International Geographical Union, Commission on Pliocene and Pleistocene Terraces*, Report no. 1, pp. 99–106.

Zittel, K. von (1901) *History of Geology and Palaeontology*, London: Scott.

PART II

Davisian Influences

CHAPTER FOUR

The Davisian Cycle in the USA and Germany

The Conflicting Roles of Landform Studies

At the end of volume 1 (p. 641) a quotation by N. M. Fenneman (1939: 355), referring to the landform studies by Powell, Gilbert, McGee and others, expressed the then conflicting role of such studies in American scholarship:

> the lives of these men were spent largely in other phases of geology, and it was left to Davis to paint the panorama of change from initial forms to their final disappearance. This is the theme of the geomorphic cycle, the most distinctive element of modern physiography.

Thus Fenneman viewed studies of landform development as a phase of *geology*; what Davis called the *geographical* cycle he termed the *geomorphic* cycle and considered this cycle to be an element of *physiography*. Even allowing for all the success associated with Davis' exposition of the cycle of erosion, which has been documented at length in volume 2, the above quotation encapsulates the ambiguities in finding scholarly accommodation for landform studies at a time of rapid change in both geology and geography.

The term *physiography*, documented in detail by Stoddart (1975, 1986), had been in intermittent use for about a century before Dana (1863: 2) defined it as beginning where geology ends; treating the final arrangements of earth features (including climates, magnetism, life etc.); and the earth's physical movements and changes (including secular variations of the atmosphere, ocean currents, heat, magnetism, etc.). However, the definitive view of physiography was given by Huxley (1877) (plate 4.1) in his book of that title which included:

> springs, rainfall and climate, water chemistry, denudation, glacial erosion, marine erosion, earth movements and volcanicity, deposition in the ocean and the formation of rocks, the geology of the Thames basin, and finally the earth as a planet, its movement and the seasons, and its place in the solar system.
>
> (Stoddart 1986: 188)

Huxley, under this title, provided an integrated approach to natural phenomena and he was followed by Powell in his definition of physiography as:

> A description of the surface features of the earth, as bodies of air, water and land. In it is usually included an explanation of their origin, for such features are not properly understood without an explanation of the processes by which they are formed.
>
> (Powell 1896: 1)

Plate 4.1 Thomas Henry Huxley (1825–95) from a painting in 1883 by his son-in-law, J. Collier

Source: Courtesy National Portrait Gallery, London

In Britain after Huxley the term physiography became synonymous with physical geography and the important textbook entitled *Physical Geography* by Philip Lake (1915) continued Huxley's tradition, being divided more or less equally between the atmosphere, the oceans and the land. Some 20 years later the American tendency to restrict physiography to the study of landforms was becoming dominant. In the United States under the influence of Davis (e.g. 1915K) the term physiography very rapidly came to mean a narrow explanatory physical geography concentrating on subaerial erosion (Russell 1958). A whole series of texts spanning half a century employed physiography to refer to regional geomorphology (Powell 1896, Bowman 1911, Fenneman 1931, 1938, Loomis 1937, Atwood 1940) (see chapter 11) before the term lost currency altogether after the Second World War in the face of 'geomorphology'. Only in the Mid-West, the bastion against Davisian attitudes, did physiography retain for some decades its connotation of a broader physical geography (Salisbury 1907, Tarr and Martin 1914). Later physical geography was often selectively concentrated on 'the background and the medium of human activities' (Sauer 1925: 27).

As has been shown in volume 1 (p. 547) the term *geomorphology* was used by McGee as 'the genetic study of topographic forms' (1888) treating the phenomena of degradation whereby 'the later history of the world growth may be read from the configuration of the hills as well as from the sediments and fossils of ancient oceans' (1891). In Britain Mackinder (1895: 373) defined geomorphology as 'the causal description of the earth's present relief' (Stoddart 1986: 204–5) but in the United States Davis (1895I, p. 8, 1915K) used the term geomorphology but preferred physiography, which he used identically. Kirk Bryan (1941) has drawn attention to Davis' attempt to divide geomorphology (the genetic description of landforms) into:

Geomorphogeny – the history of the origin, development and changes of landforms;
Geomorphography – the description of landforms.

Long afterwards Russell made a rather similar distinction between 'geological geomorphology' (1958), concerned with the history of landforms, and 'geographical geomorphology' (1949), of which he wrote:

Geographers ordinarily find difficulty in discovering useful information in the conclusions of the pure morphologist. That a particular river is a consequent stream with an obsequent extension, or that some part of a river is superimposed rather than antecedent, or that a windgap suggests a cause of stream piracy, really means little to the person working on the problems of some specific cultural landscape .... The geomorphologist may concern himself deeply with questions of structures, process, and time, but the geographer wants specific information along the lines of what, where, and how much.

(Russell 1949: 3–4)

To this distinction Strahler (1952) added that of 'dynamic geomorphology', the quantitative analysis of geomorphic processes. At any rate, by the 1940s a series of influential textbooks had acknowledged the place and scope of geomorphology in scholarship (Wooldridge and Morgan 1937, Worcester 1939, Lobeck 1939, Von Engeln 1942).

It has become clear from the foregoing that the *geological* implications of the study of landforms also have a long and complex history. In 1863 Dana distinguished between 'physiographical geology' and 'dynamical geology' and this was partly echoed some 80 years later by Bryan (1941) who considered that what he called physiography was divided into historical geology, in so far as landforms are the result of past processes, and into dynamic geology, in so far as they are the product of the interaction of endogenous and exogenous earth forces. Later, however, Bryan (1950: 198) makes it clear that the former distinction was in his view paramount:

> The large part of our present topographical forms are, however, inherited from the past ... the surface of late Pleistocene time is largely still with us and well-preserved forms of late Tertiary age are relatively common. The essence of geomorphology is the discrimination of the ancient from the modern. It is, therefore, essentially a branch of historical geology.

Dynamic geology is concerned with the processes which lead to the formation, alteration and disturbances of rocks (see volume 4) and clearly links with Strahler's recommendation of Horton's (1945) paper as:

> a document of great importance in geomorphology, not so much for the validity of its conclusions on drainage development and slopes, as for the forceful manner in which it has brought to the attention of geomorphologists the application of quantitative and dynamic methods to landform study.
>
> (Strahler 1950a: 211)

As Strahler points out, Davis' work consistently showed a lack of real concern with process (see volume 2, pp. 753–4) and Leighly (1955: 318) went as far as to assert that 'we should discard a restriction that has long been laid upon us: the prohibition of concern with process'.

Denudation chronology, considered by some to be a branch of historical geology, will be treated in chapters 7, 8 and 9 of this volume, but it is well to remind ourselves here that Davis relentlessly eschewed considerations of absolute age in favour of 'stage', except in the most vague and general terms (see volume 2, pp. 513, 527; for an exception see p. 649). Volume 1 traced the early ideas regarding the effects of variations of lithology and structure on landforms and this theme of structural geology will be pursued further in the context of twentieth-century geomorphic process studies in volume 4. Here we should note that even strongly Davisian regional physiographic treatments were cast in a dominantly structural mould (e.g. Fenneman 1931, 1938; see chapter 11) and the first half of the present century was marked by a wealth of work in which landforms were set in a secure framework of structural geology unconnected with the cycle of erosion. For example, the structural underpinning of the superficial Appalachian folds was described by Rich (1933, 1934); of certain of the Rocky Mountains by Eardley (1934) and Billings (1938); of the Basin Ranges by Bryan (1923), Blackwelder (1928), Gilluly (1937) and Sharp (1940); and of the Colorado Plateau province (Strahler 1944, 1948, Hunt *et al.* 1953). In the latter connection, special mention must be made of the superb series of Professional Papers describing the terrain of the Colorado Plateaus in terms of their lithology and structure produced by H. E. Gregory (1917, 1938, 1951, 1952; Gregory and Moore 1931) and of the geomorphic analyses of individual topographic quadrangles given by H. S. Sharp in the *Journal of Geomorphology* (1938–40).

Although from its inception Davis insisted on the term *geographical* cycle of erosion, the concept was dominantly a geomorphic one and the text on *Physical Geography* by Davis and Snyder (1898E) contained only a short section on the atmosphere. Before Huxley's *Physiography* (1877) the books on physical geography were compendious, unintegrated, poorly illustrated catalogues (Stoddart 1986: 191–2) and after Huxley the term physiography tended to replace physical geography in American usage. However, Davis' narrowly geomorphic interpretation of both terms created a legacy of conceptual problems for twentieth-century American geographers leading to a decline of interest in climatology, pedology and hydrology. Nevertheless it is of note that Strahler's excellent text on *Physical Geography* (1951) was in the best tradition of Huxley's physiography. The influence of Davis, however, was to divide geography into physiography, a study of the physical environment dominantly from a geomorphic standpoint, and ontography, the study of the human response to the physical environment (Davis 1900O, 1902E, 1903H; see volume 2, p. 735). At Harvard his courses were dominantly physiographical but he also required some ontography and encouraged his students to explore humanistic aspects of geography in a rational way. J. Russell Smith has observed in a letter to G. J. Martin in June 1982:

> No one had more to do with the un-Davising of geography than did Davis himself. He went up and down the land between 1899 and 1903 delivering addresses before various scientific groups, laying out the point that geography was a relationship between the earth and the organism that lived upon it.

Quite typical is the advice given by Davis to his former student, Isaiah Bowman, who had just taken up a position at Yale:

> The chief thing I wish to emphasize is that you should develop geography proper, physiography and ontography properly combined, and not simply physiography (as I have done too much).
>
> (Davis to Bowman, letter 18 March 1906)

Among Davis' many students who took this advice and achieved some
eminence were Isaiah Bowman, A. P. Brigham, A. H. Brooks, S. Cushing,
R. E. Dodge, H. E. Gregory, Ellsworth Huntington, Mark S. W. Jefferson,
C. F. Marbut, V. Stefannson, R. S. Tarr, W. S. Tower and R. De C. Ward.
Mark Jefferson probably did more than any other American to establish a
place for mankind in geography and to help the discipline to emerge from
'the bondage of Davisian physiography' (Martin 1968: 521). However, after
about 1908 Davis' approach to geography veered again when he began to
exalt the regional approach and to extol regional description as its ultimate
role. A. P. Brigham in 1909 summarized this tendency with considerable
acumen:

> He [Davis] sympathises fully with Hettner in regarding geography in its regional
> aspects as embodying a description of the material filling of terrestrial spaces (die
> dingliche Erfüllung der Erdräume) but gives more emphasis to the explanatory
> aspects of systematic geography, in which things of a kind are studied together,
> wherever found, and thus made ready for more appreciative treatment in their
> regional associations.
>
> (Brigham 1909: 52)

After resigning from his Chair at Harvard in 1912 Davis wrote relatively little
on methodology apart from his superb *Principles of Geographical Description*
(1915K), which, in fact, had had to wait several years for its publication.

It should be very clear from the foregoing that any approach to the study of
landforms which was proposed during the period from about 1890 to the
Second World War was faced, at a time of great changes in geography,
geology, physiography and geomorphology, with an almost impossible range
of demands. During this time the study of landforms was required to fulfil
many different roles, often conflicting.

Contemporary Dissent from Davis' Concepts in the USA

We have already demonstrated in volume 2 the success achieved by Davis'
concept of the cycle of erosion and of cyclic denudation chronology which
was fuelled by it (Beckinsale 1976, 1981), and this will be underlined by
several of the subsequent chapters in the present volume. However, it would
be wrong to assume that there was no contemporary resistance to his ideas in
his own country. Even in the north-east Davis' cyclic approach did not
achieve the full adoption which he expected. At Yale Herbert E. Gregory was
appointed instructor in physical geography in 1900 (becoming professor in
1912) and taught landform studies from a distinctly geological viewpoint.
However, in 1905 he was joined by Davis' student Isaiah Bowman (plate 4.2).
Born in 1878, Bowman was raised in Michigan, came under the influence of
Mark Jefferson at the State Normal School, Ypsilanti, and went on to study
at Harvard. The result of his teaching in the Yale Forestry School was *Forest*

Plate 4.2 Isaiah Bowman (1878–1950)

Source: From *Geographers: Biobibliographical Studies*, No. 1, Mansell, London, 1977

Physiography (1911), including within its extensive regional treatment of American landforms many cyclic and polycyclic ideas. His long letter of appreciation to Davis on the latter's retirement in 1912 has been quoted at length (volume 2, pp. 425–7) and he was clearly not happy with the replacement of Davis by W. W. Atwood (volume 2, p. 450). Bowman left Yale in 1915 to become director of the American Geographical Society, played a leading role in the American delegation to the Versailles Peace Conference and between 1933 and 1948 was President of the Johns Hopkins University. We have already noted his critical but not hostile (1926) review of

Walther Penck's work (volume 2, pp. 694–7), stimulated by his own publications on the Andes (1916) and the Atacama desert (1924) based on some 18 months working in the field and travelling by mule during the years 1907, 1911 and 1913 (Martin 1981).

Another of Davis' students, Ralph Stockman Tarr, joined Cornell University, New York State, in 1892 and published an *Elementary Physical Geography* in 1895 which evoked a critical review from Davis (Dunbar 1981). Such criticism of his former students was not uncommon in that Davis was frequently disappointed both by the manner in which the human response to the physical environment was treated and by what he considered to be an inadequate cyclic basis for landform studies. In 1898 Tarr wrote a critical article regarding peneplain identification (see volume 2, p. 242) evoking a sharp reply from Davis (1899I). Tarr's extensive *New Physical Geography* (1903) treated geomorphology, physical geology, regional physiography, the atmosphere, the oceans, plants, animals, resources and the human response, but the parts on fluvial erosion included considerations of the cycle, grade, baselevel and denudation chronology. Although Tarr had been hurt by the earlier review by Davis, he wrote in the preface:

> Doubtless the most profound influence on the author is that of his two teachers, Professors Shaler and Davis, the importance of which to him cannot be over-estimated. Together with other physiographers, the author further recognises in Professor Davis a leader in American physiography, from whom even some of the fundamental principles of the subject have been derived. An examination of the following pages would show the influence of this physiographer in many places, an influence not confined to the pure science, but extending to the pedagogy of the subject as well.
>
> (Tarr 1903: viii–ix)

However, after Tarr's early death in 1912, his *College Physiography* (1914, with Martin) appeared comprising 388 pages on processes, 110 on tectonics and volcanism, 114 on landforms and geology, 26 on the earth's interior, 72 on the oceans and 106 on the atmosphere. Strictly cyclic ideas were confined to some 40 pages. O. D. Von Engeln took over at Cornell in 1912, remained there as professor of geology until 1949, and published in 1942 the first American textbook of geomorphology to attempt a serious treatment of the views of Walther Penck.

The most sustained resistance to the adoption of the cyclic basis of the study of landforms came from the University of Chicago in the persons of Thomas Chrowder Chamberlin (1843–1928) and his former student Rollin D. Salisbury (1858–1922) (plate 4.3) (Dunbar, 1981; Pattison, 1981). Chamberlin, as will be recalled from chapter 2, was a leading worker on the glacial geology of the Mid-West and was proud of having been born on the crest of the Shelbyville terminal moraine in Illinois (Mather 1971). After becoming Chief Geologist of the Wisconsin Geological Survey (1876–82),

Plate 4.3 Original staff of the Department of Geology at the University of Chicago. Left to right: Rollin D. Salisbury, C. R. Van Hise, Thomas Chrowder Chamberlin, J. P. Iddings, R. A. F. Penrose Jr

Source: Courtesy of the University of Chicago Department of Geology

Geologist in Charge of the Glacial Division of the US Geological Survey (1881–1904) and President of the University of Wisconsin (1887–92), he became Chairman of the Geology Department of the University of Chicago in 1892. In the same year he was joined as professor of geographic geology by Salisbury who had previously been associated with Chamberlin at both Beloit College and the University of Wisconsin. In later life Chamberlin, who was described as tall, rugged, serious, severe and dignified, became a personal friend of Davis, but in 1904 the first volume of Chamberlin and Salisbury's textbook on *Geology* devoted only some 35 pages to cyclic and related ideas out of 188 pages on fluvial processes and 684 pages in all. It is reported that Chamberlin was at times critical, in a jocose manner, of some of Davis' geological views (Beckinsale 1976: 463–4). However, Salisbury, who eventually succeeded to Chamberlin's chair in 1918, was much more critical of Davis (plate 4.4). Salisbury at first accepted the cyclic model and included it in his programme of lectures in 1895, but it will be recalled from volume 2 (pp. 133–4) that Davis had probably taken his ideas of topographic youth and old age from the earlier work by Chamberlin and Salisbury in the Driftless Area of Wisconsin without acknowledgement. A further cause of antagonism was their different approach to physical geography. Davis' *Physical Geography* (with W. H. Snyder, 1898) set out to explain the origin of the physical features of the earth and to trace them to the consequences as seen in the conditions of mankind; whereas Salisbury's *Physiography* (1907) was much less deterministic (Pattison 1981). At this juncture Salisbury included cryptic statements on the stages of river-valley development, the cycle of erosion and peneplains but by the time of publication of the third edition (1919) he had almost discarded the cyclic concept. His student Stephen S. Visher wrote in 1962:

> My special teacher, R. D. Salisbury, had relatively little respect for Davis. I recall Salisbury's glee when he was employed to correct many of Davis' definitions in a new edition of *Webster's Dictionary*. Davis was 7 years older than Salisbury, and was harshly critical of him. Davis's successor at Harvard was W. Atwood, a special student of Salisbury's.
>
> (volume 2, p. 450)

Atwood (plate 4.5) was only one of a group of Mid-Western scholars who were trained in a tradition different from that of Davis. As we have seen he succeeded Davis at Harvard in 1913 and from 1920 to 1946 was President of Clark University where he worked particularly on regional physiography and built up a flourishing graduate school. However, Davis was held in much respect at Clark and lectured there several times in the 1920s. In this context the work of N. M. Fenneman of Cincinnati must also be mentioned because, although a supporter of Davisian polycyclic approaches to landforms, he did

Plate 4.4 Painting of Rollin D. Salisbury

Source: Courtesy University of Chicago

Plate 4.5 W. W. Atwood Sr in the San Juan region of Colorado with his favourite horse 'Tony'

Source: Courtesy of W. W. Atwood Jr and G. J. Martin

contribute a most important statement on non-cyclic aspects of erosion (1936; see volume 2, pp. 193–6).

In the long run, however, the influence of Carl O. Sauer (plate 4.6) was to prove especially detrimental both to Davis' geomorphic and geographical ideas. Sauer was born in Missouri in 1889 of German stock, sent to school in Württemberg for 5 years, subsequently graduated at the University of Chicago and gained a Ph.D. degree there (Leighly 1976). Shortly afterwards in 1915 he was invited by Professor W. H. Hobbs to teach geography in the Department of Geology at the University of Michigan. This department had had distinguished associations with geomorphology beginning with Israel C. Russell who held the chair there until his death in 1906. Russell was the author of the influential *River Development* (1898) which devoted some 31 pages to cyclic and denudation chronology ideas out of a total of 327, and he was succeeded as professor of geology by William Herbert Hobbs (1864–1953) (Hall 1953). Hobbs held the chair until 1934 and in 1912 published *Earth Features and their Meanings* (second edition 1931), the 517 pages of which included 82 on fluvial processes and only 8 on cyclic notions. This work, which is highly descriptive and geologically based, clearly reflected antipathy to Davis' ideas.

Carl Sauer joined the newly retitled Department of Geology and Geography at Michigan and taught there from 1915 until 1923 when he went to the University of California at Berkeley. By training and inclination Sauer was attracted to German geographical scholarship and in his Morphology of Landscape (1925) praised both Passarge's method of purely inductive or empirical description and Hettner's chorological concept of the nature of geography. In 1929 Sauer produced the first American analysis of landforms using the ideas of Walther Penck (see chapter 10). This advocacy of Germanic ideas on landscape was supported by Richard Hartshorne (1939) and together the two scholars did much to nullify the severely critical reviews of German work by Davis and his supporters. Sauer's comments on Davis have been given in volume 2 (pp. 427–8). In this connection the Penckian views of Howard Meyerhoff (1940) regarding slope development and erosion surface development independent of baselevel are also of note.

Strangely, Sauer did not visit Europe again until the age of 66 but many American scholars were more directly influenced by training received in Germany where many academics had developed themes which rendered unnecessary the insertion of deductive stages into the history of landforms. Obviously American scholars who sought further knowledge of physiography and geography in Germany would return home with an anti-Davisian bias. Salisbury spent a year (1887–8) at Heidelberg, as did Hobbs in 1888–9, and Tarr studied in Europe in the academic year 1909–10. Ellen Churchill Semple was a student under Ratzel at Leipzig in 1891–2 and 1895. Alfred Hettner's first Ph.D. student at Heidelberg was Martha Krug-Genthe who in

Plate 4.6 Carl O. Sauer

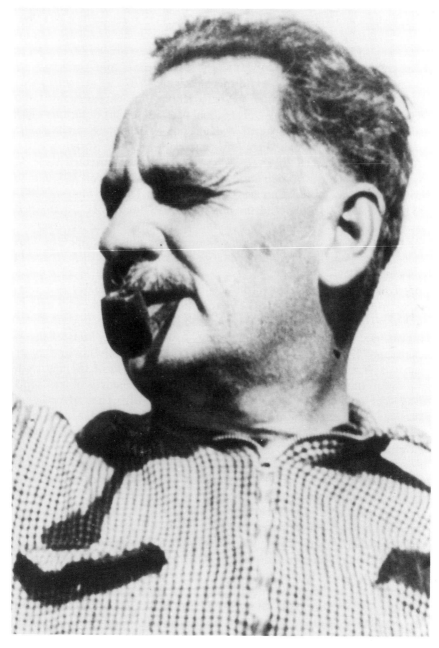

Source: Courtesy G. J. Martin

1901 accepted a teaching position at the Beacon School in Hartford, Connecticut, and stayed there until it closed in 1911 when she went back to Germany. In 1904 Semple and Krug-Genthe were the only two women among the forty-eight founder members of the Association of American Geographers. J. Russell Smith (1874–1966) worked on the Isthmian (Panama) Canal Commission until 1901 when he went to study at Heidelberg 'following the usual practice of getting a touch of German geography' (James 1981: 324). The anthropogeographical or Ratzelian ideas popularized by Semple and the explosion of commercial geography triggered off by Russell Smith must have side-tracked many geographers from Davisian physiography. In his final retrospect of American geography, Davis (1932B) suggested that the swing away from physiography towards commercial geography from about 1912 onward had been so violent that it had 'almost uprooted the Geo – from Geography'. Equally relevant is the case of W. D. Jones who, having failed to apply Davis' ideas satisfactorily to fieldwork in northern Patagonia, went as a graduate student to spend a semester with Hettner at Heidelberg in 1913, and thereafter viewed the Davis system with 'such a fishy eye' that he never used it (W. D. Jones in Martin 1950: 179).

In addition to this direct influence of tuition in Germany there was the effect of contacts made by Americans during expeditions abroad and at home. Among the thirty-two participants in the 'pilgrimage' led by Davis in Western Europe in August–September 1911 were five scholars from German universities, including Alfred Rühl from Marburg. The American Transcontinental Excursion of 1912, also managed by Davis, included forty-eight participants from Europe. Davis naturally lost no opportunity of expounding his own ideas on physical geography and one discussion ended with Dr Niermeyer telling him 'that he was no geographer but a geomorphologist'. We have already (volume 2, pp. 509–15) referred to Alfred Hettner's attack on Davisian methodology, and will return to this disagreement shortly. However, Hettner did influence Davis in the latter's attempt at an ideal regional synthesis provided by *Der Valdarno: eine Darstellungstudie* (1914K) and much later when addressing American geographers on the progress of geography in their own country Davis (1924G) discussed the 'material filling of terrestrial spaces' (*die dingliche Erfüllung der Erdräume*) and ascribed the phrase to Hettner when, as Hartshorne (1939, 1981: 146) has shown, it was the classical phrase used by Carl Ritter long before Hettner was born.

Davis and Germany

At first, Davis had considerable success in introducing his cyclic and presentational ideas into academic circles in Germany. The chief purpose of his stay at the University of Berlin, from late 1908 to early 1909, was to advertise his cyclic scheme and explanatory method of description (see volume 2, Chapter 15). He had taught himself German and sometimes conversed and lectured in that language, although he was never thoroughly at ease with it. His fullest geomorphological text, *Die Erklärende Beschreibung der Landformen* (1912J), was written in English and translated by Alfred Rühl; similarly his *Physical Geography* (1898E) was translated and adapted for German use with the collaboration of Gustav Braun (1911M; second edition 1917). Davis arrived at Berlin at a time when the Germans were enjoying a new political unity and were becoming commercial and colonial rivals of the British. Every important German university now had a professor of geography, and most of these were skilled geologists. The period from 1905 to 1914 is generally regarded as the golden age of German geography except that for some geographers the physical aspect, and especially physiography, was overemphasized. This physical bias must be attributed partly to the influence of Albrecht Penck who, since von Richthofen's death in 1904, had occupied the chair of geography at the University of Berlin. Davis himself also provided a strong impetus for landform studies and undoubtedly his excellent teaching at first proved widely popular, particularly with young German geographers. W. D. Jones, who was then in Germany, reported that Davisian ideas swept over the state like wildfire.

Davis enjoyed a cordial association with Albrecht Penck dating from around the turn of the century (volume 2, pp. 269–78) and exchanged with him in 1908–9. However, Penck did not mention the cyclic theme in his chapters on the earth's surface in Scobel's *Geographisches Handbuch* for 1895 but acknowledged it in the 1908 edition after the first major treatment of the cycle was presented in Germany by Martha Krug-Genthe in 1903. In 1911 in a review of Davis' *Physical Geography* Penck described the book as 'an excellent introduction' to the study of landforms, although somewhat inadequate in its treatment of tectonics and rock types (Tilley 1968). No doubt to Davis, however, the popularization of his ideas in Germany seemed too slow and probably he decided to speed up the process by intervening personally. His 1908–9 visit and the publication in German of his two main texts were having a decidedly beneficial effect when further expansion was hindered by the outbreak of war (1914–18) and by the antagonism which this caused (volume 2, pp. 477–8).

The change in attitude, professional and personal, between Davis and the Pencks became apparent before 1919 when Albrecht Penck published his *Die Gipfelflur der Alpen* which, in opposition to Davis' theory that summit peaks were eroded out of uplifted peneplains, explains them by a concept involving the relationship between weathering and altitude, and between rate of upheaval and rate of incision. As we have seen (volume 2, pp. 528–36), Davis wrote a severe criticism of this work (1923J). This anti-Davisian attitude in the work of Albrecht Penck must be attributed partly to his son, Walther, who like his father was a native of the Alpine fold system, and from 1912 onward became intrigued with the intense folding of the Andes. He and his father increasingly emphasized the influence that the rate of uplift exerted upon the gradient of valley-side slopes. Walther especially soon evolved a tectonically-controlled philosophy which discarded the idea of quickly up-lifted landscapes degrading progressively from youth to old age (see chapters 10 and 11). The friendship between Davis and the Pencks was strained further in 1920 when Davis (1920G) wrote a severely critical review of the *Festband* celebrating Albrecht's sixtieth birthday (1918) (see volume 2, pp. 516–19). Davis made no allowance for the disruptive effect of 4 years of war on German life. He chastised the authors of some of the chapters and generally considered the volume to be far too geological and most inaptly presented. Albrecht replied at length by letter, in which he obviously expressed serious doubts on the validity of Davis' cyclic concept. Davis answered at even greater length in a superb epistle (3 April 1921), inter-spersed with passages in German (presumably from Penck's letter) which we have already published (volume 2, pp. 519–27). Davis and Penck met for the last time in Arizona in 1927.

Walther Penck died prematurely, aged 35, in 1923 and the full impact of his theories did not seriously affect geomorphology among English-speaking nations until after the close of the Second World War (1939–45). There had been, it is true, a symposium on Penck's contribution to geomorphology held in Chicago in 1939 (Von Engeln 1940; see volume 2, pp. 715–17) and in 1942 Von Engeln devoted more than a whole chapter to a Davisian interpretation of Walther Penck's theory in his *Geomorphology*, but this attractive book was not easily available in Europe until it was reprinted in 1948. Consequently, British readers did not have ready access to the full text of Penck's *Die morphologische Analyse* (1924) until it was translated into English in 1953. In America Isaiah Bowman had discussed the original German edition and had considered that Penck's position represented 'a more explicit and detailed consideration of the *complications* of the cycle that Davis treats more briefly' (Bowman 1926: 125; see volume 2, pp. 694–9). This assessment lulled most American geomorphologists into accepting that the differences between the concepts of the two men were of degree and not of kind. However, the conflict between them took another turn in 1932 when Davis (1932G) gave an interpretation of Walther Penck's posthumous article (1925) on the piedmont-flats of the southern Black Forest. In this, as we have already discussed (volume 2, pp. 697–717), Davis mistranslated and misrepresented Penck's ideas on slope development. This misfortune was not fully exposed in print until 1957 (Tuan 1957) and not publicized widely until 1962 (Simons 1962).

The antagonism of Walther Penck must not be allowed to overshadow the continued opposition to Davis of his father who published several of his son's main writings posthumously. Between the wars geologists, such as Schuchert

(1923) and Grabau (1936), propagated ideas of a pulsating sea level to explain alternations of sedimentary sequences (see chapter 3) and inevitably alternating land uplift and depression also had its proponents. By 1928 Albrecht Penck discarded the idea of landforms degrading progressively and had replaced it with a (steady-state?) scheme developed under 'a continuous up and down surge within narrow limits' (Tilley 1968: 267). Such a scheme obviously did not accord with Davis' simple concept. Neither of the Pencks felt disposed to acknowledge that Davis had several times asserted that he was prepared to admit the possibility of different types of uplift affecting the landforms developed on them.

The Penckian attack was primarily geomorphological, whereas the opposition of Davis' two other chief German opponents, Siegfried Passarge and Alfred Hettner (plate 4.7) was all-embracing, began at an early date and continued after Davis' death. Passarge's replies to American criticisms of his own ideas on geomorphology (Passarge 1912) – that they were ponderous, too catalogic, and hyperempirical – have already been discussed (volume 2, pp. 501–9) and will be returned to again in chapter 11. His anti-Davisian writings concentrated on Davis' ignorance of process and had considerable effect in Germany, partly because they coincided with the persistent criticisms published by Hettner, a master of geographical method and no mean geomorphologist, whose opposition has been considered to be 'the most direct, forceful and effective challenge levelled against Davis' (Tilley 1968: 266) (see volume 2, pp. 509–15). Hettner, according to his own account, was unaware of Davis' concept of a cycle of erosion when he (Hettner) argued for a very similar theory in 1887–8 in order to explain striking flattenings developed in the sandstone of 'Saxon Switzerland'. However, he had discarded that argument by 1903 (Hettner 1960: 51, 64–5). After a varied career, including extensive travels in South America, Hettner accepted a professorship at Heidelberg in 1899 and held that post until he retired in 1928. He founded the *Geographische Zeitschrift* in 1895 and edited it until 1935, using it to publicize his ideas on the scope and methods of geography. From 1910 to 1924 he published in it twelve of his own articles concerned with attacking Davis' ideas on landforms or landscapes and on geographical methodology (1910: 365–84, 1911: 135–44, 1912: 665–82, 1913a, b, c: 153–61, 185–202, 435–45, 1914a, b: 129–45, 185–97, 1919: 341–52, 1920: 131–6, 1923: 37–59, 1924: 286–90). In 1921 most of these articles were elaborated and combined into *Die Oberflächenformen des Festlands* (The surface forms of the continents) which, unlike the individual articles, provoked a strong response from Davis (1923E) under the title 'The explanatory description of landforms', which was in itself provocative as Hettner was essentially hostile to Davis' 'explanatory description'. Hettner's volume strongly attacked any flaws and uncertainties in Davisian themes. Some minor landforms – caverns etc. – had been neglected; the different ways in which one kind of rock may

Plate 4.7 Alfred Hettner (1859–1941)

Source: From *Geographers: Biobibliographical Studies* No. 6, Mansell, London, 1987

react to different subaerial processes were inadequately discussed; the fact that different initial conditions and different processes may create apparently identical forms was disregarded; deduction is grossly overdone and deductively-won explanations were too often unverified; the effects of downward movements were almost ignored; some of the surfaces called 'peneplains' appeared to be merely features of a particular climate and not a penultimate stage in all climates; climatic change deserved much more attention; the whole concept of progressive landform stages was arbitrary and unreal; and so on.

Davis' reply was quick and violent (see volume 2, pp. 512–15):

> The book as a whole is not so much characterized by novel and constructive suggestions as by homilies, truisms, hesitations, obstructive misunderstandings [of Davis], and disputatious objections ... it is in reality for the most part a diatribe against the scheme of the erosion cycle .... .
>
> (Davis 1923E: 318–19)

He criticized other weaknesses and inconsistencies in detail, and many of these were corrected in Hettner's second edition (1928) which was translated into English by P. Tilley in 1972. Davis never replied to this improved edition and died probably believing that Hettner had not appreciated Davisian methods of simplification and systematization intended to make geography generally attractive. But there can be no doubt about Hettner's opinion of Davis' contribution to geomorphology and of the need to revert to pre-Davisian methods:

> I can see Davis' approach only as an episode, not as a step forward in geomorphology. Its simplicity and the energy of its advocates have rapidly won for it a wide circle of adherents; it has enlivened geomorphological research and led to a number of correct results. But as a whole it has been abortive, and studies founded upon it have produced many fallacies .... . The earlier theory now condemned as backward was in fact on the right lines; it is upon this that we must build.
>
> (Hettner 1928; see 1972: 268)

As will be shown in chapter 12, by the time of the Düsseldorf conference of 1926 the German concern for climatic (morphogenetic) geomorphology had joined that of tectonic control to provide the twin stimuli for twentieth-century geomorphology in Central Europe. However, Davis' influence was not entirely extinguished. Long after the issue of a second edition of *Grundzüge der Physiogeographie* (1917G) and of *Die Erklärende Beschreibung der Landformen* (1924L), and the publication, in collaboration with Karl Oestreich, of a manual of map reading (1918B) reference to, and criticism of, Davis' work continued in the 1930s. Undoubtedly the most durable German reference to Davis was contained in the publications of Fritz Machatschek who had studied under Albrecht Penck in Vienna and later held the chair of physical geography at that University during the period 1927–34, before transferring to Munich until his retirement in 1946. Machatschek was a prolific and well-known geomorphologist, whose *Das Relief der Erde* (1938–40) was considered by one British reviewer to be indispensable to those interested in world morphology. His earlier *Geomorphologie*, first produced in 1919, proved exceptionally popular and, successively modified and enlarged, went into a ninth edition in 1968 and was translated into English in 1969, after the death of the author in 1957. The book was a standard text in Germany for more than a generation and Machatschek represents a definite standpoint, impressed on him by the development of his subject, ... [He] emphasises the connection of the surface forms with the tectonic movements of the earth's crust, but at the same time accepts the validity, as a methodological concept, of the cyclic theories of W. M. Davis.

(H. Graul and C. Rathjens, foreword to the eighth edition)

Thus decades after the death of Davis there persisted in Germany a text which incorporated some of Davis' ideas with those of Albrecht Penck and others:

Relying on a knowledge of how exogenous processes operate, one can attempt a deductive account of the evolution of a landscape employing various simplifying hypotheses. Such is the cycle theory of W. M. Davis first proposed in 1898. . . . We will here discuss only what he called the 'normal' cycle, where the landforms are the result of flowing water and accompanying mass movement.

(Machatschek 1969: 61)

After discussion of mature and old-age forms and interruptions in the cycle, we read:

Davis's doctrine of cycles ... received considerable, often uncritical, acceptance largely because of its strict adherence to the idea of development, but, on the other hand, [received] also lively opposition from Hettner, Passarge and others. Their criticisms were directed as much against the terminology as against the method.

(Machatschek 1969: 62)

Certain difficulties in accepting the application of the Davisian method are listed, among them: separate parts of a landscape although of the same absolute age may be quite different in appearance, or appear to belong to different stages, because of differences in rock resistance or in the intensity and nature of the erosive process; the concept and application of the peneplain may be objected to because surfaces may be planated by various processes; Davis' simplified use of rapid uplift and prolonged standstill is overgeneralized, being unreal and impracticable; and account must always be taken of the relative intensity of uplift and erosion.

The expressions young, mature and old are valid as designations for stages in a development sequence only in the particular instance contemplated by Davis, when elevation does not last long and further development proceeds during tectonic rest. Therefore W. Penck replaced these expressions by neutral, flat, medium, and strong relief and spoke according to the relationship between elevation and erosion, of uniform, waxing, and waning development, and the slopes appertaining to them.

(Machatschek 1969: 64)

Machatschek, who contributed an essay on high deserts to the Düsseldorf conference (see chapter 12), deals briefly with Davis' deductive treatment of the arid cycle and suggests that it might be applicable to the deserts of east Persia and Tibet but is certainly not valid for the tableland deserts of North

Africa and Arabia. The general impression which the reader gleans from the final editions of Machatschek's *Geomorphologie* is that Davis' influence in Germany had waned and had been fragmented until it survived not as a practicable philosophy but as many specific items, mere relics amid an expanding, ever-changing geomorphological theme.

The Davisian Decline in the USA

It would be as unrealistic to assume a complete Davisian dominance in the United States during the period 1890–1950 as it would a complete Davisian eclipse in Germany. As will be shown in chapter 7, Davisian cyclic ideas manifested themselves extremely strongly in the United States through the medium of denudation chronology, but it would be quite wrong to give the impression that an overwhelming majority of American geomorphologists were Davisian in their basic outlook. This was true from the beginning of the century and support for Davis declined significantly in the later decades, particularly after his death in 1934. The present volume, dealing primarily with regional and historical interpretations of landforms, naturally tends to emphasize Davis' influence but, as will become clear in volume 4 which treats studies of process and the Quaternary aspects of geomorphology, a role-call of American geomorphologists of the period concerned with such studies impresses one with their number, intellectual stature and freedom from Davisian dogma. They include:

Fluvial processes: H. A. Einstein, H. N. Fisk, J. F. Friedkin, G. K. Gilbert, W. S. Glock, J. T. Hack, R. E. Horton, J. E. Kesseli, W. B. Langbein, J. B. Leighly, L. B. Leopold, J. H. Mackin, G. P. Merrill, C. Nevin, P. Reiche and A. N. Strahler.

Glacial processes: W. C. Alden, E. Antevs, W. W. Atwood Sr, W. W. Atwood Jr, J. H. Bretz, R. T. Chamberlin, M. Demorest, R. F. Flint, J. W. Goldthwait, L. Horberg, W. D. Johnson, F. E. Matthes, I. C. Russell, R. P. Sharp, R. S. Tarr, W. Upham and J. B. Woodworth.

Periglacial processes: C. D. Holmes, M. M. Leighton, T. L. Péwé and S. Taber.

Arid and semi-arid processes: E. Blackwelder, K. Bryan, A. C. Lawson, C. R. Longwell, WJ McGee, J. L. Rich, G. H. Smith and B. Willis.

Coastal processes: W. C. Krumbein, W. C. Putnam, R. J. Russell and F. P. Shepard.

Climatic and Karst processes: C. A. Malott, H. T. Stearns, W. D. Thornbury, S. S. Visher and C. K. Wentworth.

The symposium on Walther Penck's contribution to geomorphology (Von Engeln 1940), although it rallied support for Davis from Douglas Johnson and others, provided a forum for the anti-Davisian ideas of such authorities as Kirk Bryan. In 1950 a student of Johnson (plate 4.8) wrote:

Plate 4.8 D. W. Johnson on a Columbia University field excursion near Petersburg, West Virginia, in June 1939. Left to right: D. Babenroth, R. Mahard, H. M. Fridley, D. W. Johnson, unknown, S. Neuschel, C. W. Carlston, A. N. Strahler

Source: Photo by M. Hall Taylor Jr., courtesy A. N. Strahler

Even as recently as 1943, Douglas Johnson presented his graduate classes in geomorphology with subject matter faithfully reproducing the principles and details as written by Davis 45 years earlier.

(Strahler 1950a: 210)

In the same year Peltier's (1950) attempt to popularize the last of the Davisian-style cycles – the periglacial – was unsuccessful and served only to draw attention to continental work on climatic geomorphology (see chapter 12). Shortly afterwards John Leighly, a colleague of Carl Sauer, wrote: 'Much of Davisian geomorphology had become stale and unprofitable before its author's death' (Leighly 1955: 317).

The decline of the Davisian cycle as a basis for American geomorphology was due to a combination of drawbacks:

1 It was too theoretical (see volume 2, pp. 192–7).
2 It was required to fulfil too many different roles and this was compounded by Davis' attempts to accommodate both the equilibrium ideas of Gilbert (see volume 1, chapter 28 and volume 2, p. 196) and the tectonic ideas of the Pencks (see volume 2, chapters 22, 23 and 28).
3 It largely ignored the nuances of the effects of climate (see chapter 12) and of climatic change (see chapter 2).
4 It became increasingly associated with an ambiguous, speculative and poorly dated denudation chronology (see chapters 7 and 9).
5 It was tectonically and diastrophically naive (see chapters 1, 10 and 11, and volume 2, pp. 515–36).
6 It denied a possible unifying role for eustatism (see chapter 3).
7 It elevated 'time' to the level of a mechanism of landform production.
8 It was too wedded to denudation.
9 It provided a framework which was not capable of broad development and which ultimately became atrophied.
10 It ignored the possibility that similar forms might be produced by different means under different conditions (see Hettner 1921).
11 It was entirely qualitative and unscientific (see volume 2, pp. 753–4).
12 It was basically not concerned with process (see volume 4) and lacked any real attempt to relate form to process.

These drawbacks culminated after the Second World War in a revolution in American geomorphology which will be treated in volume 4. This revolution was spearheaded by four publications dealing with process. Horton (1945) rekindled a genuine concern for the relationship between fluvial processes and geomorphic forms (i.e. morphometry), employing a drainage network model combining both evolutionary and steady-state concepts. Mackin (1948) resolved the long-standing problem of grade by introducing thermodynamic analogies. Strahler (1950b) made the classic dichotomy between the views of Davis and W. Penck on slope development largely irrelevant with the use of quantitative methods, and introduced systems thinking to geomorphology. Finally, Leopold and Maddock (1953) illuminated the equilibrium concept of process/response in the context of hydraulic geometry.

The 1950s witnessed a new beginning for American landform studies, although Davis' *Geographical Essays* (1909C) were re-issued in 1954. In that year Preston James and Clarence Jones edited *American Geography: Inventory and Prospect* which, in discussing the influence of Davis on mid-twentieth century geomorphology, gave the impression that it was already almost of past interest only. Perhaps it was a case of 'a prophet is not without honour, save in his own country'. Strange to say, the Old World rushed in to redress the balance of the New; the Soviet Professor, I. P. Gerasimov (1956), in a long review of the volume wrote:

> At the beginning of the survey, the figure of Davis emerges in the foreground and he is characterized as 'ruler of the thoughts' of American geomorphologists at the beginning of the present century. Later, however, Davis is gradually dethroned . . . . Today this greatest of all figures in American geography is relegated almost entirely to the past. But this is hardly justified, at least so far as world geomorphology is concerned, which has every ground for setting forth its own claims to W. M. Davis . . . . Without expounding here the issue of Davis' methodology, one must acknowledge as an indisputed fact the tremendous world-wide influence of his scientific ideas on the development of geomorphology. This influence has a creative effect. In substance it amounted to the fact that Davis, by virtue of the organic wholeness of his scientific outlook, his great polemic acuity, the richness of his presentation, and his inexhaustible energy, succeeded in definitely separating geomorphology, as a special independent science, from general geology and physical geography and in creating for this new science a large and lasting authority . . . . He elevated the significance of geomorphology to a high level as a science, and from this point of view Davis achieved more than any other among the most famous European and American scholars.

References

Atwood, W. W., Sr (1940) *The Physiographic Provinces of North America*, Boston: Ginn.

Beckinsale, R. P. (1976) 'The international influence of William Morris Davis', *Geographical Review* 66: 448–66.

—— (1981) 'W. M. Davis and American Geography: 1880–1934', B. W. Blouet (ed.) *The Origins of Academic Geography in the United States*, Hamden, CT: Archon Books, pp. 107–22.

Billings, M. P. (1938) 'Physiographic relations of the Lewis overthrust in northern Montana', *American Journal of Science* 235: 260–72.

Blackwelder, E. (1928) 'The recognition of fault scarps', *Journal of Geology* 36: 289–311.

Bowman, I. (1911) *Forest Physiography*, New York: Wiley.

—— (1916) *The Andes of Southern Peru: Geographical reconnaissance along the seventy-third meridian*, New York: American Geographical Society.

—— (1924) *Desert Trails of Atacama*, special publication no. 5, New York: American Geographical Society.

—— (1926) 'The analysis of land forms', *Geographical Review* 16: 122–32.

Brigham, A. P. (1909) 'William Morris Davis', *Geographen Kalender*, year 7, Gotha: Justus Perthes, pp. 1–73.

Bryan, K. (1923) 'Erosion and sedimentation in the Papago Country, Arizona', *US Geological Survey Bulletin* 730, pp. 19–90.

—— (1941) 'Physiography 1888–1938', *50th Annual Volume of the Geological Society of America*, pp. 1–15.

—— (1950) 'The place of geomorphology in the geographic sciences', *Annals of the Association of American Geographers* 40: 196–208.

Chamberlin, T. C. and Salisbury, R. D. (1904) *Geology*, 3 vols, London: Murray, vol. 1 Geologic processes and their results (2nd edn 1909).

Dana, J. D. (1863) *Manual of Geology: Treating the principles of the science*, Philadelphia: Bliss.

Dunbar, G. S. (1981) 'Credentialism and careerism in American geography, 1890–1915', in B. W. Blouet (ed.) *The Origins of Academic Geography in the United States*, Hamden, CT: Archon Books, pp. 71–88.

Eardley, A. J. (1934) 'Structure and physiography of the southern Wasatch Mountains', *Papers of the Michigan Academy of Science* 19: 377–400.

Fenneman, N. M. (1931) *Physiography of Western United States*, New York: McGraw-Hill.

—— (1936) 'Cyclic and non-cyclic aspects of erosion', *Science* 83: 87–94.

—— (1938) *Physiography of Eastern United States*, New York: McGraw-Hill.

—— (1939) 'The rise of physiography', *Bulletin of the Geological Society of America* 50: 349–60.

Gerasimov, I. P. (1956) 'Review of "American Geography: Inventory and Prospect"', *Bulletin of the Academy of Sciences of the USSR, Geographic Series* no. 1, trans. O. A. Titelbaum, pp. 115–43.

Gilluly, J. (1937) 'The physiography of the Ajo region, Arizona', *Bulletin of the Geological Society of America* 48: 323–47.

Grabau, A. W. (1936) 'Oscillation or pulsation', *Report of the XVI International Geological Congress, Washington 1933*, vol. 1, pp. 539–53.

Gregory, H. E. (1917) 'Geology of the Navajo Country: A geographic and hydrographic reconnaissance', *US Geological Survey Professional Paper* 93.

—— (1938) 'The San Juan country', *US Geological Survey Professional Paper* 188.

—— (1951) 'The geology and geography of the Paunsaugunt region, Utah', *US Geological Survey Professional Paper* 226.

—— (1952) 'Geology and geography of the Zion Park region, Utah and Arizona', *US Geological Survey Professional Paper* 220.

Gregory, H. E. and Moore, R. C. (1931) 'The Kaiparowits region', *US Geological Survey Professional Paper* 164.

Hall, R. B. (1953) 'William Herbert Hobbs, 1864–1953', *Annals of the Association of American Geographers* 43: 284–8.

Hartshorne, R. (1939) 'The nature of geography', *Annals of the Association of American Geographers* 29: 173–469.

—— (1981) 'William Morris Davis – The course of development of his concept of geography', in B. W. Blouet (ed.) *The Origins of Academic Geography in the United States*, Hamden CT: Archon Books, pp. 139–49.

Hettner, A. (1910) 'Die Arbeit des Fliessen des Wassers', *Geographische Zeitschrift* 16: 365–84.

—— (1911) 'Die Terminologie der Oberflächenformen', *Geographische Zeitschrift* 17: 135–44.

—— (1912) 'Alter und Form der Täler', *Geographische Zeitschrift* 18: 665–82.

—— (1913a) 'Die Enstehung des Talnetges', *Geographische Zeitschrift* 19: 153–61.

—— (1913b) 'Rumpfflächen und Pseudorumpfflächen', *Geographische Zeitschrift* 19: 185–202.

—— (1913c) 'Die Abhängigkeit der Form der Landoberfläche von inneren Bau', *Geographische Zeitschrift* 19: 435–45.

—— (1914a) 'Die Entwicklung der Landoberfläche', *Geographische Zeitschrift* 20: 129–45.

—— (1914b) 'Die Vorgange der Umlagerung an der Erdoberfläche und die morphologische Korrelation', *Geographische Zeitschrift* 20: 185–97.

—— (1919) 'Die morphologische Forschung', *Geographische Zeitschrift* 25: 341–52.

—— (1920) 'Die morphologische Darstellung', *Geographische Zeitschrift* 26: 131–6.

—— (1921) *Die Oberflächenformen des Festlandes, ihre Untersuchung und Darstellung; Probleme und Methoden der Morphologie*, Leipzig: Teubner (2nd edn 1928), English translation by P. Tilley (1972) *The Surface Features of the Land*, London: Macmillan.

—— (1923) 'Methodische Zeit – und Streitfragen', *Geographische Zeitschrift* 29: 37–59.

—— (1924) 'Noch einmal die leidigen Fastbenen', *Geographische Zeitschrift* 30: 286–90.

—— (1960) 'Aus meinen Leben', *Heidelberger Geographische Arbeiten* 6: 51–65.

Hobbs, W. H. (1912) *Earth Features and their Meaning*, New York: Macmillan (2nd edn 1931).

Horton, R. E. (1945) 'Erosional development of streams and their drainage basins: Hydrophysical approach to quantitative morphology', *Bulletin of the Geological Society of America* 56: 275–370.

Hunt, C., Averitt, P. and Miller, R. L. (1953) 'Geology and geography of the Henry Mountains region, Utah', *US Geological Survey Professional Paper* 228.

Huxley, T. H. (1877) *Physiography: An introduction to the study of nature*, London: Macmillan.

James, P. E. (1981) 'Geographical ideas in America, 1890–1914', in B. W. Blouet (ed.) *The Origins of Academic Geography in the United States*, Hamden, CT: Archon Books, pp. 319–26.

Krug-Genthe, M. (1903) 'Die Geographie in den Vereinigten Staaten', *Geographische Zeitschrift* 9: 626–37, 666–85.

Lake, P. (1915) *Physical Geography*, Cambridge (new printings and editions until the 1950s).

Leighly, J. (1955) 'What has happened to physical geography?' *Annals of the Association of American Geographers* 45: 309–18.

—— (1976) 'Carl Ortwin Sauer, 1889–1975', *Annals of the Association of American Geographers* 66: 337–48.

Leopold, L. B. and Maddock, T. (1953) 'The hydraulic geometry of stream channels and some physiographic implications', *US Geological Survey Professional Paper* 252.

Lobeck, A. K. (1939) *Geomorphology: An introduction to the study of landscapes*, New York: McGraw-Hill.

Loomis, F. B. (1937) *Physiography of the United States*, New York: Doubleday, Doran and Co.

McGee, WJ (1888) 'The geology at the head of Chesapeake Bay', *7th Annual Report, 1885–86, US Geological Survey*, pp. 537–46.

—— (1891) 'The Pleistocene history of northeastern Iowa', *11th Annual Report, 1889–90, US Geological Survey*, pp. 189–577.

Machatschek, F. (1919) *Geomorphologie*, Leipzig and Berlin (2nd edn 1934, 6th edn 1954, 9th edn 1968) (English transl. by D. J. Davis, 1969, Edinburgh: Oliver & Boyd).

—— (1938–40) *Das Relief der Erde: Versuch einer regionalen Morphologie der Erdoberfläche*, 2 vols, Berlin: Bornträger (2nd edn 1955).

Mackin, J. H. (1948) 'Concept of the graded river', *Bulletin of the Geological Society of America* 59: 463–512.

Mackinder, H. J. (1895) 'Modern geography, German and English', *Geographical Journal* 6: 367–79.

Martin, G. J. (1968) *Mark Jefferson: Geographer*, Ypsilanti: East Michigan University Press.

—— (1981) 'Ontography and Davisian physiography', in B. W. Blouet (ed.) *The Origins of Academic Geography in the United States*, Hamden, CT: Archon Books, pp. 279–89.

Martin, L. (1950) 'William Morris Davis: Investigator, teacher and leader in geography', *Annals of the Association of American Geographers* 40: 172–80.

Mather, K. F. (1971) 'Thomas Chrowder Chamberlin (1843–1928)', *Dictionary of Scientific Biography*, vol. III, New York: Scribners, pp. 189–91.

Meyerhoff, H. A. (1940) 'Migration of erosion surfaces', *Annals of the Association of American Geographers* 30: 247–54.

Passarge, S. (1912) 'Physiologische Morphologie', *Mitteilungen Geographisches Gesellschaft (Hamburg)* 26: 133–337.

Pattison, W. D. (1981) 'Rollin Salisbury and the establishment of geography at the University of Chicago', in B. W. Blouet (ed.) *The Origins of Academic Geography in the United States*, Hamden CT: Archon Books, pp. 151–63.

Peltier, L. C. (1950) 'The geographic cycle in periglacial regions as it is related to climatic geomorphology', *Annals of the Association of American Geographers* 40: 214–36.

Penck, A. (1911) 'Die Physiographie von Davis und Braun', *Zeitschrift für Erdkunde zu Berlin* 46: 560–70.

—— (1919) 'Die Gipfelflur der Alpen', *Sitzungsberichte der Preussischen Academie der Wissenschaften* 17: 256–63.

Penck, W. (1924) 'Die morphologische Analyse: Ein Kapitel der physikalischen Geologie', *Geographische Abhandlungen*, 2nd series, vol. 2, pp. 1–283 (published separately in the same year by J. Engelhorn of Stuttgart and translated into English by H. Czech and K. C. Boswell (1953) *Morphological Analysis of Landforms*, London: Macmillan).

—— (1925) 'Die Piedmontflächen des südlichen Schwarzwaldes', *Zeitschrift der Gesellschaft für Erdkunde zu Berlin*, pp. 83–108.

Powell, J. W. (ed.) (1896) *The Physiography of the United States*, New York: National Geographic Society.

Rich, J. L. (1933) 'Physiography and structure at Cumberland Gap', *Bulletin of the Geological Society of America* 44: 1219–36.

—— (1934) 'Mechanics of low-angle overthrust faulting as illustrated by Cumberland thrust block, Virginia, Kentucky and Tennessee', *Bulletin of the Association of American Petroleum Geologists* 18: 1584–96.

Russell, I. C. (1898) *River Development*, London: Murray.

Russell R. J. (1949) 'Geographical geomorphology', *Annals of the Association of American Geographers* 39: 1–11.

—— (1958) 'Geological geomorphology', *Bulletin of the Geological Society of America* 69: 1–22.

Salisbury, R. D. (1907) *Physiography*, New York: Holt (3rd edn 1919).

Sauer, C. O. (1925) 'The morphology of landscape', *University of California Publications in Geography* 2: 19–53.

—— (1929) 'Landforms in the Peninsular Range of California, as developed about Warner's Hot Springs', *University of California Publications in Geography* 3: 199–290.

Schuchert, C. (1923) 'Sites and nature of North American geosynclines', *Bulletin of the Geological Society of America* 34: 157–230.

Scobel, A. (1895) *Geographisches Handbuch*, Leipzig, Velhagen and Klasing.

Sharp, H. S. (1938–40) 'Geomorphic notes on maps', *Journal of Geomorphology* 1: 67–9, 247–8, 345–7; 2: 73–4, 161–2, 258–60, 373–4; 3: 65–6, 163–5.

Sharp, R. P. (1940) 'Geomorphology of the Ruby-East Humboldt Range, Nevada', *Bulletin of the Geological Society of America* 51: 337–71.

Simons, M. (1962) 'The morphological analysis of landforms: A new review of the work of Walther Penck', *Transactions of the Institute of British Geographers* 31: 1–13.

Stoddart, D. R. (1975) '"That Victorian science": Huxley's *Physiography* and its impact on geography', *Transactions of the Institute of British Geographers* 66: 17–40.

—— (1986) 'That Victorian science', in D. R. Stoddart *On Geography and its History*, Oxford: Blackwell, pp. 180–218.

Strahler, A. N. (1944) 'Valleys and parks of the Kaibab and Coconino Plateaus, Arizona', *Journal of Geology* 52: 361–87.

—— (1948) 'Geomorphology and structure of the West Kaibab fault zone and Kaibab Plateau', *Bulletin of the Geological Society of America* 59: 513–40.

—— (1950a) 'Davis' concepts of slope development viewed in the light of recent quantitative investigations', *Annals of the Association of American Geographers* 40: 209–13.

—— (1950b) 'Equilibrium theory of erosional slopes approached by frequency distribution analysis', *American Journal of Science* 249: 673–96, 800–14.

—— (1951) *Physical Geography*, New York: Wiley.

—— (1952) 'Dynamic basis of geomorphology', *Bulletin of the Geological Society of America* 62: 923–38.

Tarr, R. S. (1895) *Elementary Physical Geography*, New York: Macmillan.

—— (1898) 'The peneplain', *American Geologist* 21: 351–70.

—— (1903) *New Physical Geography*, New York: Macmillan.

Tarr, R. S. and Martin, L. (1914) *College Physiography*, New York: Macmillan (2nd edn 1920).

Tilley, P. (1968) 'Early challenges to Davis' concept of the cycle of erosion', *Professional Geographer* 20: 265–9.

Tuan, Y-F. (1957) 'The misleading antithesis of Penckian and Davisian concepts of slope retreat in waning development', *Proceedings of the Indiana Academy of Science* 67: 212–4.

Von Engeln, O. D. (ed.) (1940) 'Symposium: Walther Penck's contribution to geomorphology', *Annals of the Association of American Geographers* 30: 219–84.

—— (1942) *Geomorphology: Systematic and regional*, New York: Macmillan.

Wooldridge, S. W. and Morgan, R. S. (1937) *The Physical Basis of Geography: Geomorphology*, London: Longmans, Green and Co.

Worcester, P. G. (1939) *A Textbook of Geomorphology*, New York: Van Nostrand.

CHAPTER FIVE

The Davisian Cycle in France and Britain

Although in eighteenth-century France there was no organized study of landforms, there was considerable interest in the subject which often appeared incidentally or as part of other work (Ellenberger 1986). We have already praised Desmarest (see volume 1, p. 17) and other eighteenth-century French scientists who made notable advances in hydrology and hydraulics. Outstanding among them were Henri Pitot, Charles Bossut (see volume 1, pp. 87–8) and Louis Gabriel Du Buat whose *Principes d' Hydraulique* (1779, 1786), verified by numerous experiments made on behalf of the government, was for long a valued text (see volume 1, pp. 86–93). In the next century this hydraulic tradition was continued notably by the civil engineer Alexandre Surell in his *Etude sur les Torrents des Hautes–Alpes* (1841), which contains a clear description of the concept of 'grade' in the long profile of mountain streams (see volume 1, pp. 283–7). The conjunction of hydraulics and geology appears in a small book entitled *Hydrogéologie* (1802) by Jean Baptiste de Lamarck (1744–1829) (see volume 1, pp. 81–2), a prolific author who is more famous for his writings on botany, invertebrate zoology, palaeontology and the theory of evolution. His book is subtitled, in the recent translation, 'Researches on the influence which waters exert on the surface of the terrestrial globe; on the reasons for the existence of ocean basins, for their displacement and successive migration over the different points of the globe's surface; finally on the changes that living bodies exercise upon the nature and state of that surface'. The approach was essentially that of Hutton and is strongly uniformitarian, the volume being merely a part of a longer planned work on terrestrial physics.

In the first half of the nineteenth century a few French geologists other than Lamarck attracted a wide attention. Among these was Elie de Beaumont whose writings included a theory on the parallel elevation of mountain chains (1829–30) which, however, he energized by catastrophic uplift (see chapter 1). More significant was his collaboration with Pierre Dufrénoy in their *Mémoires* (1838) of the geological map of France, in which they suggested that physical units such as the Paris Basin and the Central Massif should be distinguished on geological lines. This promising idea, which ran counter to Buache's scheme of regions based on drainage basins or watersheds (see chapter 11), was grossly neglected largely because academic geography was dominated in France at that time by historians. In any event, geography teaching there was generally weak or absent and, in spite of attempted reforms in 1857 and 1864, Goethe's remark made in 1808 that 'the French sport moustaches and are ignorant of geography' was still largely true when the Franco-Prussian war broke out in 1870. Lack of concerted action and neglect of opportunities degraded the French efforts into a surprisingly disastrous defeat, and at the Treaty of Frankfurt (10 May 1871) they were humiliated and suffered the loss of Alsace and Lorraine. Rather surprisingly, the poor quality of French geographical teaching was blamed as one of the chief reasons for the national débâcle.

Geomorphology and French National Security, 1870–90

The reaction to, and remedy for, the defeat of France in the Franco-Prussian war were almost as arresting as the disaster itself. The tremendous upsurge of geography has been discussed in detail by Numa Broc (1974b). Rarely, if ever, has geography figured so much in public thought, or its utility and the need to teach it more strongly recognized. After 1870 it became one of the main elements in both national revival and the urge for revenge. In the fields of primary and secondary education, geography was greatly strengthened and at least fifteen new geographical societies were founded to popularize the subject throughout France and to aid the growth of colonialism and commercialism. The chief propagator of this geographical renaissance was Ludovic Drapeyron (1839–1901) who, in an effort to turn it away from history to 'topography' (Drapeyron 1885) and natural science, founded the Société de Topographie in 1876 and the *Revue de Géographie Annuelle* in 1877. He was less successful in remoulding university teaching and his proposals to create a National School of Geography and a specialized degree course were rejected by the Sorbonne. But by the 1880s several universities had simple lectures in physical geography for their Faculty of Letters and more advanced courses (for specialists) turned towards Nature for Science students.

We, however, are more concerned with geomorphology than geography except that physical geography is likely to favour the study of landforms. The second congress of the International Geographical Union was held in Paris in 1875 and the participants included Ferdinand von Richthofen, Elie de Beaumont and August Daubrée (1814–96) whose large work, *Etude Synthétiques de Géologie Experimentale*, was published in Paris in 1879. The congress affirmed that one of the two fundamental aims of geography was 'the knowledge of the natural configuration of the earth's surface considered as an end in itself'. Thus geography must begin with 'topographie', or the official topographic map, and the direct study of the terrain.

The actual part played by geomorphological ideas in the congress was small and occupied only about 5 per cent of the printed proceedings. It consisted largely of discussion on the mobility of the earth's crust in historic time and its measurement; the origin of mountains; the relation between surface relief and geological structure; glaciation; the origin, displacements and geographical distribution of sands; and river meanders including the law controlling their curvature.

The condition of, and the attention paid to, 'geomorphology' in France before 1885 were equally insignificant and have been discussed by Numa Broc (1975). The rather elementary works by G. L. Figuier on *La Terre et les Mers* (1864) and by A. Dupaigne on *Les Montagnes* (1873) were founded on such a weak correlation between geology and geomorphology that acceptable classifications could not be proposed. Worse still they failed to recognize the fundamental role of fluvial erosion.

Avenues of progress were seen in experimental geology and in morphometry but, as will now be shown, both of these proved dead ends. In the experimental field the master experimenter was Gabriel August Daubrée (see volume 1, p. 607), a mining engineer and later professor of mineralogy and geology at Strasbourg and at Paris where after 1861 he carried out a brilliant series of experimental researches somewhat similar to those of James Hall in Scotland (Chorley 1971). By compressing glass, limestone and mastic Daubrée demonstrated nearly all types of tectonic variations – folds, faults, landslips, overlaps, etc. – and witnessed the formation of a network of fissure lines consisting of main fissures crossed more or less at right angles by secondary fissure lines. To him, this reticulation seemed reminiscent of hydrographic or stream networks and he tried to establish a constant relationship between the surface design (*modélé du sol*) and earth fissures, for example in Picardy between the Somme and Arques rivers. But he went too far in trying to explain minor relief or small relief features by faulting or earth movements, thereby subordinating systematic erosion to tectonics. In his splendid *Études Synthétiques de Géologie Experimentale* (pp. 364–5) he has folded maps on a scale of 1:80,000 of the meandering valley of the Charente south of Ruffec and these are overlaid by a transparency showing the fault network. The meanders are shown to be the points where main fault lines meet at an angle, which, he suggests, has led to the formation of *cirques d' érosion* (erosion combes or hollows) which caused the river to follow a meandering course (figure 5.1). Daubrée (1887a, b), also worked on subterranean hydrology.

Hope of geomorphological progress through morphometry or *orométrie* dates back at least to Humboldt who in 1816, by a comparison of their respective summit heights and col heights, had demonstrated that the Pyrenees were more 'massive' than the Alps. In 1842 he also calculated the mean height of the continents. About 10 years later Elie de Beaumont imagined a pentagonal network to explain the world surface pattern of continents and orographic axes. But the master of morphometry in this

Figure 5.1 Suggested influence of initial faulting (thick lines) on the formation of valleys, as exemplified by the Charente Valley, south of Ruffec, France

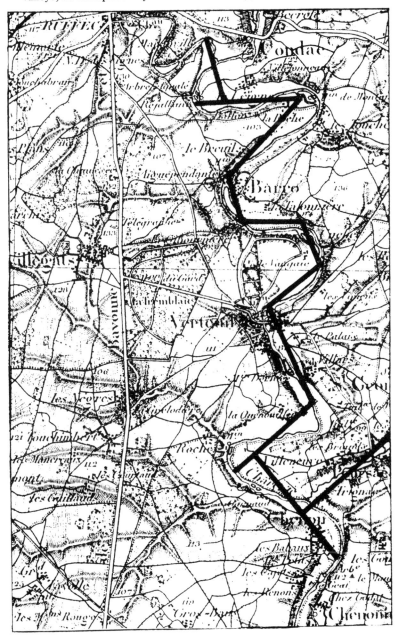

Source: From Daubrée (1879), plate V

Plate 5.1 Emmanuel J. de Margerie

Source: Courtesy of Professor Philippe Pinchemel

period was an Austrian officer, Colonel Karl von Sonklar, who worked especially in the Alps and whose *Allgemeine Orographie* (1873) is the bible of 'orométricians'. For mountain massifs he determined no less than twelve parameters such as mean height of summits, of cols and of summit line; average indentation of the chain; mean angle (gradient) of the slopes; mean height of valleys and their mean slope; and height of the base and total volume of the massif. Following Sonklar's method. L. Neumann wrote an *orométrie* of the Black Forest (1886) which he broke down hypothetically into a basal block surmounted by a series of polyhedrons representing the chains.

In Germany, Penck (1887) and others continued to look very favourably on morphometry, while in France, although Neumann's (1886) work on the Black Forest was warmly praised by E. de Margerie, it was taken up mainly by alpinists and the military. Eventually E. de Martonne in his *Traité de Géographie Physique* (1926: 517) admitted that, although morphometric calculations have an inspiring precision, morphometry cannot in itself form a science of landforms; the modern science must look in another direction.

Realization of the rather abortive nature of orometry and fracturalism happened to coincide with, or indeed probably to foster, the production of France's first great manual of landforms and its first outstanding geologist, Emmanuel de Margerie (1862–1953) (plate 5.1) whose longevity was heralded by his precocity and his remarkable facility in languages (Tobien 1974). When young, Margerie, the offspring of a cultured Parisian family, was attracted to the natural sciences by the teaching of Albert Dupaigne, and at the age of 15 he attended the lectures of Lapparent at the Institut Catholique in Paris and because of his linguistic skill took part in the international congress of geologists at Paris. In 1883 he produced the first of his prolific writings, the subject being the Colorado plateaus according to the work of American geologists. Thereafter, having added to his mastery of German and English a working knowledge of most European languages, he interpreted and assessed foreign publications on earth sciences for his countrymen.

In that same year he was introduced to Gaston de la Noë (1836–1902) (plate 5.2) who was director of the official topographic brigades and, as Margerie has recounted in his autobiographic recollections (*Critique et Géologie*, 1943: 1: 198), there quickly sprang up close mutual friendship and trust. They formed an ideal combination; the elder, erudite on traditional topographic lines, and the younger, a scholar and insatiable reader, capable of grasping the significance of landform literature and of seeing the main gist of an argument while appreciating the importance of particular details. There was, of course, by now sufficient literature to set the percipient earth scientists on the right path. They could and did draw on, for example, the researches of French topographers and hydraulic engineers, including B. Dausse's *Etudes relatives aux inundations et à l'endiguement des rivières* (1872); of famous British geologists such as Lyell, Ramsay and the Geikie brothers; of the great

Plate 5.2 General Gaston de la Noë

Source: Courtesy of Professor Philippe Pinchemel

geologists of the American West, Powell, Gilbert and Dutton; and, above all, of German-speaking scholars such as L. Rutimeyer of Basle (1869), A. Heim, Richthofen, Philippson (1886) and Penck (1887).

Margerie was already a confirmed fluvialist when he reviewed at length Richthofen's *Führer für Forschungsreisende* and, later, with A. Heim, asserted that most of the great Alpine valleys were the result not of faulting but of river erosion. Fluvialism plays a prominent part in *Les Formes du Terrain* which the Service Géographique de l'Armée issued in 1888. According to its title page it was written by General Gaston de la Noë with the collaboration of E. de Margerie, whereas the latter in his old-age reminiscences (1943: I: 198) says that he wielded the pen. It is a slim volume of 200 pages, intended especially for military engineers and has a coherence, precision, clarity and relevance that make it, for earth scientists,

> not a mere landform treatise, but a manifesto, a hymn to the new gods named actualism (realism), determinism and evolutionism.
>
> (Broc 1975: 44, Trans.)

As this important work has already been discussed by us in considerable detail (see volume 1, pp. 627–34) we must be content here with auxiliary comments. Its first part deals with the great significance of running water; the second discusses factors which determine the courses of rivers and the nature of river networks; the third and briefer part examines influences outside ordinary subaerial erosion, namely, glaciers, the sea, wind and volcanos.

From an international viewpoint, the contents – as the authors admit – contained nothing new whereas from a French point of view their collective summary was most important, as, although failing to revolutionize national thoughts on earth sciences, it provided a secure base for future progress and introduced popularly to French students concepts such as baselevel, superimposed rivers, abrasional surfaces and slope development. Inevitably it had a few flaws and uncertainties; Henri Baulig has pointed out that the authors failed to criticize weaknesses in some of the landform models proposed by German scholars, notably by Philippson. Also, there was some uncertainty about the development of plane surfaces: in the interior of continents they may be caused by streams at 'equilibrium level' and subaerial erosion, and abrasion platforms should not be attributed exclusively to the action of the sea (Noë and Margerie 1888: 187–8).

In the same year as *Les Formes du Terrain* was published, Margerie saw in a library at Zürich the first volume of Suess' *Das Antlitz der Erde* (1883) and was so impressed with it that he set out to provide a fully annotated translation of the whole work, a task which he finished 10 years later (Margerie 1914). This monumental achievement and its eustatic concepts have been discussed in chapter 3.

The reader may be surprised at the insignificant attention to W. M. Davis in *Les Formes du Terrain*. Davis' article of 1885 (D) on *geographic classification* is given in the bibliography but he is only noticed once in the main body of the text, and then in connection with glaciation. He obviously had failed to impress the authors, which may explain why they have been so diffident on the difference between platforms of marine abrasion and plains of subaerial erosion.

The Introduction of Davis to French Scholars, 1890–1910

In the decades around the turn of the century, however, Davisian concepts won a firm foothold in France (Broc, 1974a) due largely to an upsurge in physical geography and the rise of notable geologists skilled in landform technology. There had been a chair in physical geography at the Sorbonne since 1886 held in the Faculty of Sciences by Charles Vélain but the popularization of Davis' themes was the work of Margerie and his friend Albert-Auguste de Lapparent after the American scholar had produced his ingenious analysis of *the rivers and valleys of Pennsylvania* (Davis 1889 D; see volume 2, pp. 209–25). Margerie admired this intricate argument and, in drawing La Noë's attention to it, suggested that it allowed

> the topographic condition of an area at a given moment to be considered as a complex function of structure, initial height above base level and the length of time since those two factors had acquired their present character.
>
> (Margerie 1892: 2–3)

He adds that geographic landforms could now be classified within a systematic framework and the history of exposed surfaces could be reconstituted in a rational deductive way. In the light of this enthusiasm for the new method, he begins to criticize the contemporary manuals written by German scholars. Albrecht Penck's *Morphologie der Erdoberfläche* (1894) seems to him to have overdone the use of algebraic formulae and, in spite of a wise use of genetic classification, had failed to relate it to evolutionary stages by not embracing the cyclic concept and a deductive method (Margerie 1895). The fine revised edition of Alexander Supan's *Grundzüge der physischen Erdkunde* (1896), a complete physical geography rather than a geomorphology, seemed to Margerie to make use of landform classifications that were not always natural and failed regrettably to incorporate the peneplain concept and the cyclic theory which might have provided a relationship between the indefinite multitude of landforms (Margerie 1896b). Obviously, as we have already discussed in chapter 4, the leading Germans did not share the French enthusiasm for Davis.

In 1894, Davis revisited France – he had been there in 1878 – and later produced an article (1896J) on the drainage of the Paris basin, so enlightening a theme long beloved by French geologists and hydrographers. In this year there appeared prominently among Davis' supporters, Albert-Auguste de Lapparent (1839–1908) plate 5.3 a mining engineer and geologist who had produced in 1882 a large *Traité de Géologie* which went into many editions. He was the son of a cultured, noble family who became a lively demonstrator and a brilliant lecturer (1839–1908) (plate 5.3), (Lacroix, 1920, Baulig 1955; Cailleux 1973). In 1894 he was appointed to deliver a course on physical geography at the Ecole Libre des Hautes Etudes, and these lectures gave rise in 1896 to his famous *Leçons de Géographie Physique* (see also 1884–5). This

Plate 5.3 Albert-Auguste de Lapparent

Source: From A. Lacroix, *Figures de Savants*, 1932, Vol.2, Plate X, p. 200, Gauthier-Villars, Paris

treatise, despite its title, is essentially geological and geomorphic; it omits all hydrology and biogeography and dismisses climatology in barely half a chapter; it makes a clarion call for the 'new approach' recognizing that pure description and the definition and classification of landforms on external appearances only are inadequate; and it holds that 'terrestrial morphology' or morphogeny is not to be confused with geology, although it uses geology as far as internal structure controls external forms.

Lapparent's *Leçons* went far beyond *Les Formes du Terrain* of Noë and Margerie in propagating Davisian ideas, most of which are presented not with a certain amount of caution but as accepted, self-justified truths. Moreover, it goes further in adding for the first time in a French textbook an account of the cycle of erosion. We read:

> W. M. Davis uses the term life cycle to recall that the evolution of the hydrographic network comprises a series of stages, forming so many ages [life-stages] which may be distinguished from infancy to decrepitude.
>
> (Lapparent 1896: 140)

The peneplain is

> a surface brought about by the combination of all the profiles of equilibrium of the water-courses. ... It seems that the term *peneplain* coined by Davis can be incorporated almost unaltered into our language'.
>
> (Lapparent 1896: 148).

Lapparent was convinced of the dominant role of running water and completely rejected the theory of plains of marine abrasion. At the same time he took care to include the influence of rock structure and tectonics, and the existence of special morphologies, such as volcanic, karstic, aeolian, glacial and marine. In reviewing the volume, Margerie (1896a) praised its deterministic approach and prophesied its long survival. Whereas *Les Formes du Terrain* had been a flash of light for Henri Baulig and some others, it probably did not have a great effect in university circles. In contrast, the *Leçons de Géographie Physique* (1896) had a wider, more lasting and more immediate impact, and Lapparent issued in 1898 a revised, enlarged edition, making use of Margerie's comments and of the new version of Supan's *Grundzüge der physischen Erdkunde*. Noticeably he criticizes the 'closed' landform classifications of Supan and Penck (see chapter 11), so beloved by the Germans, and prefers 'open' classifications which may always be made more perfect. He popularizes in France many features now considered obvious or axiomatic, such as types of drainage patterns associated with cuestas; superimposition; river capture; and rias. He included the role of past climate; the possible double nature of slope development – maybe by retreat parallel to the first slope, maybe by progressive flattening; and the influence of faults in the basal block on the hydrographic network developed on lavas covering that base.

True, some of his explanations and his use of technical terms were sometimes weak but he enlightened and advertised most of the avenues developed later by geomorphologists (Meynier 1969: 15–6). Lapparent attended the 7th International Geographic Congress at Berlin in 1899 and stressed in discussion how strongly the French now supported the subaerial peneplain concept of Davis, as opposed to the older marine planation view (Lapparent 1899). Here Davis first delivered his *geographical cycle* (1899H).

In France itself morphogeny, although widely read, did not proliferate without opposition at least from historical and colonial geographers. For example, Louis Himly, professor of geography at the Sorbonne, continued to declaim that people were as interesting for humans as were pebbles. But in academies the new ideas made steady if slow progress, especially after general physical geography was made part of a degree course in 1896 (Maret and Pinchemel 1972: 811). The year 1904 brought particular success: at the 8th International Geographic Congress in the United States Davis chaired the section on the physiography of the land and led an excursion to Mexico which was joined by Vidal de la Blache, Albrecht Penck and Emmanuel de Martonne who became devoted to Davisian methods and principles. Davis also arranged to entertain the young Henri Baulig (1877–1962) whose first transatlantic visit in 1905 lasted for 8 years (see volume 2, pp. 256–7, 444–5).

In France, Lapparent produced in 1907 a third, further enlarged edition of his *Leçons*, now a massive tome of 728 pages which included such up-to-date topics as glacial gouging or overdeepening. The ideas introduced or popularized by himself, Margerie and Davis were now bearing fruit: numerous articles in the *Annales de Géographie* analysed the relations between river networks and geological structure; general syntheses began to appear such as O. Barré's *L'Architecture du Sol de la France* (1903); and so much use was made of Davisian peneplains (see chapter 8) that Margerie warned that they do not provide a panacea for all landform difficulties. Davis had provided for the French a coherent explanation of natural landforms that seemed to bring earth sciences into line with Darwin's evolutionary trend. So within less than two decades physical geography had been truly changed from blind empiricism to rational determinism.

The year 1909 was for geomorphology a year of great beginnings and great endings: G. Berthaut published his *Topologie: Etude du terrain*; D. W. Johnson edited a collection of Davis' earlier writings as *Geographical Essays*; (1909C) and Emmanuel de Martonne produced his *Traité de Geographie Physique* which in an ever-enlarged form was to dominate European physical geography for many decades.

Davis' French Heyday, 1910–39

Davis visited France again in 1905 and 1908 but his most prolonged and fruitful stay was in late 1911 to 1912 when his own renown coincided with a

marked change in French thought. Previously the French had remained
essentially Cartesian in outlook and René Descartes (1596–1650) was popular
especially for his *Discours de la Méthode* (1637) (Cailleux and Tricart 1971).
His philosophy encouraged the so-called traditional rationalists who held
that human reason may form a basis for *a priori* knowledge; that is, *cogito,
ergo sum* (I think, therefore I am.) In contrast, empiricism which holds
that human knowledge arises from the impact of external things on the
senses or mind, was a growing philosophical point of view because of its
affinity with the natural sciences. This latter view received support when,
during the early twentieth century, French philosophical thought was
penetrated deeply by the concepts of Henri Bergson (1859–1941), whose
writings included *Introduction to Metaphysics* (1903) and *Creative Evolution*
(1907). He reacted against Cartesianism and developed concepts in which
duration *and* intuition play a central role. For Descartes' phrase *Je suis
une chose qui pense* (I am a thing that thinks) he substituted *Je suis une chose
qui dure* (I am a thing which lasts). In other words, he substituted temporal
for non-temporal values, and values of motion and change for static
values. Thereby he restricted the role of determinism to 'selected facts',
and 'life' could be regarded as durable, mobile, continuously creative and
free.

Meynier (1969: 40–1) has suggested that most geographers probably only
remembered at most Bergson's main formulae or concepts but there seems no
doubt that his systematic and schematic ideas had a profound influence in
France particularly in higher education and government quarters. His
thoughts, almost inevitably, were distorted often to such a degree that some
of his followers began to assume that it was rather a waste of effort to prepare
plans in great detail and to attempt to foresee the outcome of all possible
eventualities. Indeed, the stage was reached when answers or solutions to
problems began to be arrived at by 'intuition'. Geographers proved especially
susceptible to this intuitive tendency. Many of them now failed to see any
utility in mathematical exactitude, and began – at least unconsciously – to
depend on intuition to find the most satisfactory interpretations of their
academic problems. This mental process helped to produce some fine
literature but only the best geographers such as Martonne, always said to
have an 'infallible intuition', had a scientific background sufficient to give
some reality to their intuition.

Many human geographers now began to embrace possibilism as opposed to
the cruder 'environmental determinism' then in vogue, and absolute determinism began to be acceptable only in physical branches of earth sciences.
The brightest contemporary star of French geography was Paul Vidal de la
Blache (1845–1918), whose regional and humanistic writings provided a
trumpet blast for intuition and possibilism. However, despite Vidal's great
influence, there was a strong contemporary upsurge of geomorphology,

Plate 5.4 Emmanuel de Martonne

Source: Courtesy of Professor Philippe Pinchemel

partly because some scholars preferred to study a subject with a more scientific (i.e. less intuitive) base. This certainly happened with Emmanuel de Martonne (1873–1955) (plate 5.4), who eventually became Vidal's son-in-law (Cholley 1956, Dresch 1956, Beckinsale 1974). He had worked in the laboratories of Richthofen at Berlin and of A. Penck at Vienna before going in 1899 to the University of Rennes where within 3 years he had established in the faculty of science a geographical laboratory complete with maps, solid models, geological specimens, photographic facilities and surveying equipment. In 1905 he moved to Lyons and 4 years later went, on the retirement of Vidal, to the Sorbonne. From 1927 until his retirement from the Sorbonne in 1944 he combined the post with the directorship of the Institut de Géographie and managed also to organize the International Congresses of Geography (1931–8) and to act as their President (1949–55). From 1904 onward he was a devoted disciple of Davis but, like Penck, was reasonably open-minded toward new concepts and techniques. His approach was usually more geographical than geophysical and, after describing and mapping the areal distribution of a natural phenomenon, he usually tried to associate that distribution with some general law and so to seek causes for it. Thus, he tended to place more importance on comparative spatial distribution than on genetic explanation and was more inclined to determine the causes of distribution than to elucidate the scientific properties of the phenomenon itself. Undoubtedly he was more interested in the patient accumulation of observed facts than in deductions (Martonne 1924).

These tendencies are shown superbly in Martonne's *Traité de Géographie Physique*. First issued in 1909 it was a great success and became the fundamental base of French physical geography for nearly half a century. Whereas Lapparent's *Leçons* were essentially landforms and geology, Martonne considered nature in all its main aspects under a quadripartite plan of climate, hydrography, landforms and biogeography, which ensured that a knowledge of weather and flowing water preceded that of landforms. Throughout, areal distributions were stressed and detailed study of elements such as the atmosphere and origin of mountains almost neglected. For example, terms such as *adiabatic* and *isostasy* did not appear in the index in the first three editions (1909, 1913 and 1921). Martonne was now in his prime and he kept the treatise up to date by careful pruning and enlargement. The fourth edition, recast in three volumes, appeared in 1925–7. Landforms (*Le Relief du Sol*) comprises the second volume, a fine work which by the seventh edition of 1947 consisted of 562 pages with over 250 illustrations, many of them drawn by the author who, like Davis, was a master depictor of block diagrams. A tenth edition was issued in 1958. The *Traité* was remarkable for its breadth of outlook, clarity of explanation, sanity in choice of examples – despite some pardonable chauvinism – and visual embellishment. It was already achieving a wide popularity and spreading Davisian concepts when Davis himself arrived in France.

As was usual with the Harvard professor, this French visit was merged with other schemes not unconnected with advertising his own concepts. He arranged a European pilgrimage that from August to October 1911 paid homage to various uplands in Ireland, England and France (see volume 2, pp. 363–75). The excursion was attended at least in parts by nine Frenchmen, including Pierre Denis of the University of Paris; Albert Demangeon (Lille University) who presented his new polycyclic ideas on the landforms of Limousin; Lucian Gallois of Paris; Antoine Vacher of Rennes; and Philippe Glangeaud of Clermont-Ferrand. The Morvan, the north-eastern prolongation of the Massif Central, was chosen by Davis as the term (morvan) for the ideal type of polycyclic doubly peneplained upland. From October 1911 to early March 1912 Davis acted as professor at the Sorbonne and then squeezed in short visits to universities at Dijon, Lyons and Grenoble before returning to the United States to lead a long international excursion there. In Paris he was received almost rapturously. Vidal de la Blache welcomed him as the person who had transformed geography from a descriptive study into an explanatory science which laid stress on 'origin, association and evolution' which he had enriched with a new terminology and fine drawings. Davis in his inaugural lecture, reproduced in the *Annales de Géographie* as *L'Esprit explicatif dans la géographie moderne* (1912A), relished the eulogic situation; he said he had striven against great odds and unpopularity, to express the principles of physical geography

> in a scientific way, as a logical discipline, and so to contribute to the systematic and serious education that best suits the students of the Sorbonne . . . . We used to say to young geographers, 'Go and see'; today we say, 'go and think!'
>
> (Davis 1912A: 2–13, trans.)

We have already described this enthusiastic welcome (see volume 2, pp. 375–9) and here must be content with its aftermath. Later in 1912 many French geographers, including Lucian Gallois, Emmanuel de Martonne, Emmanuel de Margerie and Henri Baulig went on excursions with Davis in the United States and further cemented Franco-American friendships. They were familiar with, and well disposed towards, Davisian concepts, especially of polycyclic developments. Margerie had had great influence on geomorphologists because of his fine translation and annotation of Suess' *La Face de la Terre* (1883–1908) (see chapter 3), of which 18,000 copies were sold, but he was also a keen follower and interpreter to his countrymen of Davis' ideas. Martonne (1906–7, 1914) worked on Davisian lines in Romania, and Baulig, a friend of Davis who had written on American fluvial studies in 1910, returned to France in 1912 to occupy the chair of geography at Rennes.

The Davisian honeymoon in France was brought to a brutal halt by the First World War (1914–18), infamous for its incredible massacre of millions of its direct combatants. At its close, French geographers assisted American students of Davis, such as D. W. Johnson and I. Bowman, in previewing territorial adjustments. By the Treaty of Versailles (28 June 1919) France regained Alsace–Lorraine.

After the war – which Davis as a former Quaker opposed – the drift away from his concepts was slow and his hegemony in France was scarcely dimmed by 1939. He had flourished here largely because his rise coincided with that of physical geography – a distinct contrast with affairs in Germany where he had tried to infiltrate into an already formed science. However, he had many other advantages: he acquired most powerful disciples; his brilliant literary style and exquisite draughtsmanship appealed to the French; his own insistence on his self-styled 'logicality' or 'scientific approach', and his so-called deductive method and ability to simplify a problem by subdivision, met with approval among Cartesians; and he was also helped by the Bergsonians whose intuitive and qualitative tendencies turned many geographers towards humanistic possibilist themes while breeding tolerance for Davisian concepts in physical themes. Between 1895 and 1925 Davis wrote eleven articles in French and a further seven in English on parts of France. Typically he entered their literature with their pet topic, rivers and river networks, and left it with an article on coral reefs in New Caledonia.

By the mid–1920s French geography had almost recovered from the effects of the war. There was an explosion of regional and humanistic topics whereas physical aspects, in spite of the eminence of their protagonists, moved more slowly and were taken up anew either by geologists interested in landforms or by geographers who had acquired also some training in geology. For two decades, until his retirement in 1944, Martonne was the senior professor of geography in France where the increasing precision of fieldwork was accommodated as far as possible within Davisian principles. Interruptions of the cycle or polycyclism, with its nickpoints, breaks of slope, terrace flattenings (*replats*), alluvial terraces and summit levels, became especially popular within a eustatic framework (see chapters 3 and 8). Erosional flattenings, as found by Demangeon in Limousin, Antoine Vacher (1908) in Berry and René Musset (1917) in Maine, began to be discovered almost everywhere. Even the physical application of the cycle of erosion, extended – much to Martonne's delight – to shorelines by the Americans F. P. Gulliver and D. W. Johnson, was further enlarged by Léon Aufrère's geomorphological cycle of dunes (1931).

Inevitably, signs of opposition and of modifications leading to some weakening of total adherence to Davis' general ideas began to appear as fieldwork advanced. This gentle groundswell may be glimpsed from the influence of Suess' eustatic theory upon Henri Baulig who in 1919 moved to Strasburg where he remained until his retirement in 1947. He made altimetric analyses of Brittany (1926), the Paris Basin (1928b) and the Central Massif (1928a) which led him to believe that the landmass had remained stable while eustatic changes had produced nickpoints and flattenings in the much-faulted mountain block (see chapter 8). Although most earth scientists admitted that sea-level changes had been caused by alterations of glacial and non-glacial periods, this eustatic theory did not meet with universal approval in France and Martonne (1929) attacked the idea (see chapter 3). Martonne and Baulig also disagreed over the exact nature of the profile of equilibrium in a watercourse. The former preferred a terminal profile below which deepening did not proceed; the latter considered that moving water always had some erosive power and that a long profile could not be a regular parabola because of the effect of the influx of tributaries.

Margerie remained at heart an impartial interpreter and seemed to have no great faith in eustatism nor in the results of the reconstruction of old polycyclic profiles. Martonne early recognized, as did Walther Penck in Germany, that the oversimplified concept of a sudden uplift followed by prolonged standstill was quite impossible to maintain, and from about 1925 onward considered that erosion would begin to fashion the relief before earth movements had ceased. Other scholars also sensed this tectonic weakness – as indeed had Davis himself who stressed simplicity for the sake of popularity – and the idea of the continuance of earth movements was germinating. Martonne was also troubled by Davis' overuse of 'deduction' or of his so-called logic which was too often not based on direct observation. In an obituary of his American friend he extolled him as one of the greatest contributors to modern geography, who guarded against the temptation, all too prevalent in Natural Sciences, of overestimating the value of detailed observation and missing the central objective. In the same vane, Baulig (1950: 189) pointed out that Davis' mind, like that of all great thinkers, worked with symbols, rather than 'facts'. Meynier (1969: 65) points out how imprudent this last comment was to prove in following decades.

The hold of Davis' teaching on French thought may be gleaned from volume 2 (*Le Relief du Sol*) of the seventh and eighth editions of Martonne's *Traité* (1947, 1948). Here in the seventh edition are long accounts of the *modèle* (model) of normal erosion (7th edn, pp. 549–620), in which the cycle of erosion is explained in a separate section (pp. 603–18). Details of the peneplain are included which, if we have translated correctly, say that it forms the 'dernier stade' or 'terme final' of the model of normal erosion, whereas we believe that the peneplain was the penultimate stage and that Davis never came to real grips with the final stage (see volume 2, pp. 166, 189–92 and 747). We are told that a 'cycle' cannot imply a return to its beginning (p. 605) and that Davis' attempts to apply his ideas to families of landforms outside normal erosion, such as forms of nivation and of ice, were less successful (pp. 346, 899). The cycle of littoral erosion is illuminated by a sequence of fine block diagrams characteristic of Martonne. And so the general impression is that Davis' ideas still survive but in modified and restricted form.

The Davisian Decline in France after 1939

From 1939 to 1945 Europe was plunged again into a war which brought less slaughter for France than had its predecessor but led to the occupation of the country. Baulig and René Musset were among those arrested and imprisoned and Henri Enjalbert is said to have acquired his anti-Davisian feelings in a German prisoner-of-war camp. Some of the older geomorphological stalwarts managed to survive the disaster: Margerie died in 1953 at the age of 91: Martonne in 1955 at the age of 82; Baulig in 1962 at the age of 85; and Blanchard (see 1944–56), the Alpinist, in 1965 at the age of 88. They had outlived the Davisian era in France and had witnessed the rise of new geomorphological stars, notable among them being Pierre Birot, André Cailleux and Jean Tricart. At about mid-century a number of important reviews of French geomorphology appeared (Chabot 1950, Meynier 1952, Dylik 1953, Baulig 1957b, Beaujeau-Garnier 1957, Chabot and Clozier 1957, Birot 1958).

The drift away from Davis' concepts during the two decades following the end of the Second World War in Europe was marked by an increase in the use of new techniques and in numeracy and quantification, by increasing interest in processes and tectonics, and by a great upsurge of climatic geomorphology (see chapter 12). The turn towards quantification was especially helped by laboratory investigations into the nature of materials and the mechanisms affecting their alteration. André Cholley (1886–1968), a pupil of Martonne at Lyons and his successor at the Institut de Géographie at Paris in 1944, made a detailed study of Depéret's work on sediments in the Charolais and Mâconnais and later on the silicious–calcareous deposits (*depôts meulierisés*) in the Paris Basin. In the light of these deposits and of erosional surfaces he analysed a possible sequence from a Late Miocene 'polygenetic surface', to a surface affected by tectonic movements in the Pliocene and by climatic oscillations in the ensuing Quaternary period (Cholley 1943). This research was continued by Cholley's students such as Beaujeau-Garnier, Birot, Tricart and Pinchemel and by Cailleux particularly on the fashioning of sand grains and pebbles.

The old morphometry or altimetric analysis lingered on and Martonne (1941) produced details of the mean altitudes of various areas but numeracy was now being applied in more varied ways. For instance, Charles-Pierre Péguy applied it to many geographical problems, including slope profiles, while J. Corbel expressed mathematically the amount of erosion in limestone and on the whole earth. Most pre-existing hypotheses on landform development began to be subjected to quantitative analysis and could now be published in the *Révue de Géomorphologie Dynamique*.

Attitudes to tectonic uplift changed rapidly until they were antagonistic to Davis' simplified sudden uplift followed by prolonged standstill. The work of the Grenoble school under Raoul Blanchard (1944–56) tended to show that many level parts and flattenings were caused by structure or by glaciation and lacked any cyclic significance. Some authors began to take a clearly mobilistic approach to fold mountains, a view shared widely by Soviet and German geomorphologists. In the 1950s, at the Sorbonne, Tricart and Cailleux (1954–7) taught that tectonics and erosion are capable of functioning at fairly similar rates and that the earth's surface is mobile. The Davisian idea of sudden, short phases of orogenesis followed by long stable periods was for them untenable, and they believed that the active orogenic phases lasted from 100 to 150 million years while the quieter phases were hardly one-tenth that length. They vigorously attacked the simple cyclic scheme in forthright condemnations which undermined the foundations of its already weakening popularity and, conversely, made a powerful plea for climatic geomorphology.

The influence of climate on landforms had long interested French scholars and Martonne summarized it in 1913. Later in the twentieth century much research was carried out on the effect of different climates on the disintegration of various rocks and on the fashioning of glaciated and periglacial relief. German and Soviet geomorphologists enlarged the theme of zonal geomorphology far beyond the Davisian scheme of climatic 'accidents' associated with deserts and ice. In France Martonne (1940, 1946) wrote extensively on the characteristic landforms of the tropical zone, and Pierre Birot (plate 5.5), who began his career under Martonne in 1927 and had published in the 1930s on arid granitic landforms (1931) and on the eastern Pyrenees (1937), elaborated in his *Essai sur Problèmes de la Morphologie Générale* (1949) the influence of various climates on the development of slopes and on calcareous, granitic and schistose rocks. This volume, dedicated to his masters, Martonne and Cholley, was based on lectures given at Oporto and Lisbon a few years previously. Cholley (1950, 1957) illuminated the same theme on the nature of structural and climatic geomorphology at a time when the climatic aspect climbed to predominance (see chapter 12).

Tricart, professor of geography at Strasburg (plate 5.6), and Cailleux, master of conferences at the Ecole des Hautes Etudes, singly and in collaboration set out to vindicate climatic geomorphology and to demean the Davisian concept of cycles of normal erosion to virtual insignificance. They did not mince their words: the imagined cycle does not agree with many of the facts; climates, present and past, are an essential component of land-forms; and earth scientists depend for the truth not on intuition but on the gradual accumulation of scientific observations. Tricart alone, in an anti-capitalist message meant for Communist consumption, wrote:

> The Davisian theory of the 'normal' cycle of erosion is one of the most brilliant examples of Bergsonian creative imagination applied to a scientific theme.
>
> (Tricart 1956: 9)

Plate 5.5 Pierre Birot

Source: Courtesy Mme Birot

He deplores the idealistic tone of Davis' ideas and regrets the intrusion of the cycle into what was a sound geomorphology.

Several French geographers thought that Davis' detractors had gone too far and had to some extent distorted his schemes or underestimated their depth and elasticity. The authors of some of the larger texts, such as Max Derruau's *Précis de Géomorphologie* (1956) and Birot's *Les Méthodes de la*

Plate 5.6 Jean Tricart (1985)

Source: Courtesy Professor A. Goudie, University of Oxford, Photographed by Mr Tony Lee

Morphologie (1955) tried to retain what seemed valid in the cyclic theme. The latter in an interesting chapter on landform evidence includes a section on cyclic analysis, qualitative and quantitative, and states that:

> After serving as a panacea during several decades, cyclic analysis of the Davisian kind is now subjected to increasingly violent and numerous attacks.
>
> (Birot 1955: 29)

Birot explains these defamations by the misuse of the application of the method, especially to Alpine chains, the misunderstanding of the width of Davis' original themes, and the healthy reaction of the scholars seeking new methods. Davis, he adds, did consider pedimentation and his concept on uplift merely reduced the time and place of its application. Might not standstill be valid if the cause of deformation were tangential compression or isostatic adjustment, and might not the unfinished nature of a cycle be due to the slowness of erosion rather than the persistence of surface mobility? And do not uprising mountains and slowly degrading surfaces tend under gravity towards cyclic flattening? Viewed in this light, Birot (1955: 164) suggests that Davisian concepts may appear to represent a decisive step in the history of geomorphology as they allow us to understand that the relief depends not so much on folding as on movements with a greater radius originating some kilometres beneath the surface of the earth. Birot's compromises failed to satisfy Henri Baulig who was a great admirer of both Davis and Bergson. Baulig (1957a), although 80 years old, wrote a long defence of the whole of Davisian or 'classical' geomorphology against what he seemed to regard as a defection by Birot.

Two years later Birot again revealed his attitude to Davisian ideas in his *General Physical Geography* (1959), which was published in English translation in 1966. Here he has a chapter on the evolution of slopes in homogeneous rocks towards baselevel which envisages successive stages – a first stage when accelerated stream incision exceeds slope weathering, so leading to steeper slopes, and a second stage when slower uplift is associated with flattening at the base of the slopes and parallel retreat of the upper exposed rock faces (Birot 1966: 161–4). Part Two of the volume (pp. 201–55) deals with erosion, soils and vegetation in the eight major climatic zones. In each of four major zones (tropical rain forest, tropical dry season, semi-arid and arid, tundra) considerable space is given to its cycle of erosion. In the chapter on the temperate deciduous forest zone we are merely told that any description of its cycle of erosion is partly theoretical because probably it has not existed for a long enough time and has been too disturbed by cold climate to bring about peneplanation (p. 231). Similarly, cycles are not given for subtropical forests and for coniferous forests or the taiga. In temperate grasslands the cycle of erosion on slopes is said to operate 'in very narrow limits' (p. 238). Within the volume Davis is mentioned only twice, and then only for his writings on coral reefs.

By the 1950s in France Davisian cyclic concepts had been reduced almost to insignificance; Davis' sphere of 'normal' erosion had dwindled to a rarity, while his realm of 'accidental' erosion had expanded to become the norm; the cyclic terminology was retained, although it seemed to refer to a process which in nature does not complete a cycle and today seldom shows any signs of a beginning and fewer still of an end. Climatic geomorphology was becoming dominant and the history of the relief of the world increasingly concerned the struggle between tectonics and erosion played out beneath different climates and climatic changes. Perhaps the last word should be given to Jean Dresch, professor at the Sorbonne, who always distrusted Davisian cycles:

> Since the Second World War nobody has taken Davis's ideas seriously in France; I was deeply shocked when in London for a conference last summer to see that they are still believed and applied in England. I cannot understand why England has lagged so far behind France'.
>
> (J. Dresch in conversation with C. L. Markus, March 1965)

Davisian Influences in Britain, 1890–1950

As has been documented in volume 2, Davis visited Britain on five occasions in a professional capacity before his retirement (1894, 1898, 1900, 1909 and 1911) and published seven influential articles in British journals (Davis 1895E, 1896H, 1899H, 1899M, 1906J, 1909G and 1910H). Among these were his fine essays which laid the foundations for the drainage development and denudation chronology of the English scarplands (Davis 1895E, 1899M; see volume 2, pp. 243–7), his excellent description of upland glacial sculpture (Davis 1906J; see volume 2, pp. 315–8) and, particularly, his systematic description of landforms (Davis 1909G). The latter was presented in person at the Royal Geographical Society in the same year as the *Geographical Essays* were published, and was discussed afterwards by a group of leading British geographers, geologists and biologists. Oxford geography was represented by Halford Mackinder and A. J. Herbertson. Mackinder, who had previously (1887) distinguished between physiography (Why is it?), topography (Where is it?), physical geography (Why is it there?) and geology (What riddle of the past does it help us solve), was not happy with Davis' terminology and counselled the need to 'resist the temptation of making geography into a merely supplementary chapter of geology' (Davis 1909G, discussion p. 321). Herbertson, who by 1901 was already talking of 'this great new science of geomorphology', called for a more simple descriptive method for landforms and showing his Germanic links, suggested a scheme in which 'the late mature stage of volcanic plateau denuded by running water might be 2a8'! (Davis 1909G, discussion p. 322). The geologists and others were fairly evenly divided regarding Davis' suggested terminology; for example Lamplugh was in favour whereas Hugh Robert Mill, author of the influential *Realm of Nature* (1892), believed that Davis' cyclic approach to landforms may be

> particularly dangerous when it is attempted with imperfect experience .... His method appears to me as one more for the master than the student, and I am afraid that his disciples will run away from him, and apply it in a way that will cause him anxiety at first and horror afterwards.
>
> (Davis 1909G, discussion p. 321).

During his European 'pilgrimage' in 1911 Davis was accompanied to central Wales and Snowdonia by two Cambridge geologists, O. T. Jones and J. E. Marr. Marr began lecturing on 'geo-morphology' at Cambridge in 1906 (Stoddart 1987) and Jones later achieved geomorphic prominence through his studies of the denudation chronology of Wales (see chapter 9). In 1908 the geologist Philip Lake had been appointed to a Cambridge lectureship in physical and regional geography and in 1915 he published the first of many printings of his influential *Physical Geography*. It is significant that this text, which survived in a revised form until the 1960s, contained in its first edition up to 1949 no reference to the cycle of erosion and discussed the development of a river valley in terms of the 'valley tract' and 'plain tract', although two pages were devoted to the concept of grade and two to antecedent and superimposed drainage, ideas previously accepted by British workers. Prior to the 1920s the study of landforms in Britain was dominated by geologists who were, in general, not attracted to Davisian ideas but mainly concerned with the influence of geology on landforms (Brown and Waters 1974). Early works of this type included Jukes-Browne's *The Building of the British Isles* (1888), Marr's *The Scientific Study of Scenery* (1900, ninth edition 1943) and Lake and Rastall's *Textbook of Geology* (1910). Although great geomorphic changes began in Britain in the 1920s, this geological tradition continued with Trueman's *The Scenery of England and Wales* (1938), Stamps's *Britain's Structure and Scenery* (1946), Holmes' *Principles of Physical Geology* (1944) and, latterly, Sparks' *Rocks and Relief* (1971).

Early non-structural geomorphic work in Britain, such as that by Buckman and Lake, followed Davis' approach to drainage development, but from the 1920s the major effort was directed towards denudation chronology (see chapter 9). Although some of this was conducted by geologists (e.g. O. T. Jones, A. E. Trueman, T. N. George and S. E. Hollingworth), much was carried out by geographers under the leadership of S. W. Wooldridge and D. L. Linton (Brown 1979). Indeed, British geologists largely divorced themselves from geomorphology and were particularly opposed to eustatic ideas (see chapters 3, 8 and 9) (Linton 1969). After Baulig's lectures on the changing sea level at King's College, London, in October 1933 (see chapter 8), O. T. Jones regretted having presided over one of them because of the 'mischief' it would cause. A similar distaste was evident in the second edition

Plate 5.7 Professor S. W. Wooldridge with students at Juniper Hall, Surrey, England, in the 1950s. E. H. Brown is second from the left

Source: Courtesy of King's College London, Department of Geography Archives

of W. B. Wright's *The Quaternary Ice Age* (1937), in which the author stressed isostatic controls over sea level, rather than the eustatic ones of Depéret and Lamothe (see chapter 3).

Wooldridge (1900–1963) (plate 5.7) devoted almost 40 years of his life to justifying the role of denudation chronology as the core of geomorphology (1951), as a vital part of physical geography (Wooldridge and Morgan 1937) and as a basis for geography as a whole (1949. Wooldridge and East 1951). A lay preacher, his strong views became very influential in British geomorphological work, notably:

1 His support for W. M. Davis:

> Davis towers above his predecessors and successors like a monadnock above one
> of his own peneplains.
>
>   (Wooldridge 1958:34).
>
> We continue to regard W. M. Davis as the founder of our craft and regret the
> murmurings of dispraise heard occasionally from his native land
>
>   (Wooldridge and Morgan 1959, Preface to the second edition)

He often declaimed in Davis' words against 'taking the Ge out of
Geography' (1949) or by using his own idiom of 'throwing away the baby
with the bath water'. After Davis' *Geographical Essays* was reissued in 1954
Wooldridge (1955) welcomed it rapturously, rounded on recent critics of
Davis in the United States and wished that the volume could be made
compulsory reading for university teachers of geography.

2 His unconcern for work in foreign environments. One of his favourite
Biblical texts was taken from *Proverbs 17:24*: 'The eyes of a fool are in the
ends of the earth'. In this connection, Stoddart (1987) has drawn attention
to the relative scarcity of British work in exotic locations during the first
half of the century, especially compared with that by Germans (see chapter
12).

3 His linking of denudation chronology with eustatically controlled changes
of the baselevel (see chapter 9) for which he borrowed Huxley's phrase of
using 'sea reckoning for land time' (Wooldridge 1951: 171).

> Davis, completing the work of British pioneers, first spoke clearly in terms of the
> successive *stages* of development of the southern English landscape – of cycles of
> erosion . . . . The 'chisel' had been applied not in one long uninterrupted period
> of demolition but in decipherable stages, each of which has left its mark on the
> landscape.
>
>   (Wooldridge and Linton 1939: 153)
>
> The significant development of geomorphology in the present century has been
> to attempt to link its record with that of stratigraphic geology.
>
>   (Wooldridge 1951: 170)

Despite the dominance of Wooldridge and Linton and their many students
and associates over British geomorphology between the 1920s and 1950s, as
in the United States, other kinds of mainly process-oriented studies were
being conducted in many fields (Freeman 1980, Steers 1986, Stoddart 1987):

Coastal processes: These were dominated by the work of J. A. Steers and
W. V. Lewis at the Department of Geography, Cambridge University, but
also important were E. A. N. Arber, Vaughan Cornish and C. A. M. King.
Quaternary studies: These were mainly the province of geologists such as J. K.
Charlesworth, A. R. Dwerryhouse, F. W. Harmer, S. E. Hollingworth,
P. K. Kendall, C. Lapworth, A. Raistrick, H. H. Swinnerton and L. J.
Wills.

Glacial processes: Involving W. R. V. Battle, E. J. Garwood, C. A. M. King, W. V. Lewis, L. H. McCabe, J. F. Nye, N. E. Odell and M. F. Perutz.
Periglacial processes: Involving H. G. Dines, S. E. Hollingworth and R. S. Waters.
Arid processes: Involving R. A. Bagnold, H. J. L. Beadnell, Vaughan Cornish, R. F. Peel and K. S. Sandford.
Hillslope processes: Involving P. Lake, A. Wood and A. Young.
Added to these were more isolated studies of *tectonics and geomorphology* (L. R. Wager), *palaeogeomorphology* (G. H. Dury), *karst processes* (M. M. Sweeting) and *theoretical geomorphology* (H. Jeffreys). As with the American and other process-oriented work of the period, the above contributions will be treated in volume 4.

After the Second World War, as pressures for a change of emphasis in mainstream British geomorphology increased, Wooldridge's reaction hardened (Dury 1983). In 1951 (p. 168) he deplored the recent criticism of the Davisian concept of grade, in the process completely ignoring the vital paper by Mackin (1948) published 3 years previously. As we have seen, his complaints against what he saw as deviations from Davisian orthodoxy mounted:

> There has been a recent attempt in certain quarters to devise a 'new' quasi-mathematical geomorphology. At its worst this is hardly more than a ponderous sort of cant . . . . If any 'best' is to result from the movement, we have yet to see it.
> (Wooldridge and Morgan 1959, Preface to the second edition)

Wooldridge's righteous indignation found its last public congregation at the meeting of the Institute of British Geographers in January 1958:

> I regard it as quite fundamental that geomorphology is primarily concerned with the interpretation of forms, not the study of processes. The latter can be left to Physical Geology . . . . The plea for measurement, or for morphometric work, is wholly sound and salutary providing it is remembered that the physical map is itself a quantitative statement, much of whose information only requires translating or re-casting. The urge to dabble in elementary mathematics is quite another matter. The direct attack by mathematical methods would seem to offer very limited chances of success.
> (Wooldridge 1958: 31–2)

At the corresponding meeting in Cambridge the following year the pace of change was to quicken.

References

Aufrère, L.(1931)'Le cycle morphologique des dunes', *Annales de Géographie* 40: 362–85.

Barré, O. (1903) *L'architecture du Sol de la France*, Paris: Colin.

Baulig, H. (1910) 'Écoulement fluvial et denudation d'après les travaux de l'United States Geological Survey', *Annales de Géographie* 19: 385–441.

—— (1926) 'Sur un methode altimetrique', *Bulletin de l'Association Géologique de la France* no. 10, pp. 7–9.

—— (1928a) *Le Plateau Central de la France et sa bordure Méditerranéenne: Etude morphologique*, Paris: Colin.

—— (1928b) 'Les hauts niveaux d'erosion eustatique dans le Bassin de Paris', *Annales de Géographie* 37: 385–406.

—— (1950) 'William Morris Davis: Master of method', *Annals of the Association of American Geographers* 40: 188–95.

—— (1955) 'La vie et l'oeuvre d'Emmanuel de Margerie', *Annales de Géographie* 63: 82–7.

—— (1957a) 'Les methodes de la géomorphologie d'après M. Pierre Birot', *Annales de Géographie* 66: 97–104, 221–36.

—— (1957b) 'La géomorphologie en France jusu'en 1940', in *La Géographie Française au Milieu du XXe Siècle*, Paris: Baillière, pp. 27–35.

Beaujeau-Garnier J. (1957) 'New tendencies in French geomorphology', In *La Géographie Française en Milieu du XXe Siècle*, Paris, Baillière, pp. 37–42.

Beaumont, E. de (1829–30) 'Recherches sur les quelques-unes de Révolution de la surface de la globe', *Annales des Sciences Naturelles*, 8: 5–25, 248–416; 9: 5–99, 174–240.

Beaumont, E. de and Dufrénoy, P. (1838) *Memoirs of the Geological Map of France*, 4 vols.

Beckinsale, R. P. (1974) 'Emmanuel de Martonne', *Dictionary of Scientific Biography*, vol. IX, New York: Scribners, pp. 149–151.

Bergson, H. L. (1903) *L' Introduction à la metaphysique*, Paris, Almanach Des Cahiers.

—— (1907) L' Evolution créatrice, Paris, Félix Alcan.

Berthaut, H. (1909–10) *Topologie: Etude du terrain*, Paris: Service Generale de l'Armée.

Birot, P. (1931) 'Sur les reliefs granitiques en climat sec', *Bulletin de l'Association Géographique de la France* nos 220–1, Nov.–Dec.

—— (1937) *Recherches sur la morphologie des Pyrénées orientales franco-espagnoles*, Paris: Baillière.

—— (1949) *Essai sur Quelques Problèmes de la Morphologie Generale*, Lisbon: Centro d'Estudos Geograficos.

—— (1955) *Les Méthodes de la Morphologie*, Paris: Presses Universitaire de France.

—— (1958) 'Les tendances actualles de la géomorphologie en France', *Zeitschrift für Geomorphologie* N.F. 2, pp. 123–34.

—— (1959) *Precis de Géographie Physique Generale*, Paris: Colin (English translation by M. Ledesert, 1966, London: Harrap.

Blanchard R. (1944–56) *Les Alpes Occidentales*, 7 vols, Paris: Arthaud.

Broc, N. (1974a) 'Davis et la France', *Bulletin de la Société Languedocienne de Géographie*, Montpellier, 8: 87–95.

——(1974b) 'L'establissement de la géographie en France (1870–1890)', *Annales de Géographie* 83: 545–68.

—— (1975) 'Les débuts de la géomorphologie en France: Le tournant des années 1890', *Revue d'Histoire des Sciences*, 28: (1): 31–60.

Brown, E. H. (1979) 'The shape of Britain', *Transactions of the Institute of British Geographers* N.S., 4: 449–62.

Brown, E. H. and Waters, R. S.(1974) 'Geomorphology in the United Kingdom since the First World War', *Institute of British Geographers, Special Publication* no. 7, pp. 3–9.

Cailleux, A. (1973) 'Albert–Auguste Cochon de Lapparent', *Dictionary of Scientific Biography*, vol.VIII, New York: Scribners, pp. 30–1.

Cailleux, A. and Tricart, J.(1971) 'Rappel de Descartes', *Revue de Géomorphologie Dynamique* 20: 177–8.

Chabot, G. (1950) 'French concepts of geographical science', *Norsk Geografisk Tidsskrift*, 12: 309–21.

Chabot, G. and Clozier, R. (1957) *La Géographie Française au Milieu de XXe Siècle*, Paris: Baillière.

Cholley, A. (1943) 'Recherches sur les surfaces d'érosion et la morphologie de la région parisienne', *Annales de Géographie*, 52: 1–19. 91–7, 161–89.

—— (1950) 'Morphologie structurale et morphologie climatiques', *Annales de Géographie*, 59: 321–35.

—— (1956) 'Emmanuel de Martonne', *Annales de Géographie*, 65: 1–14.

—— (1957) *Recherches Morphologiques*, Paris: Colin.

Chorley, R. J. (1971) 'Gabriel–Auguste Daubrée', *Dictionary of Scientific Biography*, vol. III, New York: Scribners, pp. 586–7.

Daubrée, A. (1879) *Etudes Synthétiques de Géologie Experimentale*, Paris.

—— (1887a) *Les eaux souterraines a l'époque actuelle*, 2 vols, Paris: Dunod.

—— (1887b) *Les eaux souterraines aux époques anciennes*, Paris: Dunod.

Dausse, B. (1872) 'Etudes relatives aux inundations et à l'endiguement des rivières', *Institut de France, Mémoirs presentés par divers savants à l'Academie des Sciences* 20: 287–507.

Demangeon, A. (1910–11) 'Le relief du Limousin', *Annales de Géographie*, 19: 120–149; 20: 316–37.

Derruau, M. O. (1956) *Précis de Géomorphologie*, (5th edn 1967).

Descartes, R. (1637) *Discours de la Méthode, pour bien conduire sa reason*, Paris: Charpentier.

Drapeyron, L. (1885) 'Que la géographie est une science grâce à la topographie', *Revue de Géographie*, 8: 401–11.

Dresch, J. (1956) 'Emmanuel de Martonne (1873–1955)', *Bulletin de la Société Géologique de la France*. 6th series, 6: 623–42.

Du Buat, L. G. (1779 and 1786) *Principes d'Hydraulique*, 2 vols, Paris, (3rd edn, 3 vols, 1816).

Dupaigne, A. (1873) *Les Montagnes*, Tours, A. Mame.

Dury, G. H. (1983) 'Geography and geomorphology: The last fifty years', *Transactions of the Institute of British Geographers*, N.S., 8: 90–9.

Dylik, J. (1953) 'Caractères du developpement de la géomorphologie moderne', *Bulletin de la Société des Sciences et des Lettres de Lödz*, classe III, vol. 4(3).

Ellenberger, F. (1986) *Histoire de la Géologie: Des anciens à la première moitie du XVIIe siècle*, vol. 1, Paris: Lavoisier.

Figuier, L. (1864) *La Terre et les Mers*, Paris, (English Translation by W. H. Davenport Adams, 1870, London: Nelson).

Freeman, T. W. (1980) *A History of Modern British Geography*, London: Longman.

Holmes, A. (1944) *Principles of Physical Geology*, London: Nelson (1st edn reprinted eighteen times; 2nd edn 1965; 3rd edn 1978).

Jukes-Browne, A. J. (1888) *The Building of the British Isles: A study in geographical evolution*, London: Bell (4th edn 1922, London: Stanford).

Lacroix, A. (1920) 'Notice historique sur Albert-Auguste de Lapparent lue de 20 décembre 1920', *Mémoires de l'Academie des Sciences (Paris)*, 2nd series, 57: 41–75.

Lake, P. (1915) *Physical Geography*, Cambridge (1st edn reprinted eleven times, 2nd edn 1949, 3rd edn 1952, 4th edn 1958).

Lake, P. and Rastall, R. H. (1910) *Textbook of Geology*, London: Edward Arnold (6th edn 1947).

Lamarck, J. B. P. de (1802) *Hydrogéologie*, Paris (English translation by A. V. Carozzi, 1964, Urbana, Illinois).

Lapparent, A. de (1885) *Traité de Géologie*, Paris: Masson (3rd edn 1900, 5th edn 1906, 3 vols).

—— (1884–5) 'Les grandes lignes de la gèographie physique', *Annales de Géographie* 4: 129–50.

—— (1896) *Leçons de Géographie Physique*, Paris: Masson, (2nd edn 1898, 3rd.edn 1907).

—— (1899) 'La question des pénéplaines envisagée à la lumière des faits géologiques', *Comptes Rendus Congrès International de Géographie*, Berlin, vol. 2, pp. 213–20.

Linton, D. L. (1969) 'The formative years in geomorphological research in south-east England', *Area*, no. 2, pp. 1–8.

Mackin, J. H. (1948) 'Concept of the graded river', *Bulletin of the Geological Society of America*, 59: 463–512.

Mackinder, H. J. (1887) 'On the scope and methods of geography', *Proceedings of the Royal Geographical Society*, N.S. 8: 698–714.

Maret, M.-P. and Pinchemel, Ph. (1972) 'L'évolution des questions de géographie aux concours d'agrégation des origins à 1914', in *La Pensée Géographique Française Contemporaine. Mélanges offert au Professeur A. Meynier*, Presse Universitaire de Bretagne, pp. 77–86.

Margerie, E. de. (1883) 'Les plateaux du Colorado, d'après les traveaux des géologiques américains', *L'Annuaire du C.A.F.*

—— (1892) 'L'évolution des formes géographiques, d'après W. M. Davis', *Les Nouvelles Géographiques*, pp. 2–3.

—— (1895) 'Review of A. Penck's 'Morphologie der Erdoberfläche', *Annales de Géographie* 4: 369–72.

—— (1896a) 'Review of de Lapparent's 'Leçons de géographie physique', *Bulletin Géographique et Historique*, pp. 394–8.

—— (1896b) 'Review of A. Supan's 'Gründzuge der Physichen Erdkunde', *Annales de Géographie* 5: 235–40.

—— (1914) 'Notice sur Eduard Suess', *Annales de Géographie* 23: (130): 371–3.

—— (1943–8) *Critique et géologie. Contribution à l'histoire des sciences de la terre*, 4 vols, Paris: Colin.

Marr, J. E. (1900) *The Scientific Study of Scenery*, London: Methuen (9th edn 1943).

Martonne, E. de (1906–7) 'Recherches sur l'évolution morphologique des Alpes de Transylvanie', *Revue de Géographie Annuelle* 1: 1–279

—— (1909) *Traité de Géographie Physique*, vol. 2, *Le Relief du Sol*, Paris: Colin (4th edn 1926, pp. 499–1057).

—— (1913) 'Le climat facteur du relief', *Scientia*, 13: 339–55.

—— (1914) 'Tendences et l'avenir de la géographie moderne', *Revue de l'Université de Bruxelles*, year 19, pp. 453–79.

—— (1924) 'Geography in France', *American Geographical Society, Research Series* 4A.

—— (1929) 'La morphologie du Plateau Central de la France et l'hypothèse eustatique', *Annales de Géographie* 38: 113–32.

—— (1940) 'Problèmes morphologiques de Brésil tropical atlantique', *Annales de Géographie* 49: 1–27, 106–29.

—— (1941) 'Hypsometrie et morphologie: Determination et interpretation des altitudes moyennes da la France et de ses grandes regions naturelles', *Annales de Géographie* 50: 241–54.

—— (1946) 'Géographie zonale', *Annales de Géographie* 55: 1–18.

Meynier, A. (1952) 'Cinquante ans de géographie Française', *Cinquantième Anniversaire du Laboratoire de Géographie (1902–1952)*, Rennes, pp. 1–19.

—— (1969) *Histoire de la Pensée Geographique en France (1872–1969)*, Paris: Presses Universitaires.

Mill, H. R. (1892) *The Realm of Nature*, London: Murray.

Musset, R. (1917) *Le Bas-Maine: Etude Géographique*, Paris: Colin.

Neumann, L. (1886) 'Orometrie des Schwarzwaldes', *Geographische Abhandlungen*, Band 1, Heft 2, 185–238.

Noë, G. D. de la and Margerie, E. de (1888) *Les Formes du Terrain*, Paris.

Penck, A. (1887) 'Ueber denudation der Erdoberfläche', *Schriften des Vereins zur Verbreitung naturwissenschaftlichen Kentnisse in Wien, in –18*, 27: 431.

—— (1894) *Morphologie der Erdoberfläche*, 2 vols, Stuttgart: Engelhorn.

Philippson, A. (1886) 'Ein Beitrag zur Erosionstheorie', *Petermann's Geographische Mitteilungen* 32: 67–79.

Richthofen, F. von (1886) *Führer für Forschungsreisende*, Hanover: Oppenheim.

Rutimeyer, L. (1869) *Ueber Thal- und See-bildung*, Basle: Schultze.

Sonklar, K. A. von (1873) *Allgemeine Orographie: Die Lehre von den Relief-Formen der Erdoberfläche*, Wien: W. Braumüller.

Sparks, B. W. (1971) *Rocks and Relief*, London; Longman.

Stamp, L. D. (1946) *Britain's Structure and Scenery*, London: Collins.

Steers, J. A. (1986) 'Physical geography in the universities 1918–1945', in R. W. Steel (ed.) *British Geography 1918–1945*, Cambridge, pp. 138–55.

Stoddart, D. R. (1987) 'Geographers and geomorphology in Britain between the wars', in R. W. Steel, (ed.) *British Geography 1918–1945*, Cambridge, pp. 156–76.

Suess, E. (1883–1908) *Das Antlitz der Erde*, 3 vols (vol. 1, 1883; vol. 2, 1888; vol. 3, 1901 and 1908), Vienna: Tempsky.

Supan, A. (1896) *Grundzüge der physischen Erdkunde*, Leipzig: Von Veit.

Surell, A. (1841) *Etude sur les Torrents des Hautes-Alpes*, Paris: Carilian-Goeury et V. Dalmont.

Tobein, H. (1974) 'Emmanuel Marie de Margerie', *Dictionary of Scientific Biography*, vol. IX, New York: Scribners, pp. 103–4.

Tricart, J. (1956) 'Le géomorphologie et la pensée Marxiste', *La Pensée* no. 69, pp. 3–24.

Tricart, J. and Cailleux, A. (1954–7) *Cours de Géomorphologie, Sorbonne*, Paris: Centre de Documentation Universitaire, Tournier and Constans.

Trueman, A. E. (1938) *The Scenery of England and Wales*, London: Gollancz (2nd edn 1949).

Vacher, A. (1908) *Le Berry: Contribution à l'étude géographique d'une région Française*, Paris: Colin.

Wooldridge, S. W. (1949) 'On taking th 'ge' out of geography', *Geography* 34: 9–18.

—— (1951) 'The progress of geomorphology', in G. Taylor, (ed.) *Geography in the Twentieth Century*, London: Methuen, pp. 165–77.

—— (1955) 'Review of 'Geographical Essays' by W. M. Davis', *Geographical Journal* 121: 89–90.

—— (1958) 'The trend of geomorphology', *Transactions of the Institute of British Geographers* no. 25, pp. 29–35.

Wooldridge, S. W. and East, W. G. (1951) *The Spirit and Purpose of Geography*, London: Hutchinson.

Wooldridge, S. W. and Linton, D. L. (1939) 'Structure, surface and drainage in south-east England', *Institute of British Geographers Publication* no. 10.

Wooldridge, S. W. and Morgan, R. S. (1937) *The Physical Basis of Geography: Geomorphology*, London: Longmans, Green and Co. (2nd edn 1959).

Wright, W. B. (1937) *The Quaternary Ice Age*, 2nd edn London: Macmillan.

CHAPTER SIX

Geomorphology Worldwide

Although by far the main volume of geomorphic work during the latter part of the nineteenth century and the first half of the twentieth was conducted in the USA, Germany, France and Britain, other work was also important. Moreover, this work was not always derivative and, in Russia and Sweden in particular, on occasion assumed a highly original character.

Russia

Up to the present Russian geomorphology has received only passing acknowledgement in our *History* (see volume 1, pp. 601–2) and we intend that the present volume will go some way to redress this omission, particularly with this chapter and chapter 11. Of course, Russian scholarship in non-mathematical subjects has tended to be undervalued in the western world, partly because of the language difficulty, but also because of the exaggerated and politically charged claims made for it by Russian reviewers during the Stalinist era (see especially Markov 1948, Borzov 1951, Dumitrashko *et al.* 1951). We shall attempt to do some justice to the imaginative and geographical sweep of Russian contributions to geomorphology. In this work we have been assisted by the above reviews, as well as those by Dylik (1953), Nikolayev (1958), Geller (1962), Kosygin (1964), Kalesnik (1968) and Gerasimov (1976). It is interesting that W. M. Davis had so little contact with Russian scholars, despite his participation in Pumpelly's Carnegie Expedition to Turkestan in 1903 which terminated in St Petersburg (see volume 2, pp. 278–82).

Mikhail Vasil'yevich Lomonosov (1711–65) (plate 6.1), the first great Russian scientist, was born near Archangel, the son of a ship owner (Kedrov 1973). He attended the Universities of St Petersburg, Marburg and Freiburg where he was exposed to the ideas of Newton, Leibniz and Descartes, and studied mathematics, chemistry, physics and, especially, mining. Returning to St Petersburg in 1741, he began an academic career in theoretical physics, experimental chemistry, mining and earth sciences. Lomonosov was made Head of the Geography Department of the St Petersburg Academy in 1758 and between 1757 and 1763 produced three books on mining and metallurgy. A supplement to his *Principles of Metallurgy and Mining* was 'On the strata of the earth' (1763) which contained his important geomorphic ideas, including

Plate 6.1 Mikhail Vasil'yevich Lomonosov (1711–65)

Source: From *Geographers: Biobibliographical Studies* No. 6, Mansell, London, 1982

those on the relations between geology and landforms. Unfortunately the whole work, intended for Russian miners, was published only in Russian and not additionally in Latin and so was neither widely read abroad nor in his home country (Markov 1948: 25, Morris 1967). Lomonosov was impressed with the reality of recent and contemporary changes of the earth's surface (1763; see 1954: 574):

> Great changes have occurred in ... [visible corporal aspects of the earth] as is shown by history and by comparing ancient geography with the geography of today, and by the changes in the surface of the earth which occur in our times.
>
> (Quoted in Gerasimov 1976: 71)

He admitted the erosive effects of wind, rain, rivers, sea and ice; recognized sedimentary layers as representing long periods of erosion; and adopted a generally uniformitarian view of earth history. Believing vertical crustal movements to be due to volcanic activity, he distinguished between mountain-building upheavals and the slower extensive movements of continental surfaces and sea beds. Lomonosov also stressed the importance of the interaction between endogenetic and exogenetic processes, with especial reference to the former which he believed to be oscillatory, as distinct from the contractional ones later in vogue (see chapter 1) (Geller 1962: 957). Lomonosov recognized river erosion but wrote little regarding it and did not connect it clearly with landforms. He had few students and the only influential one was A. T. Bolomov who described the process of headward gully erosion in 1781. Lomonosov was a true polymath – physicist, chemist, natural scientist, poet and linguist – being described as a cheerful and gentle man but also as being hot-tempered and passionate (Menshutkin 1952).

Although Peter Pallas was influenced by Lomonosov's tectonic writings during his years travelling in Russia (1768–74) (see volume 1, pp. 19–20) and incorporated them in his *Observations sur la Formation des Montagnes* (1777), Russian geomorphic studies languished until a remarkable series of explorations took place during the second half of the nineteenth century. Peter Petrovich Semenov-Tyan-Shanskiy (1827–1914), the privately educated son of a landowner and playwright, was elected a member of the Russian Geographical Society at the age of 21. In 1851 his interest in the Tian Shan Mountains was aroused by his being asked to translate Karl Ritter's *Die Erdkunde von Asien* and between 1853 and 1855 he lived in Berlin, becoming friendly with Ritter. The translation of volume 1 appeared on his return to Russia in 1855 and was recognized to contain much new material, together with a strong call for the development of Russian orographic studies (Fedchina 1975). Semenov's important expedition to central Asia during 1856–7 took him to Lake Issyk Kul, the Chu River, the Terskey Ala-Tau Range, the Tian Shan Mountains and the Tarim Basin. The results of this expedition were published in 1858, giving a description of the orography and geological structure of the mountains of central Asia and proof of the intense glaciation of the Tian Shan Mountains. In 1860 he was made President of the Physical Geography Section of the Russian Geographical Society, of which he wrote a history in 1896, and in 1864 he became Head of the Central Statistical Committee being concerned with inventories of property, crop yields and the peasant economy. Semenov was an extremely cultured man who wrote books on art and in 1910 presented some 700 paintings and 3,500 prints (many from the Dutch old masters) to the Hermitage Museum in St Petersburg.

Two other important late-nineteenth-century scholars were Petr Alekseyevich Kropotkin (1842–1921) and Ivan Cherskiy (1845–92). Kropotkin, best known for his political and social writings, joined the

Trans-Baikal Expedition to the Amur River in 1863 and in 1866 headed the Expedition to the Vitim and Olekma Basins, travelling down the Lena River. He published evidence for the former glaciation of eastern Siberia (1873), described the orography of the region (1875; see also 1904), and in 1876 produced an important book on the glacial period in which he presented evidence for the glacial transport of debris from Fenno-Scandinavia over the plains of Europe and northern Asia, giving the first comprehensive description of extensive glaciation of plains. After 1874 his political activities caused him to take refuge in Western Europe, where he became associated with the Royal Geographical Society, addressing it on the teaching of physiography in 1893. Cherskiy was also a political activist who was banished to Omsk during the period 1863–9 where his interest in geology was stimulated by his finding a copy of Lyell's *Principles* (see volume 1, p. 601). In 1871 he settled in Irkutsk and organized expeditions. Seven years later he wrote on the Lake Baikal region and proposed an embryonic form of evolutionary geomorphology based on the premise of the progressive development of erosional relief (Shchukin 1952: 558). However, as Dylik (1953: 7) has pointed out, his work was dominantly geological, lacked adequate correlation between processes and landforms, and was without a sound genetic basis. In the latter regard it is of note that Charles Darwin's ideas were not enthusiastically received in Russia partly because of the rival earlier studies of evolution by the biologist K. F. Rul'ye.

The tectonic approach to geomorphology of Lomonosov found nineteenth-century support in the work of Franz Yulevich Levinson-Lessing (1861–1939), Alexander Petrovich Karpinskiy (1847–1936) and, especially, of Vladimir Afanas'yevich Obruchev (1863–1956). Levinson-Lessing, a student of Dokuchayev, was an igneous petrologist and professor at the University of Tartu and then at Leningrad Polytechnic Institute. He believed that mountain building has been simultaneous in many different locations due to magma transfer at depth, allowing a global balance between uplift and subsidence (Levinson-Lessing 1893) (see chapter 1). His tectonic pattern for a given location involved geosynclinal sinking, slow uplift and folding at the continental margins. He recognized the impact of recent tectonic movements on landforms as did Karpinskiy who described the palaeogeography of western Russia (1888), supported the contraction theory of orogeny but believed in the importance of the vertical oscillations of the Russian Plain (1894). Obruchev took part in the Trans-Caspian Expedition of 1886–8, organized by Semenov-Tyan-Shanskiy, studying aeolian action, the origin of loess and the distinguishing features of central Asian sands which appeared mostly to be of aeolian and fluviatile origin, rather than marine as proposed by Von Richthofen (Naumov 1974). In 1888 he took up a geological post at Irkutsk, the following year travelled across the Pribaikal Mountains and in 1890–1 surveyed the goldfields of the Olekma and Vitim Basins confirming Kropotkin's discovery of 60–180 feet of superficial glacial deposits. Obruchev's report on the Irkutsky Province (1890) identified the Lake Baikal depression as massively downfaulted. Between 1892 and 1895 he worked in Mongolia and China locating the central Asian source of the loess and between 1895 and 1898 studied the geology along the route of the Trans-Siberian Railway, then under construction. As a result of these investigations he proposed that the granites and slates of the Transbaikal represented the geological nucleus of the Asian shield, a notion accepted by Suess. During the period 1905 to 1909 he worked in the Altai and Tian Shan Mountains, showing the former to be a series of step-faulted plateaus composed of horsts and grabens on a huge scale (1915), rather than the simple folds of Suess. The latest view is that the Altai are composed of a series of Palaeozoic folds which were peneplained, uparched and differentially block faulted (Naumov 1974: 168). Obruchev survived into his 93rd year and in 1936 wrote an article on the youthfulness of the faulted relief of Siberia followed by one in 1948 on what he termed 'neotectonics' in which he proposed that crustal structures created by Late Tertiary and Early Quaternary earth movements fully explain all the characteristics of the contemporary relief of the land surface of the entire globe (Geller 1962: 101). Obruchev's writings were republished in collected form between 1931 and 1949.

Of course, considerable work of a less striking nature was conducted in Russia during the latter part of the nineteenth century and prior to the First World War. Relations between geology and landforms were worked out for the Russian Plain and the Urals (by N. A. Golovkinskiy, I. F. Levakovskiy and S. N. Nikitin), for Central Asia (by P. P. Semenov-Tyan-Shanskiy and N. A. Severtsov) and for the Caucasus (by G. V. Abikh, I. T. Stebnitskiy and K. I. Podozerskiy). In the 1860s and 1870s N. A. Golovkinskiy and I. F. Levakovskiy commented on the origin of erosional river valleys. Somewhat later V. M. Lokhlin, N. S. Lelyavskiy and V. G. Glushkov wrote generally on river flow and morphology, and discharge studies were made on the rivers Chusovaya and Dniester (by V. M. Lokhlin), Volga (by N. G. Bogoslovskiy) and Dnieper (by N. I. Maksimovich) (Gerasimov 1976: 97–8).

The greatest of all the nineteenth-century Russian earth scientists was undoubtedly Vasiliy Vasil'yevich Dokuchayev (1846–1903) (plate 6.2) (Morris 1967, Yesakov 1971). Dokuchayev, the son of a village priest, was born near Smolensk and studied natural sciences at St Petersburg. His Master's degree was written on the subject of the alluvial deposits of the Katchna River, an upper tributary of the Volga near his birthplace, and shortly afterwards he was made Curator in Geology at St Petersburg University and in 1879 was teaching mineralogy and crystallography there, giving the first recorded course on Quaternary deposits. In 1877 and for his doctoral dissertation in 1878 he produced two important articles on fluvial geomorphology which:

Plate 6.2 Vasiliy Vasil'yevich Dokuchayev (1846–1903)

Source: By permission of *American Geographical Society, Occasional Publication*, No. 1, 1962, p. 16.

1 Stressed the reality of fluvial erosion, discussing river capture, migration of divides and the reasons for the lack of correspondence between the size of rivers and that of their valleys.

2 Discussed valley development from the ravine to the meandering states, introducing the terms 'senility', 'maturity', 'youth' and 'infancy' for relief (Geller 1962: 97).

3 Applied quantitative methods to the study of ravines.

4 Began the organized study of landforms in Russia based on fluvial erosion and river-valley development (Dumitrashko *et al.* 1951).

5 Described glacial deposits on the Russian Plain and criticized Murchison's theory of their 'drift' origin.

After this Dokuchayev turned more and more to the study of pedology, and in 1892 became Director of the Institute of Agriculture and Forestry at Kharkov, founding the first Department of Soil Science there. His pedological interests bore first fruit in his important *Russian Chernozem* (1883, 1949–61, vol. 3; see our volume 1, p. 602). The great Russian drought of 1891 led Dokuchayev to review farming, forestry and water management in the steppe area resulting in *Our Steppes Past and Present* (1892). Over many years, a series of scientific travels to Nizhni Novgorod, Poltava, Bessarabia, the Caucasus and the Transcaucasus stimulated his interest in the zonality of soils (1889) and his related contributions to landscape regionalization (see chapter 11). Dokuchayev believed that the study of 'genetic pedology' was one of the most important keys to the understanding of all aspects of physical geography. In this regard he distinguished five world physico-geographical zones: boreal, taiga, chernozem, arid subtropical and laterite (Gerasimov 1976: 99).

Dokuchayev's interest in the geomorphological processes of the Russian plains was paralleled by that of Aleksey Petrovich Pavlov (1854–1929), a professor at the University of Moscow. In 1889 Pavlov described the glacial landforms of the plains, identified two Pliocene cold phases and three glaciations in European Russia, studied slopes produced by the weathering of talus, and made conjectures regarding the origin of loess (which he believed to be of aqueous origin). Nine years later he described the geomorphic effects of surface and subsurface water on the plains in respect of landslides, sheetflood erosion and the eluviation of subsurface sediment by groundwater (Pavlov 1898; see also 1951).

Thus, not long after the end of the nineteenth century, nine Russian scientists – Lomonosov, Semenov-Tyan-Shanskiy, Kropotkin, Cherskiy, Levinson-Lessing, Karpinskiy, Obruchev, Dokuchayev and Pavlov – had made considerable contributions to geomorphology, although they were little known in the western world. They had also laid the foundations for Russian work in the first half of the twentieth century, in providing the impetus for research into the effects of tectonics and neotectonics on landforms, erosional development and fluvial processes, glacial landforms, landscape regionalization, orography and mapping, and practical applications. The contributions of these scientists were all the more impressive when one recalls the immense spatial scale on which they were working (dwarfing that of the American West), the problems of movement and communications (the Trans-Siberian Railway was not constructed until 1891–1905), the climatic extremes, and the social and political instability of the period. The October Revolution of 1917 had a number of geomorphic ramifications: first, it inhibited much academic work during the interwar period, after which the huge effort required of Russia during the Great Patriotic War (i.e. the Second World War) meant that full attention to scientific work could only resume in the 1950s; second, the Revolution concentrated attention on applied or 'useful' aspects of the discipline; and, third, it had a profound effect on the theoretical approach to scholarship.

After the Revolution a radical shift in the approach to the study of the earth sciences and the natural environment occurred. The ideas of Georg Hegel (1770–1831), which had been adopted by Marx and Engels, now swept through the Soviet world. The key words were *dialectic* and *materialism*. Dialectic means, crudely, the process of reaching the 'truth' through change, whereby a thesis is transformed into its opposite (antithesis) and preserved and fulfilled by it, the combination of the two being resolved in a superior form of truth (synthesis). Materialism is the theory that physical matter in its motions and modifications is the only reality and that everything (material and mental) in the universe can be explained in terms of physical laws. The theory of dialectical materialism was the Marxist interpretation of reality which viewed matter as the primary subject of change and all change as the product of a constant antagonism between opposites arising from the internal contradictions inherent in all things. These contradictions are resolved at higher levels and fresh contradictions arise.

As Markov (1948) noted, Russian geomorphology is rooted both in geology, exemplified by Levinson-Lessing, and in geography, exemplified by Semenov-Tyan-Shanskiy. Its treatment as a geological science was conducted by Bondarchuk (1949) and Yefremov (1950), whereas Shchukin (1934–8, 1952) dealt with it as a geographical discipline. In the 1930s the work of Grigor'yev (1931, 1935, 1937) provided an interesting variant to this dichotomy by combining the above approaches in a concern for the study of processes and energy flows.

However, ideological attitudes associated with the Revolution affected both geology and geography which became much more overtly applied and materialistic:

All the successes of (Soviet) geomorphological science were possible because geomorphology in our country is bound up with the practice of the socialist structure, whose development goes hand-in-hand with practice, the planning of geomorphological work depends on the needs of the national economy.

(Dumitrashko *et al.* 1951: 78)

In the case of geography, the pre-Revolution scholars, such as Lomonosov, Voyeykov, Dokuchayev and Anuchin, were considered to have held decidedly materialistic views which, around the turn of the century, had begun to be corrupted in later scholars by the 'metaphysical' attitude exemplified by Hettner:

> On the eve of the Great October Revolution, Hettnerianism had unfortunately won over a number of authoritative Russian geographers. Having cast A. Humboldt's progressive ideas into the shadows in Germany, the Hettnerian ideas also sought to gain dominance in Russia over the scientific ideology of Lomonosov and Dokuchayev. The (1917) Revolution put a stop to this.
>
> (Kalesnik 1968: 393)

Stimulated by the Marxist view of the world, a review of the theoretical positions of geography was instigated and this soon exposed a sharp theoretical division between the characteristics of physical and economic geography. The former was recognized as a natural science, the latter as a social science, and the fundamental differences in their laws and regularities made it impossible to combine them into a 'unified geography' or into a western-style 'regional geography'.

The synthesis of the interactions of the two was for the Soviets a separate geographical discipline or concept called *stranovedeniye*. Thus there developed the notion of a complex of geographical sciences, each member of which could define its own object of study. Physical (natural) geography is sometimes seen as the aggregate of all the natural sciences that enter into geography, but in practice the term is most aptly applied only to the synthetic physical–geographical disciplines, namely general earth science or general physical geography, landscape science or regional physical geography, and palaeogeography. Thus general earth science has a right to independence but remains part of the geographical family (Kalesnik 1968: 394).

The result was that, since 1917, 'Geography' and 'Geomorphology' have represented different ideas in the west and east. In the United States the influence of Davis and others caused regional geography to flourish and it grew more Germanic under the influence of Sauer and Hettner. Here geomorphology, in spite of structural regionalization, lost many adherents. In Soviet Russia, regional geography in the Davisian and Hettnerian sense was strongly obstructed and had no obvious place in concepts based on a strict adherence to 'laws'. Here geography was being smothered by geomorphology. Yet, it is clear that many of the distinctive traits of earlier Russian geography or geomorphology survived the 1917 Revolution. The pre-existing regionalization or landscape science was strengthened by economic leanings and by the establishment in 1921 of a State Planning Commission. Whereas in France and Germany the geomorphic trends began to veer towards climatic geomorphology, in Russian lands the bias turned to physico-geographical geomorphology. But, as in most other countries, the more purely geological or geotectonic approach to landform study also survived.

The encouragement given by Soviet committees to earth sciences was great and lasting, as might be expected considering the vastness of their territories ripe for further investigation and development. In 1918 a Geographic Institute was created at Petrograd and it soon had a progressive geomorphological department with a broad programme of instruction, including in addition to general geographical and general geological disciplines, special subjects such as Quaternary deposits, geochemistry and volcanology. During the 1930s the Institute of the Academy of Sciences USSR moved from Leningrad (Petrograd) to Moscow and eventually about 20 years later the Faculty of Geography at Moscow University also took up specialized research in geomorphology. Soon the three largest departments of that subject in the USSR were in the faculties of geography at Moscow, Leningrad and Kiev.

Despite the ideological and methodological furore during the period between the Revolution and mid-century, Russian work in geomorphology went ahead, largely along the lines laid out by the pre-Revolutionary scientists. We have already referred to Russian contributions to the study of crustal movements of geomorphological significance in chapter 1 and much work was directed towards the study of the epeirogenic instability of the Russian Plain (see especially Lichkov 1931, 1934a, 1941, 1949), carrying on observations in the middle of the last century by N. A. Golovkinskiy, as well as by Levinson-Lessing (1893) and Karpinskiy (1888, 1894). Tectonic studies continued to be geologically rooted, such as that by Markov (1937) on the Pamir Mountains and by Gerasimov (1946, 1959) on the major structural elements of the USSR. However, neotectonics, summarized by Nikolayev (1949), were widely believed to be responsible for the production of much of the present relief as a result of either very recent (i.e. Late Pliocene or Quaternary) vertical crustal movements or by the recent rejuvenation of existing tectonic structures. In this light Nikolayev and Polyakov (1937) studied the effects of epeirogenic movements on river courses, the Karakoram Mountains were investigated by Fedorovich (1934), the Tian Shan by Kaletskaya et al. (1945) and Shul'ts (1939), the Urals by Bashenina (1948) and the Altai by Kaletskaya (1948). During this period it was common for geomorphologists to be taken publicly to task for not taking neotectonics sufficiently into account in the explanation of existing landforms. In general, although the tectonic ideas of Walther Penck were preferred in Russia to those of Davis, both were considered inappropriate (Dumitrashko et al. 1951) in that most crustal movements were considered to be long continued, irregular, spasmodic, vertical in character, rapid in phases and contemporary in geomorphic significance. It must be remarked that western geomorphologists only became interested in neotectonics in the 1970s!

As we have seen, the study of fluvial landforms began in Russia in the nineteenth century and it is of note that S. N. Nikitin wrote the first regional geomorphology of European Russia in 1885, making a genetic connection between rivers and their valleys and commenting on the control of gullying. In the interwar period practical applications of geomorphology were paramount and B. L. Lichkov (1931, 1934b) investigated fossil hydrological networks on the Russian Plain in connection with placer mining and Yu. A. Bilibin (1938) also studied placer deposits using Davis' cyclic ideas in prospecting for them (Geller 1962: 104). It is interesting that other work based on the ideas of Davis and Penck was published after the Second World War by S. G. Sobolev (1948) on erosion and the development of ravine and gully slopes, by N. V. Dumitrashko (1948) on youthful and old relief in south-east Siberia, and by Markov (1948) on the origin of erosion surfaces (see also his review of geomorphology in honour of W. M. Davis (1952)). However, such works were invariably criticized for not taking due account of the influence of endogenetic (i.e. tectonic and neotectonic) controls over landforms. The study of channel processes in Russia was largely based on the State Hydrological Institute which was founded in Leningrad in 1919. In the 1950s research on hydraulic geometry (e.g. Makaveyev 1955) and channel processes (e.g. Fedoseyev 1957a, b) developed.

Much of the remaining Russian geomorphology during the first half of the twentieth century was concerned with the study of detailed processes which will be treated in volume 4 of our *History*. However, significant contributions were made in the fields of glacial landforms, mapping and in the production of much-needed textbooks. S. N. Nikitin (1904) studied the evolution of the glacial landforms of the Russian Plain and was one of the first scientists to describe the landforms produced by stagnant ice at the glacier margin. In 1926 N. N. Sokolov wrote on the glacial geomorphology of the northern part of European Russia; research on arctic and subarctic geomorphology was conducted by S. P. Suslov (1935) and A. A. Grigor'yev (1939–42, 1946); and I. P. Gerasimov and K. K. Markov (1939) published a general account of the Russian Pleistocene. After the Revolution the emphasis in geomorphology was increasingly concentrated on the exploitation of natural resources and on mapping (Tanfil'yev 1922–3, Obruchev 1931–49). In the 1920s A. A. Borzov directed a considerable effort in surveying and engineering geomorphology, partly concerned with the construction of the northern railways, river control, hydroelectric works and counteracting soil erosion. During the First World War a map of Russia was completed on a scale of 1:1,000,000 and subsequently attention turned especially to more detailed mapping of the Central Asian territories and Siberia in the 1930s, extending the mapping of the plateau system of Siberia which S. Obruchev and K. Salishchev produced in 1926. This practical work in geomorphology was stimulated by the founding of the national Geomorphology Institute in 1930, and later A. A. Grigor'yev and K. K. Markov prepared a map of the geomorphic regions of the USSR. Although textbooks of physical geology and physical geography were in existence from the last century (e.g. Lents 1851, Mushketov 1888–91), students had to wait until the 1930s for the first general works on geomorphology (Edel'shteyn 1932–47; and the encyclopedic Shchukin (1934–8)), until 1945 for the first Russian translation of Martonne's *Traité de Géographie Physique*, and until after the Second World War for more up-to-date texts by Markov (1948) and Bondarchuk (1949).

The geomorphic work carried on in Russia between 1917 and mid-century may appear from the foregoing to be rather limited but this is because, while admittedly there was little historically-based study of landforms (in striking contrast to the United States, France and Britain) there was considerable regional physico-geographical work which will be treated in chapter 11 and much dominantly process-oriented scholarship. The latter will be featured in our volume 4 which will include references to the following authorities:

River processes and hydrology: V. A. Dement'ev, L. K. Davydov, V. Lakhtin, M. I. Lvovich, P. I. Makaveyev and P. A. Pravoslavlev.
Soil erosion processes: P. S. Pogrebnyak, S. S. Sobolev and S. A. Yakovlev.
Glacial processes: S. V. Kalesnik and K. K. Markov.
Permafrost processes: S. P. Kachurin and A. I. Popov.
Aeolian processes: V. A. Dubyanskiy, B. A. Fedorovich, S. J. Heller, A. Ivchenko, S. N. Nikitin, V. A. Obruchev, B. A. Petrucherskiy and N. A. Sokolov.
Marine processes: O. K. Leont'yev, D. G. Panov and V. P. Zenkovich.

Above all, however, Dokuchayev's concern for soil-forming processes (disseminated abroad by translations of K. D. Glinka's *Pedology* into German in 1914 and into English in 1927) was carried on in the nineteenth century by N. M. Sibirtsev; before the Revolution by V. V. Gemmerling, P. S. Kossovich, A. I. Popov, M. Y. Tkachenko and V. R. Vil'yams; and after 1917 by K. D. Glinka, E. S. Kachinskaya, S. S. Neustruyev, B. B. Polynov, N. N. Sokolov, D. G. Vilenskiy and S. A. Zakharov (see Rode 1961).

During the 1950s and early 1960s a fascinating discussion arose among Soviet scholars which revealed strong differences of opinion on the scope and nature of geography. This had long been contentious and, in spite of official guidelines, was now brought into the open (Matley 1966). It has been ably dealt with by James and Martin (1981) and Hooson (1962) and, as it concerns geography, would be largely outside our enquiry if it had not had such a clear side-effect upon geomorphology. However, it was a great debate which did much credit to all contestants and seems to have had far-reaching consequences. In 1956, V. A. Anuchin, a distant relative of D. N. Anuchin (see

1954), wrote a well-balanced geography of Transcarpathia and went on with others to support the concept of a 'unified geography' in preference to a non-human physical geography and a non-natural economic geography, and also in preference to geographical determinism (the hallmark of bourgeois thought!) and to its opposite extreme, indeterminism. The supreme illustration of geographical method is, he says, the territorial complex or region in which physical, human and economic aspects are in balance.

In 1960 Anuchin produced his *Theoretical Problems of Geography* and immediately one group, including Baranskiy and Saushkin, welcomed it, while another, including Gerasimov (who wanted landscape science to be a practical constructive geography), Kalesnik and Konstantinov, found it unscientific and anti-Marxist. On its submission for the doctorate it was rejected by the faculty at Leningrad in 1961 and at Moscow in the following year. Eventually, after a prolonged debate, it was successful at Moscow in 1964. Possibly these discussions eased the way for the less unfavourable consideration of some aspects of western geomorphology.

Indeed, after mid-century there was a limited, but appreciable, thaw in Russian attitudes to western geomorphology from the extreme position adopted by Dumitrashko *et al.* (1951: 74):

> Today in capitalist countries the theory of geomorphological science is decadent. Authors keep rigidly to the vicious circle in the methodological treatment of the ideas of Davis or Penck and fail to put forward any new ideas .... Bourgeois geomorphology, deprived of a reliable methodological basis such as that which dialectical materialism gives to Soviet geomorphologists, is restricted to specialized tasks and is therefore unsuited to wide theoretical general investigations ... it justified itself theoretically on a premise of 'positivism' which in the West is a form of idealism.
>
> The history of foreign geomorphology is centred upon the works of W. Davis and W. Penck. Davis's cyclic concept had a false methodological base as it depended on a repetitious succession within a closed circle; W. Penck's concept was, on the whole, lop-sided and defective in so far as it was based on uninterrupted uplift whereas the formation of step-like surfaces in mountains demands intermittent uplift.

We have already described in chapter 4 how Gerasimov (1956) defended Davis against the impression given in *American Geography: Inventory and Prospect* (1954) that his work was no longer of real relevance. A further indication of interest in Davis came in 1962 when his *Geographical Essays* were translated into Russian under the title *Geomorphological Essays*. Soviet textbooks on geomorphology normally refer to the work of Davis in the introductory chapter and, for example, D. G. Panov in his *General Geomorphology* (1966) states that Davis 'creatively elaborated' earlier ideas on the evolution of relief that were expressed by G. K. Gilbert, together with those of the three Russians, V. V. Dokuchayev, I. Cherskiy and S. N. Nikitin. In 1971

A. Ye. Krivolutskiy, in his *Life on the Earth's Surface*, is reported to have discussed the Davisian cycle

> distinguishing both constructive aspects, such as the sequential approach to relief development, and negative aspects, such as an underestimation of the role of endogenous, tectonic factors in the evolution of relief. The author says that because of the many unshakable principles incorporated into Davis's concept it remains significant today and many of his propositions regarding stage and cycle continue to be used in modern geomorphology in whole or in part.
>
> (Theodore Shabad, letter dated 4 November 1976)

In 1984, in an interesting volume entitled *The Life and Explorations of W. M. Davis and W. Penck*, Mikhail Piotrovskiy of the Museum of Earth Science at Moscow State University discussed the Davis and Penck story. In 1961 he had edited and written an introduction to an edition in Russian of W. Penck's *Die morphologische Analyse*; in 1962 he translated *The Geographical Cycle* in a Russian edition of Davis' *Geographical Essays*; and he now presented his views on the concepts of Davis and Penck, which he considered

> do not represent – as many think – any past stage of science, on the contrary, they must be highly estimated anew as the first and not yet exhausted, well-elaborated concepts of self-organizing and self-regulating geomorphological (that is, inorganic) systems.

This judgement applies especially to Davis whose cycle

> is the ideal model of a physical self-organizing system in its full development from initial to final stages . . . . It is not strictly deterministic but widely flexible . . . all the complications and modifications can be explained on the basis of the general principles of the cycle.
>
> (Piotrovskiy, letter dated 12 January 1985)

As for Walther Penck, his contribution seemed to Piotrovskiy not in opposition to Davis' ideas but 'its natural, though essentially original, and enriching continuation'. In his book, Piotrovskiy also mentions the Russian and Soviet idea of the evolution of surface relief which relates weathering crusts, soils, landscapes and earth movements, as promulgated by pedologists and geobotanists such as V. V. Dokuchayev, S. S. Neustruyev, B. B. Polynov, I. M. Krasheninnikov and others. Some of them, he says, have integrated the Davisian concept into their own findings.

Finally, although we have already trespassed far outside the timescale intended for our own volume, we cannot refrain from recording that in April 1984 Piotrovsky gave an address to the Geomorphological Commission of the Moscow Branch of the USSR Geographical Society to commemorate the centenary of Davis' cyclic concept, the half-century of his death, the sixtieth year of Walther Penck's death (1923), and the posthumous publication of *Die morphologische Analyse*. It should be recalled, however, that half a century previously Borzov (1934) addressed the International Geographical Congress in Warsaw on the subject of the Davis cycle.

Despite the above examples of Russian geomorphologists taking a more benign view of the work of Davis and Penck of recent years, we must not lose sight of the fact that Russian geomorphology up to the middle of this century developed along its own very particular and distinct lines. When, in 1968, I. P. Gerasimov, Director of the Institute of Geography at the Academy of Sciences, reviewed half a century of Russian geographical thought he expressly singled out contributions in pedology, glaciology and geomorphology:

> Geomorphologists established the dynamic character of many exogenous processes (erosion, deflation, abrasion, etc.) and, on the basis of a general theory of external and internal forces and the study of recent crustal movements, developed the morphotectonic or morphostructural approach to geomorphology.
>
> (Gerasimov 1968: 242)

Scandinavia

Much of the geomorphic work carried out in Scandinavia during the first half of the twentieth century possessed a very distinctive character, presumably because of the dominant crustal, erosional and depositional legacy of Quaternary times and the consequent preoccupation with more recent geomorphic events. Such preoccupation naturally found fruitful expression in detailed studies of processes – not only glacial and isostatic, but also fluvial, aeolian and marine. In this relative emphasis, Scandinavian geomorphology contrasted strikingly with that in the United States, Britain, France and Germany.

None the less some more conventional influences were at work. Davis' contribution to the International Geographical Union meeting in Berlin in 1899 provided a powerful stimulus for all European geomorphology, not least in *Finland* (Granö 1987). The concept of the peneplain and of its utility in geomorphic explanation was adopted by J. E. Rosberg (1906), W. Ramsay (1909), J. J. Sederholm (1910, 1911), P. Eskola (1913) and J. Granö (1919). Denudation chronology was particularly explored by W. Ramsay (1917) and in later years V. Tanner (1936, 1938, 1944) applied Davis' views to Finland and Labrador. The *Swede* Hans Wilson Ahlmann (1918) discussed baselevel and peneplanation and in *Norway* K. M. Strøm (1948) described two Tertiary erosion surfaces believed to have been elevated in the Miocene and Pliocene, respectively, and dissected by later Tertiary and Pleistocene erosional processes. In this connection it is both interesting and characteristic that one of the earliest attempts to approximate an erosion surface mathematically was suggested by Svensson (1956).

Scandinavian studies of Quaternary and Recent processes and landforms will be treated in our volume 4 and reference has already been made in chapter 2 of the present volume to the very significant depositional varve studies of Gerard de Geer. The investigation of glacial erosional processes during the interwar period was led by H. W. Ahlmann, who described Norwegian geomorphology (1919) and took a leading part in the Swedish–Norwegian expeditions to the Arctic in 1931 and to Spitzbergen in 1934. During the same period A. Hellaakoski analysed the transport of material in eskers and A. G. Högbom, who had worked on drumlins in the early years of the century, discussed other glacial depositional features. The effects of post-glacial isostatic rebound were analysed by H. G. Backlund, E. Niskanen, W. Ramsay and M. Sauramo, and the earlier work of F. Nansen relating rebound to shoreline features was extended by B. Asklund's studies of the strandflat. As we have seen in chapter 2, E. Nilsson studied the effects of Pleistocene climates on the lakes of East Africa, and the relations between fossil dune forms and palaeo wind directions in Europe were examined by F. Enquist and I. Högbom. However, Scandinavian geomorphology in the first half of the twentieth century was significantly directed to the study of contemporary processes. Weathering and the production of clays concerned O. Tamm and A. Salminen, P. Eskola worked on limestone solution and G. Beskow on the effects of frost action. In *Denmark* Axel Schou related coastal processes and forms, but by far the most influential geomorphic process studies were instituted in Sweden in the 1930s by Filip Hjulstrom's classic work on debris transport and channel forms in the River Fyris. This had great influence in the United States after the Second World War and was directly instrumental in encouraging the research on stream bedload trans-port by E. Meyer-Peter in 1948 and on channel processes and forms in the 1950s by A. Sundborg for the River Klarälven and by L. Arnborg for the River Ångermanälven.

Eastern Europe

Until after the Second World War the geomorphology of Eastern Europe was dominated by German, Russian and, to a lesser extent, French influences. In *Poland* there was important nineteenth-century work on glacial, pro-glacial and fluvio-glacial forms and processes described by Rehman (1895–1904), Glodek (1952) and Czechowna (1965), as well as on general physical geography. In 1899 Eugeniusz Romer, who was to hold the chair of geography at Lwow from 1911 to 1931, published an important article on the influence of climate on landforms, very much in the German idiom. However, in spite of having been a student of Albrecht Penck in Vienna, Romer, in common with almost all leading Polish geomorphologists prior to the Second World War, accepted the Davisian approach in broad terms (Dylik 1964: 274). As late as 1960 Alfred Jahn (1961: 9) told a combined

Plate 6.3 Ludomir S. von Sawički (1884–1927)

Source: From *Geographers: Biobibliographical Studies* No. 9, Mansell,
London, 1985

Polish–American audience in Warsaw that almost all the then current
professors of geomorphology in Poland had studied originally at Lwow,
Cracow or Poznan where their work was based on the textbook by Davis
(1912J). Romer's eminent contemporary, Ludomir von Sawički (1884–1927)
(plate 6.3), was also a follower of Davis, devising a karstic cycle of erosion
(Sawički 1909a) and joining Davis' (1912G) European 'pilgrimage' in 1911.
At about this time Sawički published a number of articles on the landforms of
Italy and Britain. Despite the success of the Davisian cyclic idea in Poland
polycyclic analyses were rare there and the term denudation chronology was

virtually unknown, due mainly to the Germanic influences of Passarge and the Pencks (Dylik 1964). However, Porkorny (1911) wrote on the Carpathian Peneplain, Czyzewski (1934a) on the erosion surface in the eastern Carpathian foothills, Lencewicz (1934) on Tertiary erosion surfaces and Smolenski (1934) on polygenetic surfaces.

Other influences were also at work to moderate the impact of Davisian geomorphology in Poland (Dylik 1953, 1964). One of these was a concern for the effects of recent tectonic movements on landforms, which owed more to Russian than to German models, and which was naturally focused on the Carpathians and the extreme south of the country (Sawički 1909b, Smolenski 1910, 1922, Lencewicz 1926a, Swiderski 1932, Cehak, 1934). Of particular interest was work, such as that by Loziński (1922), specifically relating the effects of earth movements to river development. A second diversion from traditional Davisian thinking was provided by the major Polish concern for the processes and landforms associated with glacial (Glodek 1952) and periglacial (Dylik 1952) conditions. Volume 4 will be much occupied with Polish researches in the fields of lowland glaciation (e.g. by Lencewicz, Limanowski, Lomnicki, Loziński, Wollosowicz and Zaborski), upland glaciation (e.g. by Gadomski, Loziński, Pawlowski, Romer and Sawički), glaciology (e.g. by Arctowski and Dobrowolski), loess (e.g. by Sujkowski and Tokarski), Pleistocene dunes (e.g. by Lencewicz, Małkowski, Pawolwski, Przemyski and Romer) and periglaciation (e.g. by Loziński). One important offshoot of these interests was the long-continued Polish concern for the palaeomorphology of rivers, the geometry of which can be related to high Pleistocene discharges (Bakowski 1902, Sawički 1912, Pawlowski 1926). Indeed, much Polish fluvial geomorphology before the Second World War was heavily mixed with assumed Pleistocene legacies, such as the river-terrace studies of Romer (1906) and Czyzewski (1928) on the Dniester, Lencewicz (1933) on the Vistula and Dnieper, and Kondracki (1933) on the Bug. Studies of valley asymmetry (Romer 1897) and river capture (Gadomski 1934) also displayed this Pleistocene emphasis which was to give Polish geomorphology much of its increasingly distinctive flavour after the Second World War. A third feature which was to differentiate Polish geomorphology, for example from its American counterpart, was its deep rooting in university departments of geography, of which there were five by 1940 and seven in the 1960s, plus the Institute of Geography of the Polish Academy of Sciences (Dylik 1964). Building on the earlier regional studies of Kawczyński (1876), Rehman (1895–1904) and Romer (1904), morphological mapping and regional physiographic differentiation became important geomorphic pre-occupations in Poland. Among these were studies of the regions of Lwow (Pawlowski 1917), Cracow (Smolenski 1923), Poznan (Pawlowski 1927) and Pokucie (Czyzewski 1931, Zglinnicka 1931); as well as the application of morphometric methods to the analysis of lowland glacial terrain

(Zaborski 1931) and to regional differentiation (Lencewicz 1926b, Czyzewski 1934b).

After the Second World War Pleistocene and morphological work became even more dominant and Polish geomorphologists threw open their intellectual windows to outside influences, especially western ones (Dylik 1953, 1957; Brown 1964). One effect of this was to revitalize earlier Polish interest in the processes to do with weathering (e.g. by Loziński), landslides and slopes (e.g. by Sawički, Gelinski and Teisseyre), karst (e.g. by Sawički, Loziński, Friedberg and Pawlowski) and coasts (Pawlowski).

In *Czechoslovakia* before the First World War one of the two universities in Prague was German and was staffed by those who had trained either in Germany or under Albrecht Penck at Vienna. The other, Charles University, was Czech, the chair of geography at which was occupied by J. Daneš who lectured on modern geomorphic themes, including the ideas of Davis and Martonne. Daneš specialized in the study of karstic landforms, but unfortunately his considerable scholarly contribution was cut short by his premature death while on the way to the United States in 1928.

The progress of *Hungarian* geomorphology, as described by Pécsi (1964, 1985), derived partly from German influences and partly from Eastern European neotectonic ideas. Pliocene and later deposits cover the major part of the country and have been considerably acted upon by aeolian processes. Prior to the mid-1930s several authorities ascribed considerable denudation and valley cutting to desert wind action on the Pannonian (Pliocene) depositional surface, particularly in Transdanubia (i.e. west of the Danube) (Cholnoky 1910, 1918, Lóczy 1913, 1918), but the later emphasis was on fluvial action. By mid-century the picture of Hungarian landform evolution which had emerged was one of Miocene uplift of the Northern Mountains (i.e. the southern folded volcanic belt of the Carpathians) and of the Transdanubian Mountains, together with subsidence of the Great Plains (east of the Danube) and of the Little Plains in the west (Pécsi and Sárfalvi 1964). During the Pliocene, Pannonian marine deposits covered the Great Plains to a depth of more than 2,000 m while the Little Plains and parts of the Transdanubian Mountains, together with the Northern Mountains, were or had been eroded to undulating peneplains with marginal piedmont steps. As late as the beginning of the Quaternary the geomorphology of Hungary only resembled that of the present in the broadest outline. The proto-Danube flowed south across the Little Plains, west of its present position, building a large fluviatile delta, until basaltic flows and uplift along the axis of the Transdanubian Mountains diverted the river to the east where it flowed over a lower part of the range at the present location of Visegrád. Continued uplift of the Transdanubian and Northern Mountains (200–400 m) during the Quaternary formed the present gorge at Visegrád, together with many deep valleys exhibiting striking river terraces, including the Danube (Schafarzik

1918, Kěz 1933a, b) and of rivers dissecting the Vas Ridge in the west (Lőczy 1913). During the Quaternary the Great and Little Plains subsided some 400–500 m and 200–300 m, respectively, receiving up to 1,000 m of deposits from the Danube and the ever-migrating River Tisza (Cholnoky 1926a, b; Kěz 1934), the latter large river only having occupied its present course since the Holocene (Scherf 1930, Laslóffy 1982). Loess deposits (Bulla 1934) and aeolian action on the large fluviatile fans of the proto-Danube and Tisza have resulted in widespread aeolian features including wind-blown furrows and fossil dunes (Kadár 1935).

In *Romania* there was little geomorphic work during the nineteenth century, apart from the production of excellent 1:57,000 Austrian topographic maps. The predominant early-twentieth-century influence, however, was French largely through the work of Martonne who began his Romanian researches in the 1890s and produced more than seventy articles and books on Romanian geography along Davisian lines between 1899 and 1935 (Morariu 1957; Tufescu *et al.* 1966). Often accompanied by the Romanian geologist G. Murgoci (1913), Martonne worked particularly in the southern Carpathians; his classic work (1906–7), treating the morphological evolution of the Transylvanian Alps in terms of the cycle of erosion, erosion surfaces, relict glacial features, structural surfaces, slope processes, the origin of transverse valleys and river terraces, became a model for Romanian geomorphologists until the 1930s. Martonne's friend, Simon Mehedinti, founded modern geography in Romania when he began to lecture at Bucharest University in 1901, subsequently producing a number of distinguished doctoral candidates including A. Demitrescu, G. Vâlsan and V. Mihăilescu. Influenced by Martonne and Mehedinti, Romanian geography was initially dominated by Davisian-style geomorphology which concentrated on such topics as the origin of transverse valleys, the production of erosion surfaces (Mihăilescu 1932), river terraces (Ficheux 1935, Brătescu 1936), the evolution of the Danube delta, the effects of epeirogenic and isostatic movements (Brătescu 1920), morphological divisions (Mihăilescu 1931) and slope processes (Mihăilescu 1939). In 1918 physical geography was introduced at Cluj University and 3 years later interuniversity excursions were initiated. After 1930 the influence of Martonne was partly replaced by that of the geologist Mihai David of Iassy University who worked on the Subcarpathians and the Moldavian Plateau (David 1931) and interpreted landforms mainly in structural terms. This approach was followed by his student Victor Tufescu and his contemporary Nicolas Popp of Bucharest. After the Second World War, in common with that of much of Eastern Europe, Romanian geomorphology concentrated on morphological mapping and applied aspects (Sandru and Cucu 1965, Cordarcea 1966).

The *Bulgarian* geomorphologist Diniy Kanev (1957) reviewed the work of W. M. Davis indicating that the approaches to geomorphology in that country currently reflected that of the Soviet Union.

Plate 6.4 Jovann Cvijić (1865–1927)

Source: From *Geographers: Biobibliographical Studies* No. 4, Mansell, London, 1980

Southern Europe

In what is today *Yugoslavia*, the study of geomorphology was effectively introduced by Jovann Cvijić (1865–1927) (plate 6.4) who was born in western Serbia and studied in Vienna under Albrecht Penck and Eduard Suess. He became professor of geography at Belgrade in 1893, the same year as he published his first important work on karst landforms, and was strongly influenced by the teachings of Ratzel, Jean Brunhes and, especially, by Martonne. His earliest karstic publications were in German (1893) and Serbian (1895) and his full international reputation had to wait until the publication of his major cyclic work in French (1918) and its promotion in English by Sanders (1921). On his death in 1927 Cvijić left a long unpublished manuscript on the geography of the karst which Martonne later translated into French and was published by the Serbian Academy in 1960. The study of karst landforms, which Cvijić and his later compatriots did so much to advance, will be treated in detail in our volume 4. Suffice it here to note that Cvijić and Davis were friends and that the former took part in Davis' European pilgrimage in 1911 (see volume 2, p. 363). Davis himself had visited the Balkans 10 years previously, yet seems to have been uncharacteristically reticent to jump on the karstic cycle bandwagon. Rather he was content to leave the field to European scholars and not until his eightieth year did he produce a major article on karst (Davis 1930B) proposing a two-cycle development for limestone caves (see volume 2, pp. 676–86).

In most of the countries in southern and Mediterranean Europe geomorphology grew largely out of physical geography. In *Greece* the studies on landforms and landscapes were dominated by Alfred Philippson (1864–1953), a very competent geomorphologist who had trained at Leipzig under Richthofen. At the other end of southern Europe, in *Portugal*, Germanic influence also predominated at first under the inspiration of Hermann Lautensach (1866–1971) who was for some years assistant to Albrecht Penck at Berlin and, after showing skill in glacial and coastal morphology, went on to compile a superb regional synthesis (1932–7) of the Portugese homeland. Here French influence grew strong in the 1940s after Pierre Birot's lectures at Oporto and Lisbon in 1944–5 (Birot 1949).

In *Italy*, where natural sciences had an ancient and broad base, Davis made a determined effort to introduce his cyclic concepts, although his regional account of the Valdarno is decidedly Hettnerian in approach. Olinto Marinelli (1874–1926), professor of geography at the University of Florence for the first quarter of the twentieth century, attended the concluding part of Davis' pilgrimage in Europe in 1911 and with Guiseppi Ricchieri of Milan demonstrated glacial features such as morainic amphitheatres looped around the lower ends of glacial lakes. Marinelli wrote on numerous landform topics, including mountains, glaciers, rivers, deltas and karst topography.

Opposition to some of Davis' precepts began to appear in southern Europe in the 1940s, not least with an obscure work by Leuzinger (1948) which Tricart and Cailleux praised while lamenting its scarcity in France. The work is equally difficult to find in contemporary Britain, not even being held in the British Library.

Africa

The vast continent of Africa, lying athwart the equator and thus embracing a wide variety of climatic regimes including humid tropical, seasonally dry savannah, semi-arid and arid, has since the 1880s provided a fertile ground for a range of geomorphic speculation not possible in Europe or North America. These possibilities were enhanced by, on the one hand, the obvious antiquity of some African plateau landscapes and, on the other, abundant landform evidences of recent climatic changes. It was not until scientists began to investigate Africa that the full significance of wind action, of strongly seasonal streamflow and of perennial humidity was impressed upon them (detailed studies of the processes involved will be treated in volume 4). As the twentieth century progressed the slow, controversial and disquieting possibility of evolutionary landform connections on a huge hemispheric scale by way of the so-called Gondwanaland resulted largely from work on the African continent (see chapter 1). Chapter 12 describes in detail the work of German, British, French and other scholars on the geomorphic effects of climatic regimes encountered in Africa, especially by J. Walther, S. Passarge and W. Bornhardt. In the United States Davis, who had visited South Africa in 1905 (see volume 2, pp. 283–97), drew on these works to devise his arid cycle (Davis 1905L, 1906H), although placing less emphasis on the effects of wind action than did Walther, and accepting only with reservations Passarge's ideas on 'levelling without baselevelling' which he termed 'Passarge's law' (Davis 1905E). In his later years Davis made wider application of ideas regarding the formation of rock floors in arid regions (1930E; see volume 2, pp. 669–75) and published posthumously on the effects of sheetfloods and streamfloods (1938A).

Prior to the 1920s virtually all the research on African landforms was carried out by short-term European visitors to the continent, such as E. F. Gautier (1922) on the Sahara, J. D. Falconer (1911) on northern Nigeria and J. W. Gregory (1921) on the East African rifts region (see chapter 12). However, after the disruption occasioned by the First World War important indigenous geomorphic work began to be undertaken. F. Dixey's work on the physiography of Sierra Leone (1922; praised by Davis (1922M)), heralded a series of publications on the development of African semi-arid landforms (Dixey 1938, 1946, 1956). More importantly, A. L. Du Toit's (1926) work on the geology of South Africa fostered his crucial support of continental drift (1937) (see chapter 1). Important home-grown work was also conducted by E. J. Wayland (1933, 1938) on the landforms of Uganda, where he found that the flat areas were linked to abrupt inselbergs by extensive pediments. Wayland applied modified cyclic ideas to Uganda and suggested that there a planated surface formed by erosion, probably a true 'peneplain' of Cretaceous age, existed at a level above four other topographic levels which had been formed by 'etching' and were 'etched plains' rather than peneplains (Wayland 1933: 77). Etching developed as follows: a long period of relative stability of the land allowed a deep development of rotted rock (saprolite) to form at the surface; uplift of the land caused this saprolitic mantle to be removed from parts of the area, exposing the underlying rock surface as an etched plain. The repetition of uplifts and stable phases would explain the formation of four topographic levels at a lower level than the summit peneplain.

By far the most important body of geomorphic work emanating from Africa was that by Lester C. King. King studied under Charles Cotton in New Zealand before going during the Second World War to the Geology Department of the University of Natal at Durban, South Africa, where he produced, on truly Cottonian lines, his *South African Scenery: A textbook of geomorphology* (1942) which, largely rewritten and extended, was re-issued in 1951. This fine exposition is splendidly illustrated, as would be expected of anyone associated with Cotton, a close disciple of Davis, and is written with a compelling simplicity of style and explanation. Although Davis is utterly excluded from it, its author accepts a cyclic concept but replaces Davisian peneplanation by pediplanation and so substitutes Walther Penck's idea of the parallel retreat of stabilized slopes for Davis' idea of the flattening of slopes as they grow older in 'normal landscapes'. We read of the river cycle (pp. 26–44), the hillslope cycle (pp. 45–51), the landscape cycle (pp. 52–9) and the arid cycle (pp. 88–97). However, what Davis termed the 'normal' cycle is replaced by the reduction of landscapes, under assumed subhumid conditions, by the production of surfaces made up of coalescing pediments (i.e. pediplanation). As King wrote later:

> The classic account of the 'Normal Cycle of Erosion' as expounded by W. M. Davis has proved regrettably in error. With its emphasis on universal downwearing, it was a negative and obliterating conception resulting from cerebral analysis rather than from observation, and has led to sterility in geomorphologic thought and retarded progress in the subject severely. ... We accept Davis's original idea of a cycle of erosional changes in landscape while rejecting the method of landscape development which he advocated.
>
> (King 1962: 162–3)

During the next 20 years King developed this scheme of large-scale pediplanation by means of Penck-like parallel slope retreat (King 1950, 1953, 1957). Culminating with his *magnum opus, The Morphology of the Earth* in 1962, King used pediplanation to account for the extensive surfaces of low relief encountered at various elevations in the supposedly continental-drift-dismembered Gondwanaland (i.e. southern Africa, Australia, South America and Peninsular India). Pediplanation was presumed to have been produced by parallel scarp retreat under arid, semi-arid or savanna conditions leaving in its wake broadly concave coalescing surfaces (with gradients less than 6–7°) often studded with steep-sided residual inselbergs. Once formed, the pediplain surfaces were thought to be subject to only minute erosion and thus capable of persisting for very long periods until they themselves, in turn, were consumed by a new sequence of steep (15–30°) scarp retreat initiated by uplift

Figure 6.1 Schematic cross-section of the eastern part of southern Africa showing the Gondwana Pediplain (Jurassic: 4,000 ft (1,300 m)) (G), African Pediplain (650–800 m near the coast to 1,000–1,600 m inland) (A), two Late Tertiary cycles of pediplanation (C1 and C2) and Pleistocene incision (P)

Source: After King (1962), Figures 78 and 93, pp. 198 and 231

on a regional scale, perhaps of an isostatic nature caused by subcrustal metamorphic changes. In this way, one phase of pediplanation might be succeeded by uplift and marginal canyon cutting leading to a wave of scarp retreat sweeping across country for long distances. Where formed on resistant rocks, pediplains and their remnants were believed by King to achieve great antiquity, such that the highest and oldest remnants may have been formed before the break-up of Gondwanaland in the Jurassic. It is interesting that King, a compatriot of Du Toit, was one of the relatively few geomorphologists to accept the possibility of continental drift prior to 1960 (see chapter 1). For example, King correlated the highest, Gondwana (Jurassic), pediplain in southern Africa (4,000 feet (1,300 m)) with a surface of similar age in Brazil (2,300–3,300 feet (700–1,000 m)). He proposed that the subsequent break-up of Gondwanaland resulted in comparable but independent histories of pediplanation. An African pediplain surface produced in Cretaceous–Oligocene times (2,000–2,500 feet (600–750 m) near the coast and 3,000–5,000 feet (900–1,500 m) in the interior), was succeeded by a few hundred feet of uplift, two phases of Late Tertiary (Mio–Pliocene) valley-floor pediplanation, strong upwarping and, finally, the cutting of Pleistocene gorges near the coast (figures 6.1 and 6.2). King extended his model of large-scale pediplanation to try to explain the existence of residual surfaces in higher latitudes which he believed to have been formed under earlier regimes of tropical climates (see chapter 12).

Australasia

The undermining of the Davisian cause in Africa contrasts vividly with its remarkable growth in *New Zealand*. Here, in a land antipodal to Britain, Davis chanced to have his chief disciple, namely Charles Andrew Cotton (1885–1970) (plate 6.5), who from 1922 until his death poured forth a stream of Davisian literature with an approach, graphicacy and lucidity worthy of

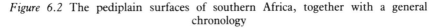

Figure 6.2 The pediplain surfaces of southern Africa, together with a general chronology

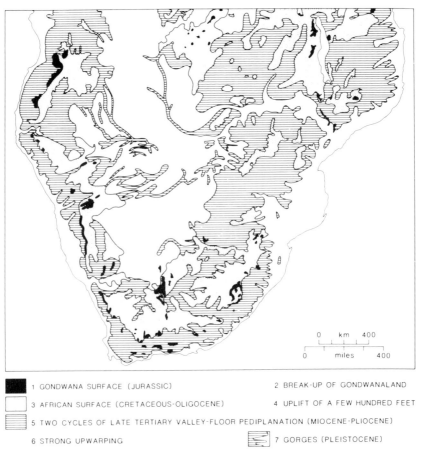

1 GONDWANA SURFACE (JURASSIC)   2 BREAK-UP OF GONDWANALAND

3 AFRICAN SURFACE (CRETACEOUS-OLIGOCENE)   4 UPLIFT OF A FEW HUNDRED FEET

5 TWO CYCLES OF LATE TERTIARY VALLEY-FLOOR PEDIPLANATION (MIOCENE-PLIOCENE)

6 STRONG UPWARPING   7 GORGES (PLEISTOCENE)

Source: After King (1962), Figure 119, facing p. 300

the master himself. Cotton when 31 years old, in a letter from Rugby, England, consulted Davis concerning the wisdom of applying for a chair of geology in New Zealand, although there 'the earth-sciences are considered of such slight importance in the syllabus' (see volume 2, p. 471). However, he went to New Zealand and as professor of geology at Wellington produced in 1922 an elementary *Geomorphology of New Zealand: Part I, Systematic* which proved popular. From 1942 onward it was revised and enlarged in several editions as *Geomorphology: An Introduction to the Study of Landforms*. Dedicated especially to Douglas Johnson and Henri Baulig, it states prefatorily that the science of geomorphology

Plate 6.5 Charles Andrew Cotton (1885–1970)

Source: From *Geographers: Biobibliographical Studies* No. 2, Mansell, London, 1978

has advanced largely as a development of the method of 'explanatory description' advocated by W. M. Davis. In the present book the treatment is intentionally Davisian in the sense that explanation is assumed to be a necessary part of landscape description. The presentation of the 'normal' cycle is Davisian also; for Davis's down-wearing theory is accepted in explanation of the origin of peneplains without reference to alternative hypotheses of slope retreat.

(Cotton 1942a, third edition, p. iii)

It contains a subsection on the geomorphic cycle (pp. 39–41), which stresses that uplift is never to be regarded as 'sudden' and that rapid uplift is a special case selected in order to simplify the introductory study; and a chapter on arid and semi-arid landscapes which mentions, rather uncertainly, pediments, plains of lateral planation, and African inselbergs.

Cotton accompanied this popular summary with three more detailed volumes: *Landscape as developed by the Processes of Normal Erosion* (1941), dedicated to the memory of the 'illustrious Gilbert, Powell, and Davis'; *Climatic Accidents in Landscape Making* (1942b), dedicated to Willard D. Johnson; and *Volcanoes as Landscape Forms* (1944). The first-named text has in its index seventy-five entries to Davis as against thirty-four to Gilbert, thirty to Douglas Johnson, twenty-four to Baulig, fifteen to Powell and fourteen to Albrecht Penck who thus may be considered the most quoted non-Davisian outsider. Its contents include a chapter on the 'Cycle of erosion; Youth of rivers' in which the problem of the rate of uplift states that (p. 38) relatively rapid uplift

> must be qualified by a free admission that the type of landscape forms developed during certain early stages of the cycle may be controlled very largely by the rate of uplift.

Further, a chapter on peneplains mentions the various hypotheses applied to the destruction of relief to a low level, such as peneplanation, lateral planation, desert pedimentation and marine erosion. Also, the author was clearly acquainted (by 1948) with King's *South African Scenery*.

Climatic Accidents in Landscape Making takes its title directly from Davis and, not surprisingly, includes subsections on the Davisian cycle of arid erosion and on the landscape cycle under semi-arid and savanna conditions. Cotton suggests that in the arid cycle the emphasis should be on deflation; under arid to semi-arid conditions the emphasis changes to pedimentation, whereas in the savanna landscape cycle much lateral river planation leaves only miniature pediment fringes and abrupt inselbergs (pp. 99–100). In his discussion of glaciated landscapes he actually queries the possibility of a cycle of glacial erosion as applied to whole regions rather than mountainous areas.

These and other writings by Cotton, although expressing reservations on Davis' simplified themes, swept through academies in the northern hemisphere where geography and geomorphology were taught. They were beautifully illustrated with skilful line drawings and for at least two decades boosted cyclic concepts as well as expounding the mature thoughts of the aged Davis. However, the popular dominance of Cotton's texts should not obscure the importance of other New Zealand work which was mainly concerned with the mechanisms and topographic effects of processes to do with mountain glaciation, periglaciation and mass movements.

In *Australia* the rise of geomorphology was less dominated by adherence to Davisian themes and, particularly from the late 1920s, detailed studies of the

Figure 6.3 The physiographic evolution of south-western Australia: Gondwana
Pediplain (G); Australian Pediplain (A); Mio–Pliocene two phases of pediplanation
(C); Pleistocene incision (P)

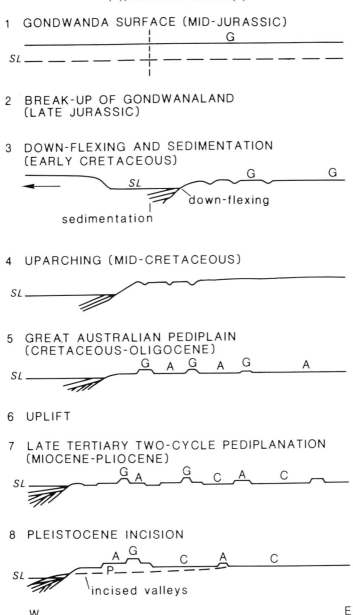

1 GONDWANDA SURFACE (MID-JURASSIC)

2 BREAK-UP OF GONDWANALAND
   (LATE JURASSIC)

3 DOWN-FLEXING AND SEDIMENTATION
   (EARLY CRETACEOUS)

4 UPARCHING (MID-CRETACEOUS)

5 GREAT AUSTRALIAN PEDIPLAIN
   (CRETACEOUS-OLIGOCENE)

6 UPLIFT

7 LATE TERTIARY TWO-CYCLE PEDIPLANATION
   (MIOCENE-PLIOCENE)

8 PLEISTOCENE INCISION

Source: After King (1962), Figure 133, p. 333

morphological effects of processes relating to 'non-normal' climates (in the Davisian sense) began to dominate. Notable among Australian geographers in the early part of the century was Griffith Taylor, who was born at Walthamstow near London in 1880 and emigrated to New South Wales with his parents in 1893. He graduated in geology at the University of Sydney in 1904 and was employed by the Australian government as a physiographer. After a 3-year return to England to work in Cambridge he went back to the southern hemisphere in 1910 to serve for 3 years as senior geologist on the Scott expedition to Antarctica. Thereafter, mainly under the influences of Davis, Herbertson and Ellsworth Huntington, he became a human and cultural geographer with a strong leaning towards 'scientific determinism'. In his early work on the physiography of eastern Australia, Taylor (1911) adopted the Davisian approach in his use of stream patterns to infer the migration of the Great Divide in the Hunter Valley region by capture and this was followed by a number of similar conventional works by others. For example, Ward (1925) proposed that drainage discordances in central Australia were due to superimposition from a Cretaceous cover; Madigan (1931) used antecedence to explain other drainage discordances in the western Macdonnell Ranges; Hills (1940) believed that the eastern highlands of Victoria preserve part of a Cretaceous peneplain uplifted and tilted to the west mainly in Pleistocene times; and Voisey (1942) identified a Tertiary erosion surface preserved beneath basalt flows in southern New England, New South Wales. It was not long, however, before Australian denudation chronology was given its own distinctive flavour by Woolnough (1918, 1927a, b) who used the existence of duricrust to identify an extensive former peneplain. We have already referred to the later Gondwanaland studies of L. C. King which were specifically designed to replace Davisian-style polygenetic landform studies under subhumid conditions. At mid-century King (1950, 1951) extended his theory of pediplanation to Australia and in 1962 summarized a proposed history for central and western Australia (figure 6.3) resulting in Gondwana Pediplain remnants (G), the Australian Pediplain (400–500 m) (A), two phases of Mio–Pliocene pediplanation (C) and Pleistocene incision (P).

A particularly Australian approach to landform evolution under tropical conditions was heralded by Marks' (1913) study of the humid tropical coastlands of north-eastern Queensland in which he showed that chemical weathering was more significant than mechanical erosion and that local relief differences arose from the different weathering rates of the component rocks, rather than from the times of denudation. These percipient observations were utterly lost on the Davisian supporters then entrenched in the Department of Geology at Sydney University. However, the years between the First World War and mid-century did see the production of a growing body of indigenous geomorphic research concentrating on the study of processes, commonly within the context of climatic change, which was so very different from that generated in the Davisian northern mid-latitudes. F. W. Whitehouse, R. L. Crocker and W. R. Browne investigated climatic changes; W. G. Woolnough and R. L. Crocker the effects of climatic changes on landforms and soil types; J. A. Prescott, L. J. H. Teakle, K. H. Northcote and C. G. Stephens the processes of soil formation; W. G. Woolnough and F. W. Whitehouse laterite formation; E. R. Stanley, A. J. Marshall and F. W. Whitehouse humid tropical processes; and W. G. Woolnough, J. T. Jutson and W. R. Browne the ancient river systems in Western Australia. The Simpson Desert Expedition of 1939 focused the continued attention given by Australian geomorphologists to aeolian land-forms (C. T. Madigan and D. Carroll) and the extent and character of the Australian Desert (J. A. Prescott and A. J. Marshall). In keeping with the expansive southern hemisphere vision exemplified by L. C. King, E. S. Hills studied the morphological effects of large-scale tectonics in Australia. The publication in 1967 of *Landform Studies from Australia and New Guinea* edited by J. N. Jennings and J. A. Mabbutt, with an introduction by E. S. Hills, marked a milestone in Australian geomorphology.

Elsewhere

Although geomorphic research was conducted in Asia and Latin America (in addition to the climatic geomorphology referred to in chapter 12), we will restrict our comments here to Japan and Brazil.

Japan's landscape and civilization were much admired by W. M. Davis, who spent some months in that country at the age of 27 (see volume 2, pp. 81–94). Physical geography began to appear in Japanese academic courses in about 1890 under Bunjiro Koto (1856–1935), who had studied geology at Munich and Leipzig. An early professor of geography was Takuji Ogawa (1870–1941), also a trained geologist, who had worked under Eduard Suess in Vienna and who took a chair at the University of Kyoto in 1907 (James and Martin 1981: 275–8). The German influence presumably began to weaken after 1911 when Naomasa Yamasaki (1870–1929), another graduate geologist who had worked with Albrecht Penck in Vienna, travelled with Davis in Europe and became professor of geography at Tokyo. Yamasaki was especially interested in the landforms of glaciated mountains and in faulted terrains. However, it is generally held that Davis' cyclic scheme was introduced into Tokyo by his successor, Taro Tsujimura. By the 1930s Japanese students had ready access to translations of Hettner and of certain elementary works of Davis, whose *Elementary Physical Geography* (1926C) and *Practical Exercises in Physical Geography* (1901E) were translated in 1930 (Davis 1930I, 1930J). Thereafter research in Japanese universities continued to emphasize the physical aspects of geography.

In *Brazil* geomorphological advances were, as in Romania, accompanied by considerable assistance from France (James and Martin 1981: 256–8), particularly during and after the 1930s. Martonne was, once again, especially important, in 1940 discussing the morphological problems of tropical Atlantic Brazil (see chapter 12). A few years later, under his direct influence, strong physical aspects were introduced into Brazilian geographical training. On his return to France, Martonne created two institutes, one at Paris and the other at Bordeaux, specifically to study Latin American problems. It should also be remembered that some time later L. C. King (1956, 1962) linked the geomorphic evolution of South America with that of Africa by identifying the Gondwana Pediplain (700–1,000 m) in Brazil, so further cementing modern approval of the tradition of the ancient conjunction of the two continents.

References

Ahlmann, H. W. (1918) 'Erosionsbas, peneplain och toppkonstans', *Geologiska Föreningens i Stockholm Förhandlingar* 40: 627–44.
—— (1919) 'Geomorphological studies in Norway', *Geografiska Annaler* 1: 1–148, 193–252.
Anuchin, D. N. (1954) *Geographical Works*, Moscow: Geografgiz (in Russian).
Anuchin, V. A. (1960) *Theoretical Problems of Geography*, Moscow, trans. R. J. Fuchs and G. J. Demko (eds), Ohio State University, 1967.
Bakowski, K. (1902) 'Former river beds near Cracow' (in Polish), *Rocznik Krakowski* 5: 138–72.
Bashenina, N. V. (1948) *The Origin of the Relief of the Southern Urals*, Moscow: Geografgiz (in Russian).
Bilibin, Yu., A. (1938) *Fundamentals of the Geology of Placers*, Moscow: AN SSSR (2nd edn 1955) (in Russian).
Birot, P. (1949) *Essai sur Quelques Problèmes de Morphologie Générale*, Lisbon: Centro de Estudos Geográficos.
Bondarchuk, V. G. (1949) *Fundamentals of Geomorphology*, Moscow: Uchpedgiz (in Russian).
Borzov, A. A., (1934) 'Supplement to the study of the normal cycle', *Lecture delivered at the International Geographical Congress, Warsaw* (in Russian).
—— (1951) *Geographical Works: Fifteen Years of Geomorphology in the USSR*, Moscow (in Russian).
Brătescu, C. (1920) 'Mouvements épirogénétiques et caractèrs morphologiques dans le Basin du Bas Danube' (in Romanian), *Bulletin de la Société Roumaine de Géographie*, vol. 39.
—— (1936) 'Critériums pour la détermination de l'âge des terrasses quaternàires' (in Romanian)', *Bulletin de la Société Roumaine de Géographie* vol. 52.
Brown, E. H. (1964) 'Geomorphology in Poland', *Professional Geographer* 16:22–5.
Bulla, B. (1934) 'Problems of the loesses and the river terraces of Hungary' (in Hungarian; German summary), *Földrajzi Közlemények* pp. 136–49, 166–8.
Cehak, A. (1934) 'Détermination quantitative des déformations des profils longitudinaux dans la région du Pokucie', *Bulletin International de l'Académie*

Polonaise des Sciences et des Lettres, Classe des Sciences Mathématiques et Naturelles série A, (Cracovie), anné 1933, pp. 226–38.

Cholnoky, J. (1910) 'The surface of the Great Hungarian Plain' (in Hungarian), Földrajzi Közlemények pp. 413–36.

—— (1918) 'Hydrography of Lake Balaton' (in Hungarian), A Balaton Tud. Tan. Eredm. vol. 1, part 2.

—— (1926a) 'On river valleys' (in Hungarian; German summary), MTA Mat. ěs Term. tud. Ert. 42: 101–10.

—— (1926b) The Knowledge of the Features of the Earth's Surface: Morphology, Budapest (in Hungarian).

Cordarcea, A. L. (1966) 'Cent années de recherches géologiques en Roumainie' (in Romanian), Revue Roumaine de Géologie, Géophysique et Géographie, Série de Géologie 10: (2):129–48.

Cotton, C. A. (1922) Geomorphology of New Zealand, Part I: Systematic, Wellington: Dominion Museum.

—— (1941) Landscape as developed by the Processes of Normal Erosion, London: Cambridge University Press, (2nd edn 1948).

—— (1942a) Geomorphology: An introduction to the study of landforms, Christchurch: Whitcombe & Tombs (2nd edn 1942, 6th edn 1958).

—— (1942b) Climatic Accidents in Landscape Making, Christchurch: Whitcombe & Tombs.

—— (1944) Volcanoes as Landscape Forms, Christchurch: Whitcombe & Tombs.

Cvijić, J. (1893) 'Das Karstphänomen: Versuch einer morphologischen Monographie', Geographische Abhandlungen von A. Penck 5: 215–319.

—— (1895) Karst: Geografska Monografija, Beograd.

—— (1918) 'L'hydrographie souterraine et l'évolution morphologique du karst', Révue de Géographie Alpine 6: 375–426.

—— (1960) 'La géographie des terrains calcaires', Academie Serbe des Sciences et des Arts, Classe des Sciences Mathématiques et Naturelles, Monographies, tome 341, no. 26, Belgrade.

Czechowna, L. (1965) History of Polish Geomorphology in the Period 1840–1939 (in Polish), Ph.D. thesis, Poznan.

Czyzewski, J. (1928) 'Histoire d'une partie de la vallée du Dniestr', Prace Geograficzne, Lwow, 10: 33–66.

—— (1931) 'Physiographie du Pokucie occidental', Prace Geograficzne Wydzialu prozez E. Romera 12: 5–32.

—— (1934a) 'Le problème de la surface d'aplanissement dans l'avant-pays des Karpates Orientales Polanaises', Proceedings of the International Geographical Congress (Warsaw), p. 68.

—— (1934b) 'Les aplatissements des versants et les plaines superieures des crêtes commes base de la division regionale', Kosmos 57: 321–30.

David, M. (1931) 'Le relief de la région subcarpatique des départements Neamt et Bacău' (in Romanian), Bulletin de la Société Roumaine de Géographie vol. 50.

Dixey, F. (1922) 'Physiography of Sierra Leone', Geographical Journal 60: 41–61.

—— (1938) 'Some observations of the physiographical development of Central and Southern Africa', Transactions of the Geological Society of South Africa 41: 113–70.

—— (1946) 'Erosion and tectonics in the East Africa rift system', Quarterly Journal of the Geological Society of London 102: 339–88.

—— (1956) 'Some aspects of the geomorphology of Central and Southern Africa', Transactions of the Geological Society of South Africa vol. 58, annex.

Dokuchayev, V. V. (1877) 'Ravines and their importance', Trudy Volnogo Ekonom. Obšč. t. 3 (in Russian).

—— (1878) 'Means of formation of river valleys in European Russia', *Transactions of the St Petersburg Naturalists' Society* vol. 9 (in Russian).

—— (1883) *Russian Chernozem*, St Petersburg, (see 1949–61, vol. 3) (in Russian).

—— (1889) *The Study of Natural Zones: Horizontal and vertical soil zones*, St Petersburg, (see 1949–61, vol. 6) (in Russian).

—— (1892) *Our Steppes Past and Present*, St Petersburg, (see 1949–61, vol. 6) (in Russian).

—— (1949–61) *Collected Works*, 9 vols, Moscow: Academy of Sciences of the USSR (in Russian).

Dumitrashko, N. V. (1948) 'Youth and senility of relief in south-east Siberia', *Trudy Inst. Geogr., vyp. 39, Probl. Geomorfologii* (in Russian).

Dumitrashko, N. V., Kamanin, L. G. and Meščheriakov, J. A. (1951) 'The present state of geomorphology and its problems', *Izviestiia Akademii Nauk SSSR, Ser. Geog.* no. 5, pp. 71–82 (in Russian).

Du Toit, A. L. (1926) *Geology of South Africa*, Edinburgh: Oliver & Boyd (3rd edn 1954).

—— (1937) *Our Wandering Continents*, Edinburgh: Oliver & Boyd.

Dylik, J. (1952) 'The concept of the periglacial cycle in middle Poland', *Bulletin de la Société des Sciences et des Lettres de Lödz* classe III, vol. 3 (5).

—— (1953) 'Caractères du développement de la géomorphologie moderne', *Bulletin de la Société des Sciences et des Lettres de Lödz* classe III, 4: (3): 1–40.

—— (1957) 'Dynamical geomorphology, its nature and methods', *Bulletin de la Société des Sciences et des Lettres de Lödz* classe III, vol. 8 (12).

—— (1964) 'Some remarks on the development of modern geomorphology in Poland' (in Polish), *Czasopismo Geograficzne* 35: 259–77.

Edel'shteyn, I. S. (1932–47) *Fundamentals of Geomorphology*, Moscow: Gosgeolizdat (2nd edn 1949) (in Russian).

Eskola, P. (1913) 'On the phenomena of solution in Finnish limestones and on sandstone filling cavities', *Fennia* 33: (4): 1–50.

Falconer, J. D. (1911) *The Geology and Geography of Northern Nigeria*, London: Macmillan.

Fedchina, V. N. (1975) 'Petyr Petrovich Semyonov-Tyan-Shansky', *Dictionary of Scientific Biography*, vol. 12, New York: Scribners, pp. 299–302.

Fedorovich, B. A. (1934) 'The Karakoroms of Turkomenia, their geomorphology and origin', *Trudy Inst. Geogr., vyp. 39, Probl. Geomorfologii* (in Russian).

Fedoseyev, I. A. (1957a) 'The development of the study of river-bed processes', *Trudy In-ta istorii estestvoznaniya i tekhniki* 9: 207–34 (in Russian).

—— (1957b) 'On the history of the study of transverse flow in rivers', *Voprosy istorii estestvoznaniya i tekhniki* no. 3, pp. 130–41 (in Russian).

Ficheux, R. (1935) 'Terrasses et niveaux d'érosion dans les vallées des 'Munţii Apuseni', *Comptes Rendus de l'Institut de Géologie Roumaine* vol. 21.

Gadomski, A. (1934) 'Le captage du Poprad des Beskidy', *Wiadomsci Sluzby Geograficznej*, Warszawa, 8: 37–64.

Gautier, E.-F. (1922) *Structure de l'Algérie*, Paris: Librarie de la Société de Géographie.

Geller, S. Yu. (1962) 'Geomorphology', in C. D. Harris (ed.) *Soviet Geography: Accomplishments and Tasks*, American Geographical Society, Occasional Publication no. 1, pp. 96–107.

Gerasimov, I. P. (1946) *Experiment in Interpreting Geomorphologically the General Pattern of Geological Structure in the USSR*, Moscow (in Russian).

—— (1956) 'Review of "American Geography: Inventory and Prospect"', *Bulletin of the Academy of Sciences of the USSR, Geographical Series* no. 1, pp. 115–43 (trans. O. A. Titelbaum).

—— (1959) *The Structural Features of the Relief of the Earth's Crust on the Territory of the USSR and their Origin*, Moscow: Izd-vo AN SSSR (in Russian).

—— (1968) 'Fifty years of development of Soviet geographic thought', *Soviet Geography* IX: (4): 238–52.

—— (1976) *A Short History of Geographical Science in the Soviet Union*, Moscow: Progress.

Gerasimov, I. P. and Markov, K. K. (1939) 'The Ice Age in the USSR' *Transactions of the Geographical Institute, Academy of Sciences of the USSR* no. 33 (in Russian).

Glinka, K. D. (1927) *Pedology*, 3rd edn Moscow (German edn 1914; English translation by C. F. Marbut, 1927).

Glodek, J. (1952) 'History of investigations of the Polish Quaternary' (in Polish), *Biuletyn P.I.G.* no. 66, pp. 589–673.

Granö, J. G. (1919) 'Les formes de relief dans l'Altai russe et ler genèse', *Fennia* 40: (2): 1–128.

Granö, O. (1987) 'On the Spread of the Peneplain Concept in Finland', *Striae* 31: 37–42.

Gregory, J. W. (1921) *The Rift Valleys and Geology of East Africa*, London: Seeley, Service and Co.

Grigor'yev, A. A. (1931) *Physico-Geographical Study of the Union*, Moscow (in Russian).

—— (1935) 'In search of regularity in the morphological structure of the globe', *Problems in Physical Geography*, vol. 2, Leningrad (in Russian).

—— (1937) *Experiment in Describing Analytically the Composition and Structure of the Earth's Physical Geographical Envelope*, Moscow (in Russian).

—— (1939–42) 'Fundamental environments in physical geography: The arctic and the subarctic', *Probl. Fiziceskoi Geografii*, vols 7, 11, Academy of Sciences of the USSR (in Russian).

—— (1946) *The Subarctic*, Moscow: Akad. Nauk SSSR (in Russian).

Hills, E. S. (1940) *Physiography of Victoria*, Melbourne and Christchurch (2nd edn 1946, 3rd edn 1951).

Hooson, D. J. M. (1962) 'Methodological clashes in Moscow', *Annals of the Association of American Geographers* 52: 469–75.

Jahn, A. (1961) 'Basic problems of geomorphological research in Poland', *The Review of the Polish Academy of Sciences* 6: (2): 9–14.

James, P. E. and Martin, F. G. (1981) *The Association of American Geographers 1904–1979*, Association of American Geographers.

Jennings, J. N. and Mabbutt, J. A. (eds) (1967) *Landform Studies from Australia and New Guinea*, London: Cambridge University Press.

Kadár, L. (1935) 'Studies on running sands in the area between the Danube and the Tisza' (in Hungarian), *Földrajzi Közlemények* pp. 4–15.

Kalesnik, S. V. (1968) 'The development of general earth science in the USSR during the Soviet period', *Soviet Geography* 9: 393–407 (from (1967) *Izvestiya Vnesoyuznago Geograficheskogo Obshchestva* no. 5, pp. 376–83).

Kaletskaya, M. S. (1948) 'Development of relief in the north-east Altai Mountains', *Trudy Inst. Geogr., vyp. 39, Probl. Geomorfologii* (in Russian).

Kaletskaya, M. S., Avsyuk, G. A. and Matveyev, S. N. (1945) *The Mountains of Southeastern Kazakhstan*, Alma-Ata: AN KazSSR (in Russian).

Kanev, D. (1957) 'William Morris Davis' (in Bulgarian), *Geographia*, Sofia, 10: 20–1.

Karpinskiy, A.P. (1888) 'The regularity of outline, the distribution and the structure of the continents', *Mining Journal* vol. 1 (see also (1939) *Collected Works*, vol. 2, Moscow and Leningrad) (in Russian).

—— (1894) 'The general character of movements of the earth's crust within the boundaries of European Russia', *Proceedings of the Academy of Sciences of the USSR*, 5th series, vol. 1, pp. 1–19 (see also (1939) *Collected Works*, vol. 2, Moscow and Leningrad) (in Russian).

Kawczyński, M. (1876) 'An introduction to the study of general geography and the geography of the Lwow district' (in Polish), *Kosmos* 1: 295.

Kedrov, B. M. (1973) 'Mikhail Vasilievich Lomonosov', *Dictionary of Scientific Biography*, vol. VIII, New York: Scribners, pp. 467–72.

Kěz, A. (1933a) 'Terrace gravels of the Vár Hill in Buda' (in Hungarian), *Földrajzi Közlemények* pp. 266–8.

—— (1933b) 'Cutting through by the Danube at Visegrád' (in Hungarian; German summary), *MTA Mat. és Term. tud. Ert.* 49: 713–51.

—— (1934) 'On the formation of the Györ–Budapest reaches of the Danube' (in Hungarian; German summary), *Földrajzi Közleményk* pp. 175–93.

King, L. C. (1942) *South African Scenery: A textbook of geomorphology*, Edinburgh: Oliver & Boyd (2nd edn 1951).

—— (1950) 'The study of the world's plainlands', *Quarterly Journal of the Geological Society of London* 106: 1–32.

—— (1951) 'The cyclic land-surfaces of Australia', *Proceedings of the Royal Society of Victoria* 62: 79–95.

—— (1953) 'Canons of landscape evolution', *Bulletin of the Geological Society of America* 64: 721–52.

—— (1956) 'A geomorphological comparison between Brazil and South Africa', *Quarterly Journal of the Geological Society of London* 112: 445–74.

—— (1957) 'The uniformitarian nature of hillslopes', *Transactions of the Geological Society of Edinburgh* 17: 81–102.

—— (1962) *The Morphology of the Earth: A study and synthesis of world scenery*, Edinburgh: Oliver & Boyd (2nd edn 1967).

Kondracki, J. (1933) 'Die Terrassen des unteren Bug', *Przeglag Geograficzny, Polskie Towarzystwo Geograficzne*, Warszawa, 13: 104–26.

Kosygin, Yu. A. (1964) 'The position of geology among the sciences and the principal problems of modern geology', *National Lending Library for Science and Technology*, vol. 6, no. 4, pp. 349–70.

Krivolutskiy, A. Ye. (1971) *Life on the Earth's Surface*, Moscow (in Russian).

Kropotkin, P. A. (1873) 'Account of the Olekma–Vitim Expedition', *Zapiski Russkogo geograficheskogo obshchestva po obshchey geografi* 3: 1–482 (in Russian).

—— (1875) 'A general essay on the orography of eastern Siberia', *Zapiski Russkogo geograficheskogo obshchestva po obshchey geografi* 5: 1–91 (in Russian).

—— (1876) 'An investigation of the Glacial Period', 2 parts, *Zapiski Russkogo geograficheskogo obshchestva po obshchey geografi* vol. 7 (in Russian).

—— (1893) 'On the teaching of physiography', *Geographical Journal* 2: 350–9.

—— (1904) 'The orography of Asia', *Geographical Journal* 23: 176–207, 331–61.

Laslóffy, W. (1982) *A Tisza*, Budapest: Akadémiai Kiadó.

Lencewicz, S. (1926a) 'Quartäre epirogenetische Bewegungen und Veränderungen im Flussnetz Mittelpolens', *Przeglag Geograficzne. Polskie Towarzystwo Geograficzne, Warszawa* 6: 99–125.

—— (1926b) 'Morphological differences between central and western Poland' (in Polish), *Ksiega Pamiatkowa XII Zjazdu Lekarzy i Przyrodnikow Polskich, Warszawa* 1: 113–4.

—— (1933) 'Relations entre les terraces de la Vistule et du Dniepr', *Proceedings of the International Geographical Congress, Paris, 1931, Comptes Rendus* II(1): 134–7.

—— (1934) 'Surfaces d'aplanissement tertiaire dans les monts Lysigóry', *Proceedings of the International Geographical Congress*, Warsaw, vol. 2, pp. 492–6.

Lents, E. (1851) *Physical Geography*, St Petersburg (in Russian).

Leuzinger, V. R. (1948) *Controversias geomorfologicas*.

Levinson-Lessing, F. Yu. (1893) 'Ancient movements of land and sea', *Reports of Tartu University* (in Russian).

Lichkov, B. L. (1931) 'Some features of the geomorphology of the European part of the USSR', *Tr. Geol. In-ta AN SSSR* no. 1, Leningrad (in Russian).

—— (1934a) 'On the epeirogenic movements of the earth's crust in the Russian Plain', *Transactions of the Geomorphological Institute, Academy of Sciences of the USSR* no. 10 (in Russian).

—— (1934b) 'Rivers and the genesis of caustobiolites', *Transactions of the Geomorphological Institute, Academy of Sciences of the USSR* no. 10 (in Russian).

—— (1941) 'On the rhythm of changes in the earth's surface in the course of geological time', *Priroda (Nature)* no. 4 (in Russian).

—— (1949) *Fundamental Laws of the Development of the Relief of the Earth's Sphere*, Moscow (in Russian).

Lóczy, L. S. (1913) 'Geology and morphology of the region of Lake Balaton: Part I' (in Hungarian), *A Balaton Tud. Tan. Eredm.*, vol. 1, part 1, section 1.

—— (1918) *Description of the countries of the Hungarian Holy Crown*, Budapest (in Hungarian).

Lomonosov, M. V. (1763) 'On the strata of the earth; 2nd supplement to the *Principles of Metallurgy and Mining*, St Petersburg (see also vol. 5 of the (1954) *Collected Works of M. V. Lomonosov*, Academy of Sciences of the USSR (in Russian)).

Łoziński, W. (1922) 'Sur les conditions tectoniques du développment des rivières dans les Carpates de Flysch', *Kosmos* 46: 525–38.

Madigan, C. T. (1931) 'The physiography of the Western Macdonnell Ranges', *Geographical Journal* 78: 417–31.

Makaveyev, N. I. (1955) *The Bed of a River and the Erosion in its Basin*, Moscow: AN SSSR (in Russian).

Markov, K. K. (1937) 'A geomorphological essay on the Pamir Mountains', *Transactions of the Institute of Geography, Academy of Sciences of the USSR* no. 17 (in Russian).

—— (1948) *Basic Problems of Geomorphology*, Moscow: Geografyiz, (in Russian).

—— (1952) 'Review of geomorphology in honour of the centenary of the birth of William Morris Davis', *Proceedings of the Academy of Sciences of the USSR, Geographical Series* (in Russian).

Marks, E. O. (1913) 'Notes on a portion of the Burdekin Valley with some queries as to the universal applicability of certain physiographic theories', *Proceedings of the Royal Society of Queensland* 24: 93–102.

Martonne, E. de (1906–7) 'Recherches sur l'évolution morphologique des Alpes de Transylvanie', *Revue de Géographie Annuelle*, vol. 1.

—— (1940) 'Problèmes morphologiques de Brésil tropical atlantique', *Annales de Géographie* 49: 1–27, 106–29.

Matley, I. M. (1966) 'The Marxist approach to the geographical environment', *Annals of the Association of American Geographers* 56: 97–111.

Menshutkin, B. N. (1952) *Russia's Lomonosov: Chemist, courtier, physicist, poet*, Princeton, NJ.

Mihăilescu, V. (1931) 'Les grandes divisions morphologiques de la Roumainie' (in Romanian), *Bulletin de la Société Roumaine de Géographie* vol. 48.

—— (1932) 'Les divisions des Carpates Orientales' (in Romanian), *Bulletin de la Société Roumaine de Géographie* vol. 51.

—— (1939) 'Glissements de terrain et éboulements: Leur classification' (in Romanian), *Revue Géographie Roumaine*, II, fasc. II–III (Bucharest).

Morariu, T. (1957) 'Emmanuel de Martonne: Son oeuvre de géographie regionale et sa contribution au développement de l'école géographique de Cluj' (in Romanian), *Proceedings of the Romanian Academy* pp. 421–35.

Morris, A. S. (1967) 'Mikhail Lomonosov and the study of landforms', *Transactions of the Institute of British Geographers* 41: 59–64.

Murgoci, G. (1913) 'Etudes de géographie physique dans la Dobrogea du nord' (in Romanian), *Bulletin de la Sociéte Roumaine de Géographie* vol. 33.

Mushketov, I. V. (1888–91) *Physical Geology*, St Petersburg, (2nd edn 1905) (in Russian).

Naumov, G. V. (1974) 'Vladimir Afanasievich Obruchev', *Dictionary of Scientific Biography*, vol. 10, New York: Scribners, pp. 166–70.

Nikitin, S. N. (1885) *Social Geological Maps of Russia*, Lublin (in Russian).

—— (1904) 'The Oka Basin', *Research of the Hydrological Department, 1894–98* no. 2 (St Petersburg) (in Russian).

Nikolayev, N. I. (1949) 'Recent tectonics of the USSR', *Trudy Komissii po Izucheniyu Chetvertichnogo Perioda*, vol. 8, Moscow: AN SSSR (in Russian).

—— (1958) 'The history of the development of the basic ideas in geomorphology, *Ocherki po Istorii Geologicheskikh Znaniy, No. 6, Moscow, AN SSSR* (in Russian).

Nikolayev, N. I. and Polyakov, B. V. (1937) 'Epeirogenic movements in the northern pre-Caspian region and their significance to the curves of river beds', *Problems of Soviet Geology* no. 3 (in Russian).

Obruchev, V. A. (1890) *A Geological Sketch of Irkutsky Province* (in Russian).

—— (1915) *Altai Studies* (in Russian).

—— (1931–49) *History of Geological Research in Siberia*, 5 vols, Moscow–Leningrad (in Russian).

Pallas, P. (1777) *Observations sur la Formation des Montagnes et les Changements Arrivés au Globe, Particulièrement de l'Empire Russe*, St Petersburg.

Panov, D. G. (1966) *General Geomorphology*, Moscow (in Russian).

Pavlov, A. P. (1889) 'Genetic types of continental formations of glacial and post-glacial epochs', *Izvestiya Geologicheskago komiteta* 7: 243–61 (in Russian).

—— (1898) 'On the topography of plains and its changes under the influence of underground and surface waters', *Zemlevedenie* 5: 91–147 (in Russian).

—— (1951) 'Geological influences over the relief of the plains and the differences between the forms of fluvial valley slopes', *Stati po geomorfolgii i prikladnoi geologii*, Moscow (in Russian).

Pawlowski, S. (1917) 'The morphology of the Lwow District' (in Polish), *Kosmos* 41: 211–2.

—— (1926) 'On the system of diluvial and post-diluvial drainage in Greater Poland' (in Polish), *Ksiega Pamiatkowa Zjazdu Lekarzy i Przyrodnikow Polskich* pp. 111–13.

—— (1927) 'La géologie et la morphologie des environs de Poznan', *Przewodnik Kongresowy II. Zjazdu Geogr. słow. Krakow* pp. 116–8.

Pécsi, M. (1964) *Ten Years of Physico Geographic Research in Hungary*, Budapest: Akadēmiai Kaidŏ (in Hungarian).

—— (1985) 'Environmental geomorphology in Hungary', in M. Pécsi (ed.) *Environmental Geomorphology*, Budapest, pp. 1–15.

Pécsi, M. and Sárfalvi, B. (1964) *The Geography of Hungary*, London: Collet's.

Piotrovskiy, M. (1984) *The Life and Explorations of W. M. Davis and W. Penck*, Moscow (in Russian).

Porkorny, W. (1911) 'Ein Beitrag zur ehemaligen Karpatischen Peneplaine in der Umgebend von Chyrów', *Kosmos* 36: 549–58.

Ramsay, W. (1909) *Geologins grunder*, Helsingfors.

—— (1917) 'Fennoskandias ålder', *Fennia* 40: (5): 1–21.

Rehman, A. (1895–1904) *The Former Territories of Poland and Neighbouring Slavonic Countries Described from a Physico-Geographical Standpoint*, Lwow: Nizowa Polska, (in Polish).

Rode, A. A. (1961) *The Soil Forming Processes and Soil Evolution*, Jerusalem: Israel Program for Scientific Translations.

Romer, E. (1897) 'Studies on the asymmetry of valleys' (in Polish), *Sprowozdania c.k. wyzszej szkoly realnej za rok 1897*, Lwow, pp. 1–45.

—— (1899) 'Climatic influence on the forms of the earth's surface' (in Polish), *Kosmos* 24: 243–71.

—— (1904) *The Land: Physical Geography of the Polish Territories*, Lwow: Macierz Polska (in Polish).

—— (1906) 'Contribution sur le développement de la vallée du Dniestr', *Kosmos* 31: 363–86.

Rosberg, J. E. (1906) 'Anteckningar om Sibbo dalen', *Medd. Geogr. För. Finl.* 7: 1–16.

Sanders, E. M. (1921) 'The cycle of erosion in a karst region (after Cvijić)', *Geographical Review* 11: 593–604.

Sandru, I. and Cucu, V. (1965) 'Le développement de la géographie en Roumainie' (in Romanian), *Revue Roumaine de Géologie, Géophysique et Géographie, Série de Géographie* tome 9, no. 1.

Sawički, L. (1909a) 'Ein Beitrag zum geographischen Zyklus im Karst', *Geographische Zeitschrift* 15: 185–204, 259–81.

—— (1909b) 'Die jüngeren Krustenbewegungen in den Karpathen', *Kosmos* 34: 361–401.

—— (1912) 'Dead landscapes in Poland' (in Polish), *Wszechswiat*, Warszawa, 31: 275–81.

Schafarzik, F. (1918) 'Palaeohydrology of the Danube in its Budapest reaches' (in Hungarian; German summary), *Hidrológiai Közlöny* pp. 184–205.

Scherf, E. (1930) 'Geology and geomorphology of the Pleistocene and Holocene beds of the Great Hungarian Plain and their relationship with evolution of the soils, particularly with the formation of alkaline soils' (in German; Hungarian summary), *Földtani Intězet Evi Jelentěse* pp. 269–301.

Sederholm, J. J. (1910) 'Sur la géomorphologie de la Finlande', *Fennia* 28: No. 1, 22–30.

—— (1911) 'Lignes de fracture: Leur importance dans la géomorphologie de la Fennoscandia', *Atlas de Finlande 1910*, *Fennia* 30: (5, 6a): 72–87.

Semeonov-Tyan-Shanskiy P. P. (1858) 'First trip to the Tian Shan or Heavenly Range as far as the upper reaches of the Jaxartes or Syr Darya River in 1857', *Vestnik Russkogo geograficheskogo obshchestva* no. 23, pp. 7–25 (in Russian).

—— (1896) *History of Fifty Years Activity of the Russian Geographical Society 1845–1895*, vols 1–3, St Petersburg (in Russian).

Shchukin, I. S. (1934–8) *General Morphology of the Land*, vols 1 and 2, Moscow: ONTI (in Russian).

—— (1952) 'Geomorphology', *Bolšaia Sovetskaia Encyklopedia* tome 10 (in Russian).

Shul'ts, S. S. (1939) 'Recent tectonics of the Tian-Shan', *Transactions of the 17th Session of the International Geological Congress (1937)* vol. 2, Moscow (in Russian).

Smolenski, J. (1910) 'The genesis of the northern Podolian edge and the morphological role of the younger movements in Podolia (in Polish), *Rozprawy Wydzialu Matematyczno-Przyrodniczego Polskiej Akademii Umiejetności*, series A, B, Kraków, 10: 31–67.

—— (1922) 'Influence des perturbations du cours inferieur des rivières sur la travail d'erosion de leur cours superieur', *Sprawozdania Polskiego Instytutu Geologicznego, Warszawa* 1: 489–507.

—— (1923) 'The landscape of the Cracow area' (in Polish), *Ziemia. Polskie Towarzystwo Krajoznawcze, Warsawa* 8: 102–8.

—— (1934) 'Les surfaces polygéniques et les conditions de leur genèse', *Kosmos* series A, 57: 273–8.

Sobolev, S. G. (1948) *Development of erosional processes in the regions of the European USSR*, Moscow (in Russian).

Sokolov, N. N. (1926) 'A geomorphological sketch of the region of the Volkhov River and Lake Il'men', *Materialy po Issledovaniyu Reki Volkhova i yego Basseyna* vol. 7 (in Russian).

Strøm, K. M. (1948) 'The geomorphology of Norway', *Geographical Journal* 112: 19–23.

Suslov, S. P. (1935) 'Geographical observations in the forest and tundra of the Yenesi', *Trudy Inst. Fiz. Geogr.* vyp. 14 (in Russian).

Svensson, H. (1956) 'Method for exact characterizing of denudation surfaces, especially peneplains, as to their position in space, *Lund Studies in Geography* series A, no. 8, pp. 1–5.

Swiderski, B. (1932) 'Sur l'évolution tectonique et morphologique des Karpates polonaises au tertiaire et au quaternaire', *Rocznik Polskiego Towarzystwa Geologicznego, Kraków* 8: 239–65.

Tanfil'yev, G. I., (1922–3) *Geography of Russia, the Ukraine, and the Territories Adjacent to them on the West within the Russia of 1914*, Odessa (in Russian).

Tanner, V. (1936) 'Om peneplanet i Finland', *Arsbok-Vuosikirja. Societas Scientiarum Fennica* 14B: No. 3.

—— (1938) 'Die Oberflächengestaltung Finnlands', *Bidr. Känned. Finl. Nat. o. Folk.* 86: 1–762.

—— (1944) 'Outlines of the geography, life and customs of Newfoundland–Labrador', *Acta Geographica* 8: (1): 1–906.

Taylor, G. (1911) *The Physiography of Eastern Australia*, Bulletin no. 8 of the Bureau of Meteorology, Australia.

Tufescu, V. *et al.* (1966) 'Le développement de la géographie en Roumainie et l'activité de l'Institut de Géographie' (in Romanian), *Revue Roumainie de Géologie, Géophysique et Géographie, Série de Géographie* tome 10, no. 2, pp. 115–38.

Voisey, A. H. (1942) 'The Tertiary land surface in southern New England', *Journal and Proceedings of the Royal Society of New South Wales* 76: 82–5.

Ward, L. K. (1925) 'Notes on the geological structure of central Australia', *Transactions of the Royal Society of South Australia* 49: 61–84.

Wayland, E. J. (1933) 'Peneplains and some other erosional platforms', *Annual Report and Bulletin, Protectorate of Uganda, Geological Survey* no. 1, pp. 77–9.

—— (1938) 'Outlines of the physiography of Karamoia', *Annual Report and Bulletin, Protectorate of Uganda, Geological Survey* no. 3, pp. 145–53.

Woolnough, W. G. (1918) 'The Darling peneplain of Western Australia', *Journal and Proceedings of the Royal Society of New South Wales* 52: 385–96.

—— (1927a) 'The chemical criteria of peneplanation', *Journal and Proceedings of the Royal Society of New South Wales* 61: 17–23.

—— (1927b) 'The duricrust of Australia', *Journal and Proceedings of the Royal Society of New South Wales* 61: 24–53.

Yefremov, G. K. (1950) *Status of Geomorphology in the Geological Sciences* (in Russian).

Yesakov, V. A. (1971) Vasiliy Vasil'yevich Dokuchayev, *Dictionary of Scientific Biography*, Vol. 4, (New York: Scribners, pp. 143–6.

Zaborski, B. (1931) 'The morphometric analysis of the relief of a lowland territory' (in Polish), *Wiadomosci Sluzby Geograficznej*, Warszawa, 5: 177–203.

Zglinnicka, A. (1931) 'Les régions morphologiques du Pokucie', *Prace Geograficzne*, Lwow, 12: 81–97.

PART III

Historical Geomorphology

CHAPTER SEVEN

American Polycyclic Geomorphology[1]

The Cyclic Interpretation of Erosion Surfaces

There can be little doubt that one of the most important results of Davis' formulation of the concept of the cycle of erosion was that after about 1890 landform assemblages were increasingly viewed as the result of the super-imposition of erosional cycles and successive parts of cycles:

> The simplicity and beauty of the conception of allotting all parts of an area to their respective cycles is alluring. So much so that we are prone to think in terms of diagrams, in which each higher level gives way visibly to a lower and younger surface, a newer peneplain which is constantly enlarging at the expense of the older and constantly losing by the spread of still newer and lower surfaces. The conception embodied in such diagrams is so simple, so illuminating, so useful, in many cases so true, and it burst so suddenly upon the science, hitherto without it, explaining so many things, and introducing order where chance had reigned, that it can not be wondered at if its application was, for a time, made too broad.
>
> (Fenneman 1936: 90)

Although polycyclic interpretations of landforms had been attempted pre-viously (volume 1, pp. 610–20), especially by McGee (1888), there is no doubt that the major impetus was provided by Davis (1889D; volume 2, chapter 11) in his monumental *The Rivers and Valleys of Pennsylvania* which made the Appalachian region the archetype for polycyclic geomorphology.

In chapter 11 of volume 2 Davis' approach to denudation chronology was analysed:

> I have repeatedly insisted that it was only by recognizing the existence of a peneplain that uplift or deformation could be determined in certain cases; and that only in this way could certain stages of geological history be discovered, in the absence of what might be called orthodox geological evidence in the form of marine deposits.
>
> (Davis 1899I: 212)

It was in the Appalachian region that this reasoning was to be pressed to extremes by Davis and his associates in an environment which had no datable marine or fluviatile deposits between the youngest folded rocks (Lower Permian) and the higher Pleistocene terraces of the Susquehanna River – a period of some 250 million years. Within this large speculative time interval the fertile imagination was free to range relatively unfettered and it was not surprising that the historical interpretation of landforms came to be associated with the recognition of supposed baselevelled surfaces of which the lower, most extensive and continuous were assumed to be younger in origin than the higher and more fragmentary. Thus the identification of supposed erosion surfaces, their association with assumed baselevels and their linkages with postulated drainage changes came to be at the heart of American denudation chronology. The interpretation of the concept of baselevel was a particularly thorny one and it will be recalled that before Davis' (1889D) great analysis of the drainage of Pennsylvania McGee (1888) had already made considerable use of the baselevel concept to suggest the polycyclic character of the 'Middle Atlantic slope' (see volume 1, pp. 613–20).

In naming baselevel, Powell (1875: 203; see volume 1, pp. 529–31) defined three quite separate meanings:

1 Grand (ultimate or general) baselevel – the plane surface or geoid formed by the landward projection of sea level. This usage was supported by Davis (1902A) and Johnson (1929) and was termed 'ultimate baselevel' by Malott (1928). It was associated with the production of peneplains and plains of marine denudation.
2 Temporary baselevel – the plane surface, commonly imposed headward of a resistant outcrop below which surrounding erosion cannot take place until its longer-term lowering. Davis (1902A; volume 2, pp. 173–4) included lakes and resistant outcrops as forming local or temporary baselevels. This was later called 'structural baselevel' (confusingly termed 'temporary local baselevel' by Johnson (1929)), and its extreme influence was associated with the introduction of structural or stripped erosion surfaces supported by near-horizontal resistant strata.
3 Local baselevel – a concept which presented particular difficulties to Davis (1902A). This was defined by Powell (1875: 203) as one of a class of baselevels of erosion playing a 'local and temporary' role and defined by

> the levels of the beds of the principal streams which carry away the products of erosion. (. . . What I have called the base level would, in fact, be an imaginary surface, inclining slightly in all its parts toward the lower end of the principal stream draining the area through which the level is supposed to extend, or having the inclination of its parts varied in direction as determined by tributary streams).

To Davis this was clearly a definition of the peneplain and to Chamberlin (1930) a definition of grade. As Cotton (1948: 97) later stated, this interpretation implies 'the level of every point on a river may be regarded as a local base-level for the river above that point with all its tributaries'.

Close to the heart of the polycyclic interpretation of landforms lay the belief that significant parts of topographic surfaces could be identified which

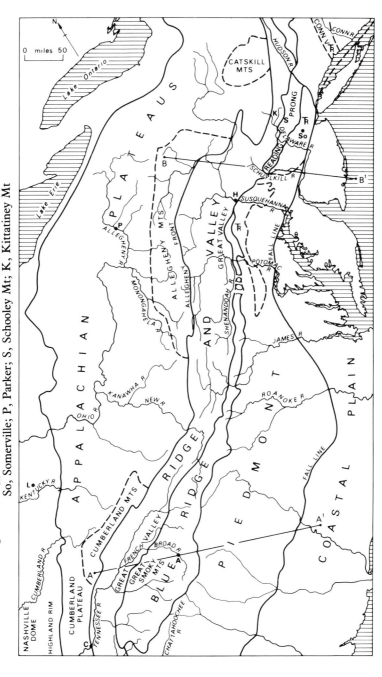

Figure 7.1 The physiographic provinces of the Appalachians and adjacent areas. The lines A–A¹ and B–B¹ indicate the locations of cross-sections shown in figure 7.2: T_R, infaulted Triassic sandstones; L, Lexington; A, Asheville; C, Chattanooga; H, Harrisburg; So, Somerville; P, Parker; S, Schooley Mt; K, Kittatiney Mt

Source: After Fenneman (1938)

had formerly been eroded with reference to identifiable grand baselevels. Nowhere else has this interpretation been put to such a searching test as in the Appalachian region of the eastern United States.

The Appalachians

The Appalachian Mountain system, generated in the last instance by an Early Permian deformation of a complex Palaeozoic geosyncline, dominates the eastern quarter of the United States, extending from the Gulf Coastal Plain in the south to New England and eastern Canada in the north (figure 7.1). It is a feature of the preserved Palaeozoic sedimentary rocks that facies generally coarsen towards the east, giving rise to the late-nineteenth-century idea that an 'oldland' (Appalachia) had existed to the east prior to and during the orogeny, directing the original drainage across the region to the west into the proto Ohio and Mississippi river systems. Accepting this thesis, it is not surprising that Davis and his contemporaries, employing the excellent topographic maps becoming available at the end of the last century, were impressed by what they saw as major drainage anomalies in the present Appalachians. Chief among these was the contrast between the major Atlantic/Mississippi drainage divides of the southern and central sectors. In the south major drainage systems (e.g. the French Broad and the New Rivers) originate near the topographic highs in metamorphics and granites near the tectonic axis of the mountains and flow west into the Mississippi. Further north the James, Potomac, Susquehanna, Schuylkill and Delaware Rivers rise well to the west close to the Allegheny Front and flow eastward crossing the tectonic axis. Given this anomaly, combined with the assumption of original westward drainage from Appalachia, it is little wonder that Davis (1889D) was so concerned to derive a plausible explanation for the assumed drainage reversal of the central Appalachians.

The Appalachian tectonic axis is located along the belt of the Older Appalachians, roughly dividing the Inner (western) Piedmont from the Blue Ridge province (figure 7.2). The dominantly metamorphic rocks (e.g. gneiss and schist) of the Blue Ridge support the highest elevations in the Appalachians. In its broad southern part there are extensive subsummit elevations of about 4,000 feet in the Carolinas and of 3,100–3,200 feet in northern Georgia and southern Virginia, above which isolated higher summits are located. Further north in Pennsylvania the province narrows and the general summit level is about 2,000 feet. Below the subsummit level in the south there is an extensive stretch of terrain along the French Broad River near Asheville, North Carolina, at 2,100–2,200 feet. The complexity of some of the drainage history is suggested by the existence of a striking series of wind gaps east of the Shenandoah River in Virginia.

To the east of the Blue Ridge province lies the Piedmont, mainly composed

Figure 7.2 Appalachian cross-sections indicated in figure 7.1: A–A¹, A general tectonic cross-section of the southern Appalachians; B–B¹, A structural cross-section of the Pennsylvania and New Jersey Appalachians

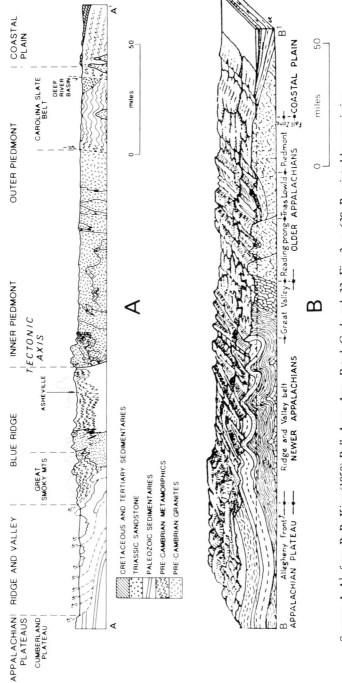

Sources: A–A¹, from P. B. King (1950) *Bull. Amer. Assn. Petrol. Geol.*, vol. 32, Fig. 2, p. 639. Reprinted by permission. B–B¹, from D. W. Johnson (1931) *Stream Sculpture on the Atlantic Slope*, Copyright © 1931, Columbia University Press. Used by permission.

of granites and metamorphics (including slates) with tracts of infaulted Triassic sandstones (with intruded diabases). The general level of the Piedmont is about 500–700 feet in Pennsylvania (with residuals 300–400 feet higher), rising to more than 800 feet further south (residuals 300–500 feet higher). Although in places the Blue Ridge forms a striking east-facing escarpment, the Inner Piedmont in the south rises to a general level of 1,400–1,800 feet in Georgia and even higher in the Carolinas, with higher summits above this. The older rocks of the Piedmont generally slope gently towards the south-east but at the Fall Line (or Fall Zone) (figure 7.1), there is a break of slope marked by rapids and the offsetting of rivers. Seaward of this the older rocks dip more steeply under the Cretaceous and Tertiary overlapping sedimentary rocks (figure 7.2 bottom) forming the Coastal Plain province.

To the north-west of the Blue Ridge province the earlier Palaeozoic sedimentary rocks (Cambrian to Silurian shales, limestones and sandstones (quartzites)) have been symmetrically folded to produce the features of the Ridge and Valley province first analysed by Lesley (volume 1, pp. 346–54). These gently folded rocks are now known to occur in low-angle thrust sheets (figure 7.2 top) and the relief of the province is mainly provided by the resistant quartzite ridges which rise 1,000 feet up to about 1,500 feet above the less resistant shale and limestone valley floors. The plunging ends of the folds produce the characteristic zig-zag ridges in plan with 'cigar-shaped' anticlines and 'canoe-shaped' synclines. The topography of the Ridge and Valley province is characterized by two apparently conflicting tendencies:

1 A very strong structural control over terrain, including striking local accordance of the majority of ridge crests.
2 Features which appear to depart markedly from structural control, including aligned water gaps through several ridges (e.g. of the Susquehanna, Delaware and Potomac, respectively), a multiplicity of wind gaps suggesting a complex drainage history dominated by capture, and a tendency for some ridge crests to decline towards the main water gaps.

The main valley floors lie at about 600–800 feet in the vicinity of the Susquehanna River, rising to 1,100–1,200 feet north-eastward and south-westward to 1,200 feet near the Roanoke River and to 2,200 feet in the New River Basin. Towards the south-eastern margin of the province broad outcrops of shale and limestone abut giving the Great Valley (or Shenandoah Valley in Virginia) a general elevation of about 600 feet in the north, rising significantly towards the south-west. The terrains of the valley floors are complicated by elevations on the limestones being as much as 200 feet lower than those on the shales, by elevations differing markedly depending on location in the erosional drainage basin (e.g. being about 400 feet near the major rivers and 2,400 feet near the major divide of the same basin), and because the major rivers themselves may be incised 500–600 feet below the main valley-floor levels.

To the north-west of the Ridge and Valley province lies the Appalachian Plateau province, separated from it by the Allegheny Front (figure 7.2 bottom). This is a sharp south-east-facing erosional escarpment, having a relief of more than 1,000 feet, formed where the Palaeozoic rocks assume generally low dips in a north-westerly direction and are capped by the very resistant Carboniferous (Pocono and Pottsville) sandstones. In the Allegheny Mountains section where gentle folds persist the terrain is somewhat more complicated but the province as a whole presents the features of sandstone hills with linear skylines declining gently to the north-west and dissected by steep-sided valleys many hundreds of feet deep. Such a surface naturally gave rise to suggestions that it had been formed by structurally controlled stripping but Fenneman (1936) pointed out that it was extremely unlikely that such extensive stripping could occur significantly above grand baselevel and that even summits of the Appalachian Plateau are indicative of near-sea-level peneplanation. These summits are at about 2,300–2,500 feet in northern Pennsylvania near the Allegheny Front, rising to 2,700 feet in Maryland, to 3,500–4,000 feet in West Virginia, and even higher further south where the plateau narrows to become the Cumberland Plateau in Kentucky and Tennessee. To the north-west the summits decline to 1,400 feet in south-west Pennsylvania and 1,200–1,300 feet in Kentucky, reaching only 900–1,000 feet where the Nashville and Cincinnati domes of the Interior Low Plateau province take over (Fennemann 1938, Thornbury 1965).

The Davis Polycyclic Interpretation and its Aftermath

As we have already noted (volume 1, p. 624; volume 2, pp. 208–9), Davis' (1889D) composite theory of the development of Appalachian drainage and topography was preceded by earlier work. Tietze (1878: 600) preferred an antecedent origin for the major drainage which appeared to be most applicable to the west-flowing southern rivers like the French Broad and the New. Löwl (1882: 405) believed the major drainage of the middle Appalachians to be due to a combination of lake overflow behind rising mountain folds and, particularly, of headward erosion by fingertip tributaries (Davis 1883D). Powell had previously thought that the courses of rivers such as the Delaware, Susquehanna and Potomac had been guided by previous fractures and Peschel (1880) believed in the fracture control of wind and water gaps (see later; Hobbs 1904).

Almost simultaneously with the development of the concept of the geographical cycle (the term 'geomorphic cycle' of Lawson (1894: 253) did not catch on), Davis (1889D; volume 2, pp. 208–33) caused it to blossom into a complex polycyclic model. He proposed that, as in the south, the original

Permian drainage of the middle Appalachians was north-westward consequent on the original gradient from the highlands to the east and antecedent
on the rising folds. The major drainage line was that of the assumed
Anthracite River located in Pennsylvania. The piecemeal reversal of the
middle Appalachian drainage was believed by Davis to have been initiated in
the Triassic period by the downsinking of the Appalachian oldland, the
associated eastward warping of the region, and the downfaulting of the
Triassic basins. This caused the new, steeper and more aggressive east-
flowing rivers to encroach westward by a complicated pattern of drainage
reversal, headward erosion and capture involving the present middle
Susquehanna, Schuylkill and Potomac drainage. Davis believed that the
present westward-flowing drainage of the Appalachian Plateaus is the
beheaded remains of the original Permian drainage, although further east the
disappearance of the postulated major Anthracite River continued to present
him with problems. The assumed general accordance of summit levels on the
Appalachian ridges, in the Allegheny Mountains and on the narrow northern
extension of the Blue Ridge province (e.g. Reading Prong) suggested to
Davis a Jurassic–Cretaceous (Schooley) peneplanation, correlated with the
sub-Upper Cretaceous unconformity beneath the Coastal Plain sedimentary
rocks. Davis' idea was supported later by Hayes and Campbell (1894), but
doubted by Shaw (1918b) on an age basis and by Ashley (1931) on the basis of
lack of continuity and the probable amount of lowering which the Appalachian
folds must have suffered. Following a verbal suggestion by R. S. Tarr
regarding the possible superimposition of the lower Connecticut River in
New England from previously more extensively overlapping Cretaceous
rocks, Davis proposed that some of the most easterly parts of the
Appalachian drainage, notably the lower and middle Susquehanna, might
have been superimposed from a modest overlap of Cretaceous marine rocks
or from fluvial floodplain or deltaic deposits (volume 2, pp. 223–4). Willis
(1895) gave support to this view, believing that the Delaware–Roanoke
drainage systems had been reversed by the eastward tilting of a flat
Cretaceous peneplain on which the drainage originally developed and from
which it was superimposed from its own floodplain deposits. Doubts were
later cast on this possible mechanism of superimposition, despite the work of
Crickmay (1933) proposing lateral floodplain development as a mechanism
for erosion surface production (panplanation). Tertiary uplift was assumed to
have resulted in the destruction of the Schooley surface on the softer outcrops
where a widespread lower partial peneplain developed (Davis 1889D, 1890K,
1891H; see volume 2, pp. 226–32). On the Triassic rocks of New Jersey a
local surface was named the Somerville by Davis and Wood (1889J). A
further phase of post-Tertiary uplift was held by Davis to have produced the
present narrowly incised river valleys.

The problem of attempting to formulate an overall polycyclic theory to embrace the topography and drainage of an area so geographically extensive and geologically varied as the Appalachians was very challenging. The absence of datable deposits for an approximate 250 million year period between the Early Permian and the Early Pleistocene meant that dating had to be approached by the assumed previous extension of peripheral outcrops (e.g. the Cretaceous of the Coastal Plain) or by purely morphological evidence in which terrain elements were related to different baselevels. The former method was extremely speculative and the latter, while being of necessity geared to the 'higher older, lower younger' assumption, faced the following severe difficulties:

1 A given peneplain, if sufficiently extensive, can exhibit considerable relief. Even Davis (1922I: 588; volume 2, pp. 490–7) allowed the interior parts to lie 3,000 feet or more above sea level.
2 Different rock types can be eroded to different levels during partial peneplanation (Fenneman 1936). In this way, particularly on less resistant outcrops, a 'berm' is the evidence of a former mature valley (Bascom 1931) and a 'strath' is that of a laterally cut fluvial valley (Bucher 1932). The Somerville surface of the Ridge and Valley limestones and the Black Belt of Alabama (Cleland 1928) were held to be examples of straths. Thus a number of different partial peneplains can be formed at different levels on outcrops of differing erosional resistance all related to one baselevel of erosion.
3 The trenching of rivers may not, as Davis assumed, be uniquely caused by baselevel changes. Malott (1920) emphasized 'static rejuvenation' of streams by decrease of load, increase of runoff or extension of the drainage basin. Although Rich (1914) ascribed 'entrenched' and 'ingrown' meanders to rapid and slow uplift, respectively, Moore (1926) thought that hard and soft rock, respectively, provided the dominant control during uplift. However, Cole (1930) did not believe that all incised meanders were necessarily two cycle in origin and Mahard (1942) similarly thought that they could develop during a single cycle.
4 All erosion surfaces are capable of being lowered more or less uniformly without necessarily involving valley cutting (Fenneman 1908). Chamberlin (1910) calculated from structural reconstructions that the upper surface of the originally folded Pottsville rocks of the Ridge and Valley was some 3 miles above the present terrain. Hayes (1899) proposed that such downwearing of peneplain surfaces was greatest on less resistant, narrow outcrops. Various researchers have estimated surface reduction in the Appalachians as at least 300 feet (Hayes 1899) several hundred feet (Wright 1931) and 100 feet per million years (Ashley 1935). This subsequent erosion would be able to transform one extensive, ancient, uplifted peneplain into many surfaces at different levels in the course of time.

5 In an extensive region like the Appalachians, particularly where the central parts are drained both by short, east-flowing, aggressive Atlantic rivers and by longer rivers flowing westward to the Gulf of Mexico, different drainage basins can exhibit different elevations during erosion to the same baselevel. Davis (1903M) believed that surfaces of unequal elevation east (lower) and west (higher) of the Blue Ridge were both of equivalent (Harrisburg) age due to this reason.
6 Subsequent deformation can locate a given peneplain at different vertical elevations.
7 The effect of temporary (structural) baselevels on rivers crossing outcrops of hard rock (e.g. the Blue Ridge metamorphics) would be to create or sustain different baselevel effects in a single drainage system.
8 The erosional effects of small baselevel movements may be blurred or coalesce as they are translated inland (Davis 1891H).

Thus, it soon became clear that purely morphological evidence for the postulation of old baselevels was open to considerable ambiguity and question.

The morphological evidence for Appalachian landform evolution was especially marked in the Ridge and Valley province where the assumed older (Schooley?) surface showed up to 2,000 feet of elevation variation and the lower (Harrisburg?) several hundred feet. Campbell (1903) suggested that the Harrisburg surface had been warped and Ward (1930) that the Somerville 'partial peneplain' was simply the result of differential lowering of the major limestone outcrop by solution below its original Harrisburg level. In its upstream reaches the Harrisburg valley-floor surface appeared as a system of high-level graded valleys; downstream it sloped towards the major rivers and was believed to merge with the Somerville surface (Clark *et al.* 1906). To Ver Steeg (1930) the faint lowering of the ridge levels towards the main water gaps indicated that they were remnants of a dissected fluvial peneplain. It is therefore not surprising that figure 7.3 shows a general proliferation of the number of suggested erosion surfaces recognized in the Ridge and Valley province and adjacent areas during the half-century following Davis' original synthesis (see especially Bascom 1921, Knopf 1924 and later Hickok 1933). The general view of drainage evolution was that it developed on an upper surface, the uplift of which led to the extension of longitudinal consequents, aligned water gaps, multiple captures and the production of wind gaps.

However, it was not only the direction of the main drainage in the north which presented problems; that of the Tennessee system in the south also appeared anomalous. Hayes and Campbell (1894), Campbell (1896) and Hayes (1899) interpreted the changes of direction of the Tennessee from south-west to north-west across Walden Ridge and the Cumberland Plateau as due to Late Tertiary capture of the original south-westerly flowing strike

Figure 7.3 Various interpretations of the number of Appalachian erosion surfaces: [1]correlated with the Cumberland surface in the Southern Appalachians; (Adams 1928); [2]renamed Harrisburg by Campbell (1903); [3]later assumed to be uparched Schooley Peneplain (2,600–4,600 feet); [4]correlated with Harrisburg surface by Fridley (1929); [5]later assumed to be the equivalent of the Asheville and Piedmont Peneplains; also of the Tertiary and valley-floor surfaces (Stose 1928)

DAVIS (1889D, 1889J)	HAYES & CAMPBELL (1894)	WILLIS (1895)	KEITH (1894, 1896)	HAYES (1899)	BASCOM (1921)	KNOPF (1924)	FRIDLEY-NÖLTING (1931)	JOHNSON (1931)	VER STEEG (1940)	COLE (1941)
		BALDS KITTATINNY	CATOCTIN-ALLEGHENY	CUMBERLAND	KITTATINNY	KITTATINNY[3]	KITTATINNY		KITTATINNY-SCHOOLEY	UPLAND
CRETACEOUS[1] (SCHOOLEY)	CRETACEOUS				SCHOOLEY	SCHOOLEY		SCHOOLEY		SCHOOLEY
			WEVERTON				ALLEGHENY-WEVERTON			
					HONEYBROOK	MINE RIDGE HONEYBROOK SUNBURY[4]				
TERTIARY	TERTIARY	SHENANDOAH[2]	UPPER TERTIARY	HIGHLAND RIM	HARRISBURG-BRYN MAWR	HARRISBURG[5]	HARRISBURG	HARRISBURG	HARRISBURG-WORTHINGTON-LEXINGTON	HARRISBURG
			LOWER TERTIARY		LANCASTER	LANCASTER				
SOMERVILLE				COOSA	EARLY BRANDYWINE			SOMERVILLE	PARKER STRATH-SOMERVILLE	SOMERVILLE
					LATE BRANDYWINE					
			UPPER PLEISTOCENE		SUNDERLAND	SUNDERLAND				
					WICOMICO	WICOMICO				
			LOWER PLEISTOCENE		TALBOT	TALBOT				

Source: Mostly after Thornbury (1965), Table 4.1, p. 79 and Table 8.2, p. 150 reprinted by permission of John Wiley and Sons, Inc., all rights reserved.

river to the Gulf by a west-flowing river. This was supported by Simpson
(1900) who used faunal evidence to show that the Tennessee River once
continued to the south-west along the, now beheaded, Coosa River.
However, Johnson (1905) thought that the present course of the Tennessee
developed on a summit peneplain (named the Cumberland and dated as
Cretaceous by Hayes (1899)) and was incised and superimposed after uplift.
Adams (1928) thought the north-west lower course of the Tennessee developed due to a westward tilt associated with the Tertiary emergence of the
Coastal Plain but Wright (1936) used the north-east slope of the Harrisburg
surface from the Tennessee–Coosa divide in the southern Appalachians to
suggest that the original drainage direction here must have been north-
easterly, not south-west to the Gulf via the Coosa River.

To the west of the Allegheny Front on the Appalachian Plateaus a high-
level surface corresponded with the westward-dipping sandstones, but
declining in places much more steeply (e.g. up to 30 feet/mile). The major
levels lie at 2,300–2,500 feet in northern Pennsylvania (Campbell 1903),
2,700 feet in western Maryland (Martin 1908) and at 4,200–4,600 feet in
West Virginia (Wright 1925). In West Virginia and Pennsylvania Fridley and
Nölting (1931) described two peneplain remnants, the upper, older
Kittatinny and the lower, younger Allegheny, both Tertiary in age, the
Allegheny correlating with the Schooley (Fridley 1929) or with the Weverton
of Keith (1894). Other workers correlated the Kittatinny with the Schooley
and the Allegheny with the Harrisburg (Ver Steeg 1931a, 1940), or assumed
that the Allegheny existed only as a differentially lowered part of the
Schooley–Kittatinny Peneplain (Ashley 1935). Further south the summit
peneplain, termed the Cumberland, was assumed to have been uplifted to
initiate the lower Lexington (Campbell 1898) or Highland Rim Peneplain
(Hayes 1899). The latter surface, dated as early Tertiary (Eocene?) was
thought to lie at 1,000–1,100 feet around the Nashville Basin, sloping north-
west to 1,000–900 feet in north-central Kentucky and to 800 feet in southern
Illinois (Galloway 1919, Jillson 1928, Wilson 1935). In western Pennsylvania
a lower surface, the Worthington, lying about 100 feet below the Harrisburg
level, was identified (Butts 1904) but the assumed same surface in the valleys
of the New and Kanawha Rivers was thought to be intermediate between the
Schooley and Harrisburg (Stone 1908). Later Ver Steeg (1940) correlated the
Worthington with the Lexington and Harrisburg surfaces. A still lower
surface appeared as a dominant former valley-floor terrace along the
Allegheny River at about 1,000 feet (later possibly correlated with the
Somerville level; Ver Steeg 1940) and as a level on shales and limestones in
Kentucky and Indiana dissected by valleys filled with Pleistocene outwash
(Ver Steeg 1936). Although these are the earliest datable deposits involved in
the Appalachian denudation chronology, a very interesting and widespread
gravel deposit, the Lafayette, occurs at higher levels all over the Interior Low

Figure 7.4 The possible origin of wind gaps in the Blue Ridge as the result of capture and beheading by the south-westward-extending Shenandoah River. The original drainage (A) was assumed to have developed on the Schooley Peneplain

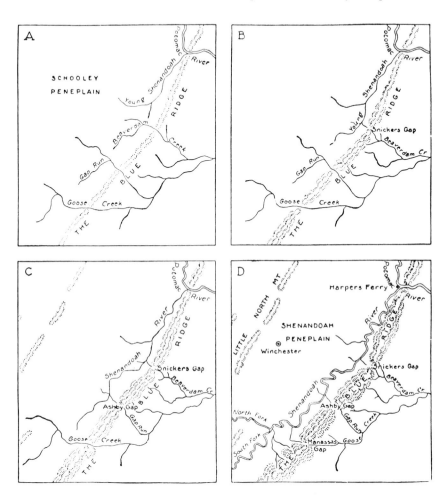

Source: From Willis (1895) and Fenneman (1938), Fig. 47, p. 169

Plateaus to the west of the Appalachian Plateaus several hundred feet above the present rivers (McGee 1893, Campbell 1898). Lusk (1928) found similar fluvial gravels on the Highland Rim Peneplain in Tennessee, decreasing in level in the major downstream direction and suggesting river deposition near baselevel.

Moving eastward, north of the Potomac the Blue Ridge province has a

general level of more than 2,000 feet which Davis and Hayes and Campbell (1894) correlated with the summit peneplain in the Ridge and Valley province. In the south, as the province widens, elevations rise and become less regular, especially as the capture of drainage by the southern extension of the Shenandoah River has left a series of wind gaps (Willis 1895, Davis 1903M) (figure 7.4). In the main mass of the Blue Ridge Mountains Willis (1889) suggested that an ancient summit peneplain was located at approaching 6,000 feet and that an Asheville surface lay at 2,100–2,400 feet along the French Broad and adjacent rivers (see volume 1, p. 620). The latter was believed to be older than the valley-floor surface of the Great Valley (e.g. the Shenandoah Peneplain at some 1,200 feet), but Wright (1931) later correlated them, assuming that they were both the differentially lowered Harrisburg surface. Wright (1928) also postulated a subsummit surface (below 3,800–4,000 feet) in the upper basins of the New and French Broad Rivers which he assumed to be Schooley in age. The higher Blue Ridge was thus an area where the assumed effects of local baselevels and different distances to the coast seemed to present especial difficulties in correlating supposed erosion levels.

The east-facing Blue Ridge escarpment (relief 1,500–2,500 feet) was originally thought to be due to an eroded monoclinal flexure (Hayes and Campbell 1894) but Davis (1903M) thought that it was the result of the intersecting of two peneplains, possibly related to the same baselevel but eroded on rocks of differing resistance with respect to differing distances from the Gulf and Atlantic coasts. Wright (1928, 1931) partly agreed with Davis, proposing that the two surfaces (the Blue Ridge subsummit and the Piedmont) were both originally Schooley in age but that the Piedmont had been later lowered during the Harrisburg cycle by the erosion of softer rocks. This association of the Piedmont surface with the Harrisburg was popular in the 1920s (Bascom 1921, Knopf 1924; see also Campbell 1933) and it was thought that a Pliocene gravel-covered part of the Harrisburg in the Pennsylvania–Maryland Piedmont (the Bryn Mawr surface) had been later upwarped (Stose (1930) disagreed). Towards the outer edge of the Piedmont the Fall Line was marked by the offsetting of rivers. This feature was originally thought by McGee to be due to faulting or to the effect of Pliocene offshore bars and later coastal emergence (Darton 1894). Davis (1899I) supported Darton (1894: 570) in the belief that the Appalachian summit peneplain had been bent down under the Coastal Plain sediments as an unconformity (see also Clark et al. 1906) and Shaw (1918b) also identified the uncomformity as a buried peneplain. However, Renner (1927) proposed that the Fall Line was a 'morvan' formed by the intersection of the older (Jurassic) erosion surface forming the Coastal Plain unconformity (later called the Fall Zone Peneplain by Sharp (1929)) and the Harrisburg Peneplain (figure 7.5). Davis' lower Somerville surface was recognized as

Figure 7.5 The Fall Line interpreted as the intersection between two peneplains, a Tertiary (?) and a pre-Cretaceous

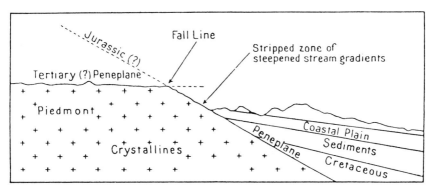

Source: From Renner (1927) Figure 13, p. 285. By permission American Geographical Society.

younger and possibly correlated with Cooke's (1931) Sunderland terrace on the Coastal Plain sediments (Bascom 1921, Knopf 1924).

The depression of the Fall Zone (Jurassic?) Peneplain in the Coastal Plain was followed by Cretaceous to Miocene deposition (Salisbury 1895). The present drainage appeared to have been superimposed either from the Miocene marine rocks (Stephenson 1928) or from 'Lafayette gravels' (Davis 1907C) and a Pliocene erosion surface formed. In a series of publications Cooke (1931, 1935, 1936, 1943, 1945) proposed seven Pleistocene marine terraces along the Atlantic Coast (Brandywine, 270 feet; Coharie, 215 feet; Sunderland, 170 feet; Wicomico 100 feet; Penholoway, 70 feet; Talbot, 42 feet; Pamlico, 25 feet), associated with Pleistocene eustatic or isostatic variations. However, Campbell (1931) thought the highest two were of subaerial alluvial origin and Flint (1940, 1941) only recognized two marine terraces, the Suffolk (20–30 feet) and the Surry (90–100 feet). In the eastern Gulf Coastal Plain Shaw (1918a) found evidence of four cycles; the highest he correlated with the Coosa Peneplain although a higher (Highland Rim?) surface was possibly thought to exist. To the west of the Mississippi delta Stephenson (1926) found evidence for several erosional cycles on the Coastal Plain, dominated by a Pliocene surface.

In the northern Appalachians the largely metamorphic and igneous rocks of New England and adjacent areas fitted less well into the schemes proposed for the central Appalachians. Davis (1895K) proposed the existence of a single peneplain, equivalent to the Schooley, sloping seaward from some 2,000 feet around the Green Mountains. He accepted that such a surface could exhibit a considerable range of elevations (Davis 1899I, 1902A). Keith (1916) proposed the existence of five peneplain remnants in western Massachusetts (figure 7.6), although Lobeck (1917) identified a single New

Figure 7.6 Altitudes of accordant summit levels and terraces in western New England. All names by Barrell except asterisked.

SURFACE	S NEW ENGLAND (Barrell, 1913,1920A, 1920B)	TACONIC MOUNTAINS (Pond,1929)	E and C VERMONT (Meyerhoff and Hubbell, 1928)	W MASSACHUSETTS (Keith, 1916)
DORSET*	......	3200	(?)3000(?)-3200	......
BRAINTREE*	......	2700	2700 -2800	(?)2800
BECKET	2450	2500	2300(?)-2500	......
CANAAN	2000	2000	2000 -2150±	2050-2200
HAWLEY*	......	......	(?)1820 -1920	......
CORNWALL	1720	1700	1660(?)-1720	1600-1700
GOSHEN	1380	......	1360 -1420	......
LITCHFIELD	1140	......	1180 -1240	1100-1200
PROSPECT	940	900±	(?) 980 -1080(?)	......
TOWANTIC	740	700+	780 - 900(?)	......
APPOMATTOX	540	......	540 - 620(+)	500
NEW CANAAN	450	......	420 - 480	......
SUNDERLAND	240	......	(?) 260 - 300	......
WICOMICO	120	......	(?) 160 - 180	......

Source: After Barrell, Pond, Meyerhoff and Hubbell, and Keith; from Fenneman (1938), p. 351. By permission McGraw-Hill Publishing Co., New York.

Figure 7.7 Cross-sections of the New England Upland south of the White Mountains: section I NW–SE; section II WNW–ESE; section III NE–SW; section IV W–E

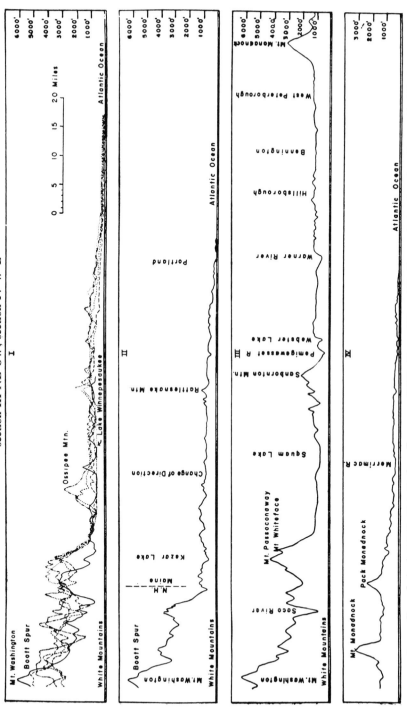

Source: From Lobeck (1917), Plate II; from Fenneman (1938), Figure 103, p. 365. By permission American Geographical Society.

England Peneplain at about 1,000–1,100 feet (figure 7.7) above which groups of monadnocks rise. However, the higher areas of the White Mountains were thought to be the remains of a higher and older erosion surface (J. W. Goldthwait (1914), R. P. Goldthwait (1940)); a lower surface was identified there at 2,600–3,000 feet by Lane (1921); and Cushing (1905) and Miller (1917) proposed two peneplains in the Adirondacks, differentially warped or faulted. An important work by Meyerhoff and Hubbell (1928) proposed the existence of as many as fourteen fluvial terraces (incipient or partial peneplains or straths) in eastern and central Vermont produced by successive, intermittent uplifts and separated by fluvial nickpoints. Each terrace had an average vertical range of 115 feet, an average vertical interval of only 127 feet between terraces and an average width of little more than 1 mile. Doubts were expressed as to whether this degree of polycyclic refinement could be preserved. Davis (1932G) did not believe that this Penck-like flight of fluvial levels could be maintained and Rich (1938) believed that such supposed breaks of topographic slope would soon disappear under the action of subaerial erosion. The important earlier work by Barrell (1913, 1920a, b), using the method of projected profiles (figure 7.8), also employed by Meyerhoff and Olmsted (1934), identified a much more widespread flight of terraces across southern New England and possibly as far south as the Potomac River. They were interpreted as marine in origin each averaging some 9 miles wide backed by a wavecut but fluvially degraded slope averaging some 200 feet high. This work reviewed the controversy on the relative efficacy of fluvial and marine processes to produce erosion surfaces and was criticized because of the alleged subjectivity of the method of projected profiles, the existence of erosional residuals on the supposed marine surfaces and the complex sequence of depression and wave cutting followed by emergence necessary to produce Barrell's surfaces. Nevertheless, this interpretation was extended into south-east Connecticut by Hatch (1917) and by Lane (1921) into southern New Hampshire and south-west Maine. As late as 1946 Olmsted and Little proposed the existence of a Miocene marine terrace at less than 600 feet extending across New England. In contrast to studies of possible erosion surfaces, research relating to the evolution of drainage in the northern Appalachians/New England region was more integrated with that farther south. Building on the work of Fairchild (1925), Ruedemann (1931, 1932) proposed that the headwaters of the ancestral Susquehanna were located in the Adirondacks, and that the river flowed south-west across a Permian surface before being broken up by captures involving the Hudson and Delaware Rivers. The original Hudson River was naturally thought by Davis (1891H) to have developed on the Schooley surface, whereas Tarr (1902) held that the Hudson Valley was originally the site of Triassic rocks on which the river developed. The anomalous southern-most south-east course of the Connecticut River was first attributed by Davis

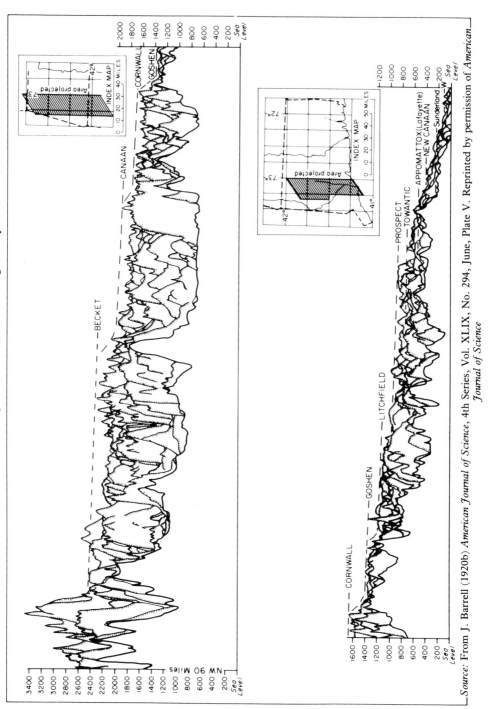

Figure 7.8 Projected profiles across the New England Upland

Source: From J. Barrell (1920b) *American Journal of Science*, 4th Series, Vol. XLIX, No. 294, June, Plate V. Reprinted by permission of *American Journal of Science*

to faulting but later to superimposition from a limited landward extension of marine sediments. Sharp (1929) postulated that the New England (Upland) Peneplain was a younger erosion surface which intersected an older Fall Zone Peneplain (figure 7.9). It was on such bases that Johnson (1931b) developed his idea of regional superimposition of drainage for both the middle and northern Appalachians extending inland to the Green Mountains.

Figure 7.9 Generalized profile showing the suggested intersection of the Upland and Fall Zone Peneplains in Connecticut

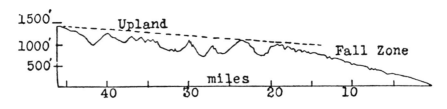

Source: From Sharp (1929); and Fenneman (1938), Figure 105, p. 372

Johnson's Synthesis

By the early 1930s the gamut of denudational possibilities in the Appalachians appeared to have been explored as far as the erosionally dominated landscape permitted. At that time Douglas Wilson Johnson (1931b, previewed shortly before in Paris (1931a)) made a significant breakthrough by, first, accepting the long-held idea that the Schooley Peneplain was different from, and younger than, the Fall Zone Peneplain and, more important, that regional superimposition had taken place from a widespread Cretaceous marine overlap which covered the Jurassic Fall Zone Peneplain (figure 7.10).

Johnson was born in 1878 at Parkersburg, West Virginia, the son of a lawyer, who was an ardent prohibitionist and crusading newspaper owner. Not surprisingly, Douglas Johnson developed into a man of self-discipline, with a legalistic mind and emotional austerity which in later years made him extremely formal, socially inflexible and vain (Bucher 1946, Chorley 1973). In his university days he was greatly influenced by the clarity of mind, deductive reasoning and power of exposition of his Harvard teacher W. M. Davis, as well as by the mercurial Nathaniel Southgate Shaler. In 1907 he became assistant professor at Harvard, where he organized field trips with military precision, and in 1909 edited Davis' *Geographical Essays* (1909C). Moving to Columbia University, New York, in 1912, Johnson showed himself to be Davis' most effective and distinguished geomorphological pupil, producing many Ph.D. students (notably Armin K. Lobeck in 1917)

Figure 7.10 Cross-section of northern New Jersey suggesting a geomorphic evolution by regional superimposition; W and P indicate the Watchung and Palisade sills; X the intersection of the Schooley and Fall Zone Peneplains; and M the Musconetcong and other adjusted rivers of the highlands.

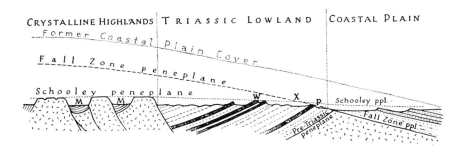

Source: From D. W. Johnson (1931b) Stream Sculpture on the Atlantic Slope, Fig. 15, p. 77. Copyright © 1931, Columbia University Press. Used by permission

and continuing until the 1940s to follow the methodology of the Master. Until the mid-1920s he concentrated largely on military aspects of geomorphology and in applying cyclic ideas to coastal physiography. His major work *Stream Sculpture on the Atlantic Slope* (1931b) was a masterpiece of which Davis approved and to which he wrote a preface. One of his last Ph.D. students, Arthur Strahler, began his scholarly career by pursing research on the denudation chronology of the Appalachians (Strahler 1945) and thus Johnson provided the link between the two most influential American geomorphologists of the twentieth century. In later years Johnson (1938–42) became especially interested in the application of scientific methodology to geomorphology, but in 1942 he suffered a severe heart attack. In 1943, 3 months after marrying for the second time, he narrowly escaped injury and death in a notorious rail crash on the way to Florida in which more than eighty passengers were killed. Two months later in 1944 he died of another heart attack in Sebring, Florida.

Johnson's (1931b) thesis cut the Gordian knot of the Appalachian problem by employing a supposed extensive and significantly pre-Schooley superimposition of the main drainage lines and was supported by a series of strikingly persuasive block diagrams (figure 7.11). He thus avoided the problems encountered by Davis and the other researchers who assumed great antiquity (e.g. Permian) for the present drainage system (Thornbury 1954) and which led them into a maze of supposed drainage modifications, involving piecemeal capture and stream dismemberment. Johnson (1931b) assumed that superimposition of the dominant, consequent, east-flowing drainage had taken place from a Cretaceous cover, extending as far west as the present Appalachian Plateaus, on to the Fall Zone Peneplain. This lead to

Figure 7.11 Johnson's theory of the development of Appalachian landforms. – see *Notes* opposite.

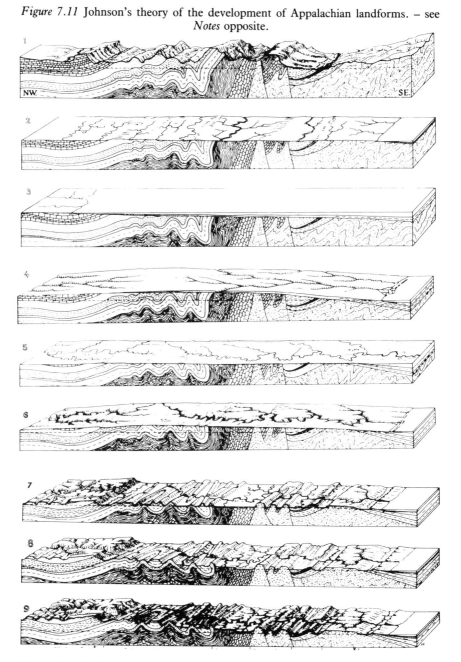

Source: From D. W. Johnson (1931b) *Stream Sculpture on the Atlantic Slope*, Figs. 96, 97 and 98. Copyright © 1931, Columbia University Press. Used by permission

Notes:
1.Rejuvenated Appalachians in Post-Triassic time; 2. Formation of Fall Zone Peneplain; 3. Marine overlap and deposition of Cretaceous beds; 4. Arching of Fall Zone Peneplain and superimposition of south-east flowing rivers; 5. Formation of Schooley Peneplain; 6. Arching of the Schooley Peneplain; 7. Dissection of the Schooley Peneplain and erosion of the (partial) Harrisburg Peneplain on less-resistant limestones and shales; 8. Further uplift and erosion of the (partial) Somerville Peneplain on least-resistant limestones; 9. Latest uplift and trenching of the present inner valleys

the major observed discordant features of the drainage such as aligned water gaps and wind gaps which predated the Schooley cycle sufficiently to give time for the present-day drainage adjustments to structure to have occurred. The proposed stages in this sequence of events are shown in Figure 7.11:

1 The original Mississippian–Permian relief produced by the Appalachian orogeny together with later Jurassic faulting and deposition. Johnson follows Davis (1889D) in postulating a major original drainage divide well to the east in Appalachia.
2 The subsidence of Appalachia, Jurassic(?) denudation and the development of the Fall Zone Peneplain extending far inland.
3 Eastward tilting, extensive Cretaceous transgression and the deposition of a marine cover on the Fall Zone Peneplain across the whole of the present Ridge and Valley province and to the west of the Allegheny Front.
4 The broad uparching of the Fall Zone Peneplain, together with its Cretaceous cover, with the crest of the arch extending from the present Adirondack Mountains through south-west Pennsylvania close to the present Allegheny Front, from which west- and east-flowing drainage was superimposed. Thus the dominantly east-flowing drainage was super-imposed on the buried folds bevelled by the Fall Zone Peneplain. The easternmost part of the buried Fall Zone Peneplain was further depressed and continued to receive (Tertiary) sediments eroded from the Cretaceous cover further inland.
5 The continued removal of the Cretaceous cover, except for a near-coastal strip, was part of the erosional cycle leading to the production of the Schooley Peneplain (mid-Tertiary?) which intersected the buried coastal remains of the Fall Zone Peneplain along the future Fall Line.
6 Uparching and the commencement of the fluvial dissection of the Schooley Peneplain with the major drainage lines running transverse to the structural strike.
7 The production of the Harrisburg partial peneplain (Late Tertiary) at a new grand baselevel on the less resistant shale and limestone outcrops of the future Ridge and Valley province and the Triassic sandstones, leaving remnants of the eroded Schooley surface bevelling the resistant quartzite ridge crests as well as on the Reading Prong of the Blue Ridge province and

the surface of the Appalachian Plateaus. At the same time the eastward stripping back of the Cretaceous and Tertiary rocks covering the Fall Zone Peneplain remnant in the east began to reveal the Fall Zone.

8 Further limited uplift initiated a new cycle of erosion affecting particularly the less resistant Ridge and Valley limestones and Triassic sandstones and producing the Somerville strath.

9 Renewed minor uplift led to the initiation of the current cycle of erosion during which present rivers were incised below the Somerville level.

There were clearly major problems which remained or which had been created by Johnson's synthesis:

1 There is no evidence for the extensive Fall Zone Peneplain except in the narrow Fall Zone and as the sub-Cretaceous Coastal Plain unconformity.

2 There is, in particular, no evidence for the extensive Cretaceous cover required by the theory, despite the fact that Johnson had neatly circumvented the absence of Cretaceous deposits inland of the Coastal Plain by associating the assumed Cretaceous deposits with the now almost totally destroyed Fall Zone Peneplain.

3 The present evidence for the Schooley Peneplain lies in a very wide range of ridge-top and plateau elevations which could have other non-cyclic explanations (e.g. structural control in the Appalachians; see later).

4 The Harrisburg and Somerville surfaces in the Ridge and Valley province are restricted to separate geological outcrops of differing erosional resistance and the differing topographic levels could be the result of differing rates of erosion rather than of different cycles of erosion.

Nevertheless, despite these objections, Johnson's masterful synthesis rapidly achieved very considerable support, as well as the approbation of his master, W. M. Davis, in a preface to the work.

The years following Johnson's synthesis were marked, first, by efforts to show how different erosional levels could have developed without the need for a complex denudation chronology involving baselevel changes; second, by attempts to introduce a more complicated baselevel change chronology; and, third, by attempts to modify the Fall Zone Peneplain superimposition theory. Ashley (1935) believed that no remains of any general Appalachian peneplain exist and that the surfaces of even the most resistant outcrops are at least 800 feet below the original erosion surface, because these outcrops have been eroded at a rate of at least 100 feet per million years, with less resistant outcrops being reduced probably ten times faster. Thus there may be evidence of only one previous erosion cycle in the Ridge and Valley province in that a single previous erosion surface might be expected to exist currently as accordances at different levels due to local baselevelling, stripping or to parallel lowering of differing amounts on different rock types (see also

Fenneman 1936). Ashley (1931, 1935) also believed in the youthfulness of much Appalachian scenery with the oldest surface being possibly only of the Pliocene age. Thompson (1939: 1330) similarly believed that the decline of the ridge-top (Schooley) surfaces towards the major transverse rivers was due to post-Schooley erosion and therefore is not diagnostic of superimposition, and Cole (1941) proposed that the Schooley 'surface' was merely a subsummit surface on existing ridge tops. In the southern Appalachian valleys Wright (1942) applied similar reasoning to explain the different levels to differential lowering of the Harrisburg surface. At the other extreme, Hickok (1933) correlated stream nickpoints in southern Pennsylvania and proposed fourteen–eighteen partial peneplains between elevations of 320 and 1,660 feet produced by intermittent uplift. Attempts to use quantitative techniques such as altimetric frequency curves (Thompson 1939), modified projected profiles (Bates 1939, White 1939) and summit distributions (Thompson 1941) were increasingly attacked as being subjective and inaccurate (e.g. Rich 1938). Suggested modifications of Johnson's superimposition theory were provided by Stose (1940) who proposed a Jurassic age for the Schooley Peneplain which was then covered by Cretaceous–Oligocene rocks, uplifted in the post-mid-Tertiary and further dissected, and by Von Engeln (1942) who believed in superimposition from a Pliocene covering of the Schooley. These proposals for a superimposition more recent than from the covered Fall Zone surface were countered by the lack of outliers of the younger sediments and by the apparently well-adjusted mature and antique impression of the drainage system. Bethune (1948) held a different view that drainage superimposition was from the floodplain deposits of the major rivers developed on transverse downwarps in the Schooley surface.

To the west in the Appalachian Plateaus Sharp (1932) (three or four surfaces) (figure 7.12), Ashley (1933), Cole (1935) (four surfaces) and Fennemann (1938) (five surfaces) proposed various chronological combinations of the Upland, Kittatinny, Allegheny, Schooley, Worthington, Lexington, Harrisburg and Parker surfaces. In New England Meyerhoff (1938) proposed superimposition from a Miocene marine cover, Atwood (1940) supported a main Schooley surface with monadnocks above and Harrisburg valley-side benches below, and Adams (1945) agreed that the range of general summit elevations was not too great to be ascribed to a single uplifted peneplain.

Non-cyclic Interpretations

In the face of all the foregoing work, it should not be assumed that explanations of Appalachian landforms of a non-cyclic character were lacking or even unimportant. Beginning in the nineteenth century, attempts were made to account for topographic accordances by a variety of means not involving peneplanation. Some of these were:

Figure 7.12 Three assumptions of the ages and correlation of the peneplains believed to exist as remnants in the Allegheny Mountains (east), the Appalachian Plateau (centre) and the Central Lowland (west): A, Allegheny Mountains = Schooley; Plateau = Central Lowland = Harrisburg; B, Allegheny Mountains = Schooley (oldest); Plateau = Allegheny (middle); Central Lowland and Plateau Valleys = Harrisburg (youngest); C, Allegheny Mountains = Schooley (oldest); Plateau = Harrisburg (middle); Central Lowland = Somerville (youngest)

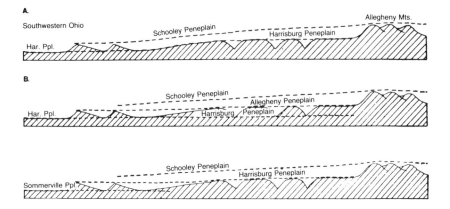

Source: After Sharp (1932); from Fenneman (1938), Fig. 82, p. 296. Courtesy Denison Univ. Bull.

1 A variant of the *Gipfelflur* concept which assumed greater efficiency of erosion at higher altitudes leading to more or less uniform summit elevations (see chapter 10). In this connection Tarr (1898) believed that the rapidity of erosion above the general tree line might produce terrain accordances.

2 Summit accordance due to uniform isostatic equilibrium of individual upland masses (Daly 1905).

3 Accordance of hilltops due to dissection by uniformly-spaced streams (Shaler 1899, Smith 1899). Glock (1932) elaborated this notion by showing that if the available relief (i.e. the vertical elevation between the uplifted surface and the general level of the major valley floors) exceeds some 300 feet, the expected stream spacing and valley-side slope will allow continued erosion to produce hilltop and ridge accordance unrelated to the original uplifted surface.

4 Dissected marine planation surfaces (Barrell 1913, 1920a, b, Johnson 1916).

5 Stripped or structural surfaces, such as possibly the Appalachian Plateaus. However, Fenneman (1936) believed that large-scale stripping could only occur close to grand baselevel and therefore was difficult to distinguish from true peneplanation.

6 Surfaces of lateral stream planation (i.e. panplains) (Crickmay 1933). Again such a surface could be viewed as a variant of the peneplain.

The most interesting applications of non-cyclic notions occurred after Johnson's (1931b) synthesis and were intended to obviate the complexities which it embodied. In a series of articles Meyerhoff, with Olmsted (1934, 1936, Meyerhoff 1938) proposed that the present south-east-flowing rivers of the central Appalachians were established during the original Permian orogeny on low-angle thrust sheets or recumbent folds (figure 7.13) along structural sags (Thornbury 1954) or fracture zones. The original drainage divide was the culminating elevation of these structures which lay to the west of the present Ridge and Valley province. This drainage (e.g. the Delaware, Potomac and Schuylkill) was then superimposed on the underlying fold or thrust structures and was further guided by the developing Triassic fault troughs across which rivers now flow preferentially and whose sediments were primarily derived from these pre-existing rivers. However, Mackin (1938) and Strahler (1945) objected to this structural attempt to cut the polycyclic Gordian knot in the following terms (Thornbury 1954):

1 It is difficult to envisage the geometry of the structures required;
2 There is no clear relation between Permian structures and present stream courses;
3 The position of the major divide was not far enough west to account for all the existing drainage features;
4 There is no general evidence for the Triassic sediments having been deposited by major south-east-flowing rivers;
5 There is no clear evidence of origin of Permian drainage;
6 Too much has been made statistically of the correlation of the locations of major rivers and that of the Triassic basins.

Another non-cyclic mechanism which was suggested as a means of simplifying cyclic explanations was that of progressive piracy (Thompson 1939). It was proposed that the original drainage divide lay along the central axis of the Blue Ridge province and that the steeper and more aggressive eastward-flowing rivers had been able to extend their headwaters westward by a slow process of headwater piracy (figure 7.14) and drainage reversal, and possibly by subterranean solution piracy in limestone (Fridley 1929). Of course, this idea was not new (Löwl 1882, Tarr and Martin 1914: 566) and had formed a part of Davis' (1889D) thesis. Thompson believed that westward drainage extension operated very selectively along lines of structural weakness and avoided particularly wide resistant outcrops. This process explained why the divide had only shifted westward 10–20 miles in the south, 40–70 miles in the Roanoke and James basins, and 100 miles further north in the Potomac basin. This differential shifting, requiring only an average rate of 0.7 miles

Figure 7.13 Postulated Appalachian drainage superimposition from an asymmetric fold: A. d_1 and d_2 show successive positions of the erosion surface and drainage divide.

B–D. Block diagrams showing the development of superimposed drainage.

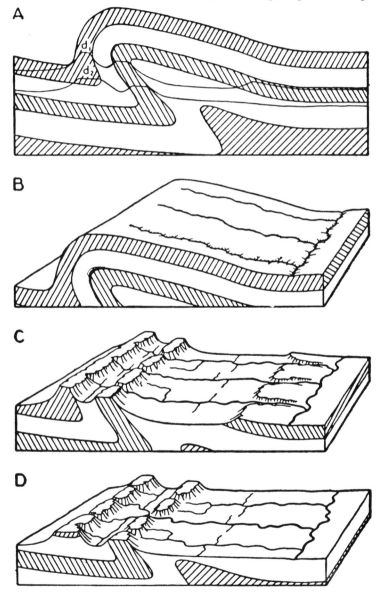

Source: From H. A. Meyerhoff and E. W. Olmsted (1936) *American Journal of Science*, 5th Series, vol. XXXII, no. 187, July, Fig. 2, p. 26. Reprinted by permission of *American Journal of Science*

Figure 7.14 The possible westward migration of the drainage divide on the folded Appalachians by progressive piracy resulting from the elevation of stream b being less than that of stream a. The drainage divide is shown to migrate from D_1 in Block A to D_2 in Block B

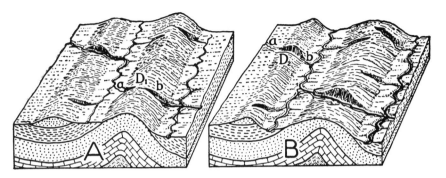

Source: From H. D. Thompson (1939) *Bulletin of the Geological Society of America*, Vol. 50, Fig. 2, p. 1332

per million years, was believed to be due to the decreasing width of the resistant Blue Ridge outcrop towards the north and to possibly greater eastward tilting there. However ingenious this theory, it clearly could not satisfactorily explain aligned water gaps nor the general independence of the Appalachian drainage from structural control. By the Second World War cyclic theories, dominated by that of Johnson, were paramount in the eastern United States.

The Western United States

Between the Appalachian Plateaus and the Missouri River polycyclic notions were early applied (Hershey 1896, Marbut 1896) with, particularly, the mid-Tertiary Lancaster Peneplain of the Driftless Area of Wisconsin (Trowbridge 1921) being correlated with the Highland Rim surface in Tennessee (Hayes 1899), the Lexington in Kentucky (Campbell 1898) and the Ozark Peneplain further south (Hershey 1902a, b). The latter author postulated two peneplains in the Ozark region: the lower Ozark Peneplain in the Springfield Plateau which had partly replaced a higher peneplain, remnants of which form the summits of the Boston Mountains to the south. This older surface was estimated to be Jurassic–Cretaceous in age by Hershey (1896) and correlated with the Schooley by Fenneman (1938), although Purdue (1901) believed this surface to be predominantly of structural origin. The Ozark Peneplain was held to have been uplifted in late Tertiary time (Purdue 1901) and dissected by Pliocene valley straths, correlated with the Lafayette Gravels (Fenneman 1938). The warped Boston Mountains Peneplain was believed to correlate with the upper (Ouachita) peneplain of the Ouachita Mountains (Hershey 1902a, b). This surface was proposed as Early Cretaceous in age (Purdue and Miser 1923, Miser and Purdue 1929) and to decline southward passing as an unconformity below the Coastal Plain sediments (Melton and McGuigan 1928). The lower (Hot Springs) peneplain was considered to be Eocene.

In contrast to the Ozarks and the region lying east of the Mississippi–Missouri, the Great Plains, a large expanse of low relief, gentle gradients and dominantly depositional surface presented different problems to the denudation chronologer. It was clear from the stratigraphic record that by the Late Oligocene or Early Pliocene there had been a complex history of erosional baselevelling and deposition culminating in the production of an extensive surface sloping gently eastward and capped by gravels and sands (notably the Pliocene Ogallala Formation) largely derived from the Rocky Mountain granites. This sloping surface, termed the Cypress Plain Peneplain in Alberta and Montana (Alden 1932, Collier and Thom 1918) and the Great Plains Peneplain further south, when projected westward was held to be continuous with the high erosion surface in the Raton and Pecos areas of the southwestern Great Plains (Ray and Smith 1941, Morgan and Sayer 1942), the Subsummit peneplains of the Front Range, Beartooth and Bighorn Mountains (Alden 1932) and the Blackfoot Peneplain of the Lewis Front Range in Montana (Willis 1902). The general view emerged that by the Oligocene–Pliocene the High Plains and the adjacent Rocky Mountains which had been broadly uplifted in Late Cretaceous (Laramide) time had been fashioned into a complex sloping surface, part erosional and part depositional (Alden 1924, Fenneman 1931). Major eastward drainage developed on this surface often controlled by ancient sand-dune orientations (Russell 1929, Price 1944). Intermittent uplift led to the excavation of much of the thicker depositional material close to the mountains in the west; produced a series of polycyclic valley-in-valley forms in Montana (Collier and Thom 1918), the Black Hills (Fillman 1929), the Raton Section (Ray and Smith 1941) and the Pecos Valley (Morgan and Sayer 1942); and produced a series of peneplain remnants and terraces converging in level towards the east. These erosional forms were thought to date from the Pliocene to Late Pleistocene (Alden 1932) and regional differences in texture and patterns of erosion were considered to be mainly due to geological or vegetational differences (Johnson 1900).

The mountains of the Middle Rockies and the Front Range were those which attracted the major interest from denudation chronologers, being far enough south for their higher parts not to have been extensively destroyed by cirque glaciation. This series of individual ranges was produced during the Larimide (Late Cretaceous) orogeny as broad anticlinal arches with marginal thrusts and fractures. Their major common geomorphic features appear to be a vestigial high-level older peneplain (Summit or Flattop), a lower and

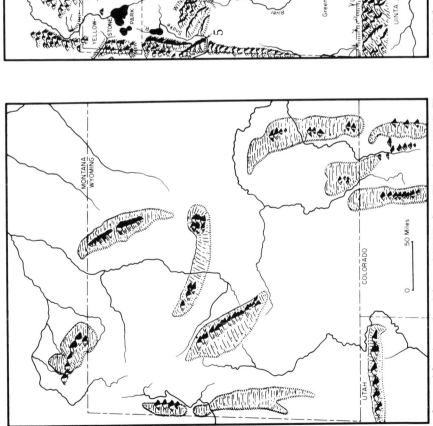

Figure 7.15 The assumed development of the drainage of the central Rockies: left, partially buried ranges in Late Tertiary time; right, present topography with nine numbered superimposed gorges

Source: After Atwood and Arwood (1938); from Thornbury (1965), Figures 17.4 and 17.5, pp. 327 and 331. Reprinted by permission of John Wiley and Sons Inc., all rights reserved.

younger peneplain (Subsummit or Rocky Mountain) and a series of gorges where they are crossed by the main rivers which were originally ascribed by Powell (volume 1, p. 531) to antecedence but came to be viewed as superimposed (figure 7.15). Although by mid-century the geomorphic inter-pretation of the evolution of the mountains of the western United States had strongly diverged from that in the east, it was natural that ideas of Appalachian denudation chronology should have been applied west of the Mississippi. Davis (1911I; volume 2, pp. 356–60) naturally adopted a chronological sequence for the Colorado Front Range similar to that of the Appalachians involving peneplanation, a Cretaceous sedimentary cover, uparching, imperfect peneplanation, followed by renewed uplift and valley erosion. The major work by Lee (1922) proposed the production of two peneplains in the region at a relatively steep angle due to the distant location from the coast: the restricted, older and higher Flattop and the more extensive, younger and lower Rocky Mountain Peneplain which was corre-lated with the Pliocene deposits of the High Plains. A final phase of uplift caused canyon cutting and the erosion of the flanking sediments. Lovering (1929) believed the Flattop Peneplain to be Davis' partly exhumed pre-Cretaceous peneplain, Fenneman (1931) correlated the Rocky Mountain Peneplain with the extensive South Park surface further south, and Love (1939) believed that North Park had been filled with Pliocene sediment to make a surface continuous with the Subsummit surface of the Medicine Bow Mountains further north on which the upper part of the North Platte River was developed. Long before, Blackwelder (1909) had proposed an Eocene summit peneplain in the Medicine Bow higher than the Sherman Peneplain of the Laramie Range. The same author (1915) later recognized a (Pliocene?) summit peneplain at above 12,000 feet in the Wind River Range with evidence of three younger cycles (see also Baker 1912, 1946, Westgate and Branson 1913) and Fryxell et al. (1941) extended the summit surface into the adjacent Grand Teton Range. Further north geomorphic interpretations were confused by possible effects of glaciation and by Tertiary volcanism. The assumed late Tertiary Subsummit Peneplain of the Bighorn Mountains (Fenneman 1931) was tentatively ascribed by Eakin (1916) to high-level periglacial planation (altiplanation) and to similar cryoplanation by Bryan (1946). In the Beartooth Range Bevan (1925) assigned the 12,000 feet summits to a tilted (Oligocene?) peneplain buried to the east below the Miocene volcanics of the Absaroka Range, the summits of which were believed to correlate with the Subsummit (Pliocene?) surface of the Beartooths (10,000–10,500 feet). However, the record here is less obvious than that further south, where Hughes (1933) postulated that the Subsummit surface is an exhumed plain of marine erosion and Sharp (1938) failed to recognize any peneplains at all. On the north flank of the Uinta Mountains Bradley (1936) recognized two major pre-Pleistocene surfaces:

1 The Gilbert Peak surface (Oligocene or Early Miocene?), steeply dipping (up to 400 feet/mile) and forming the summit of the range, which is fretted with cirques like the other ranges of the Middle Rockies. Bradley believed this to be the remains of a pediment which was at one time continuous with the Bishop Conglomerate of the Green River basin to the north (Rich 1910). Subsequent erosion had removed many of the sediments forming this part-erosional and part-depositional surface (Atwood 1909) (figure 7.16).
2 This erosion gave rise to a younger and lower pediment surface (the Bear Mountain) graded northward on to basin fill.

Two lower pediments, the Tipperary and the Lyman, were believed to be of Pleistocene age. Further west in the Wasatch Range Eardley (1944) identified two surfaces which he correlated with the Gilbert Peak and Bear Mountain. Thus by the Second World War a general theory of Rocky Mountain geomorphology had evolved which postulated the erosion of the ranges and the filling of the basins to produce compound erosional/depositional surfaces. Renewed episodic erosion exposed the eroded stumps of the mountains giving three or four partial-erosion surfaces (Westgate and Branson 1913, Blackwelder 1914) and allowing the superimposition of drainage over the exposed lower parts of the eroded ranges. Atwood and Atwood (1938) produced the following overall scheme (Figure 7.15):

1 Late Cretaceous (Larimide) uplift of large anticlinal arches;
2 Erosion to form an Early Tertiary peneplain (Flattop or Summit);
3 Renewed uplift;
4 Further erosion and basin filling giving an Oligocene–Pliocene (Rocky Mountain) peneplain correlated with the Ogallala deposition;
5 Uplift of several thousand feet, superimposition of gorges from the sedimentary infill and canyon cutting;
6 Minor renewed Pleistocene uplift.

Mackin (1947) believed the foregoing thesis was too dependent upon the peneplanation concept of the Appalachians and followed Bradley (1936), Rich (1938) and Howard (1941) in stressing pedimentation (to graded alluvial surfaces) as a means of producing steeply formed surfaces at inland elevations high enough to avoid some 6,000 feet of Pliocene and Pleistocene uplift required by the peneplanation thesis of Atwood and Atwood (1938). He also drew on the ideas of Eakin (1916) and Bryan (1946) to suggest that the Summit surfaces of the Rockies might be *younger* than the Subsummit surfaces in that the former had been produced at a relatively late (Pleistocene) date by high-level periglacial planation processes. Except for exhumed surfaces, he believed all the Rocky Mountain erosion surfaces to be post-Eocene in age.

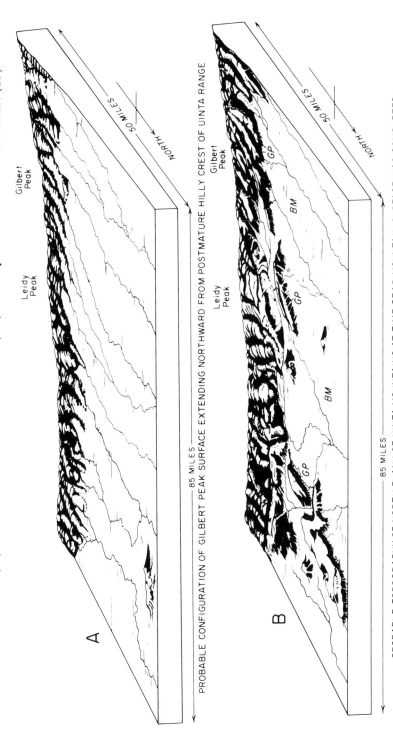

Figure 7.16 Possible stages in the development of the topography of the north flank of the Uinta Mountains: A, formation of the Gilbert Peak surface; B, formation of the Bear Mountain surface (BM) at the expense of the Gilbert Peak surface (GP)

Source: After Bradley (1936), Plate 42; from Thornbury (1965), Figure 20.2, p. 364. Reprinted by permission of John Wiley and Sons Inc., all rights reserved.

Thus the central Rockies provided the stimulus for the development of the most coherent Rocky Mountain denudation chronology. Further north, the mountains of Idaho presented a more complex picture because of the prevalence of Tertiary faulting and lava flows and the greater destruction by glacial erosion of the crests of the ranges. Nevertheless, a widespread Idaho (Summit) peneplain was recognized (Atwood 1916, Mansfield 1924, Anderson 1947), which was variously dated as Eocene (Umpleby 1912, Atwood 1916, Kirkham 1930) or Miocene or later (Blackwelder 1912, Rich 1918). Anderson (1929) proposed three erosion surfaces in northern Idaho of Cretaceous, Eocene and Miocene age, respectively. However, the complexity of the region was reflected in Blackwelder's (1912) sequence – erosion, dissection, deposition, erosion, deposition, faulting, renewed erosion – and in Mansfield's (1927) sequence of peneplanation, uplift, erosion, lava flows in valleys, renewed uplift and erosion. In the Lewis and Clark Range some flat-topped ridges in the east were thought to be peneplain remnants (Willis 1902, Alden 1932). In the southern Rockies the uplifted San Juan Peneplain was recognized and correlated with the Rocky Mountain surface in a region affected by volcanism (Atwood 1911, Atwood and Mather 1915, Atwood and Mather 1932) and in the Jemez Mountains there appeared to be a high-level (Rocky Mountain?) surface and a flight of some half-dozen terraces below (Behre 1933, Powers 1935).

To the west of the central and southern Rockies lies the Colorado Plateau, an extensive tract of dominantly flat-lying Palaeozoic and Mesozoic sedimentary rocks broken by faults and monoclines, pierced by intrusive igneous rock bodies and flooded by Tertiary lavas particularly on its southern and western margins. This region was made classic by Powell's (1875) work on the Colorado River (see volume 1, chapter 27) and, especially, by Dutton's monographs of 1880 and 1882 (see volume 1, pp. 573–86). Dutton was impressed by the average erosion of some 10,000 feet of rocks during the early and middle Tertiary which he termed the 'Great Denudation'. This period of erosion was equated with the plateau cycle by Davis (1901H) which he believed to have been preceded by the faulting and monoclinal flexures and to have been complicated by later intermittent uplifts. The broad valleys or 'Parks' of the Kaibab and Coconino Plateaus were thought by Davis to be mature features of the upper parts of the Plateau Cycle erosion surface as yet unconsumed by the later cycle but Robinson (1910) ascribed these to a separate partial cycle of erosion younger than the Plateau but older than the later Canyon Cycle. Later a theory of Strahler (1944) ascribed the Parks to the abandonment of fluvial valleys when they were eroded down to the level of the major limestone formation, being subsequently deepened by solution. Davis (1901H) thought that the course of the Colorado River was consequent on the mid-Tertiary Plateau Cycle peneplain, although Powell and Dutton believed it to be older and antecedent with regard to the Tertiary uplift. Later work by Blackwelder (1934) and Longwell (1946) dated the present course of the river much later than did Davis, proposing Pleistocene and Late Miocene–Early Pliocene, respectively. Moore (1926) assigned the drainage of the Colorado Plateau to superimposition from a more extensive Eocene sedimentary cover. The peneplain or stripped plain produced during the Plateau Cycle was dissected and the Grand Canyon and other canyons produced following several thousand feet of uplift (Pliocene?) and the initiation of the Canyon Cycle (Davis 1901H). This latest cycle may have been punctuated by further baselevel changes during which, for example, the stripped bench of the Grand Canyon Esplanade was formed. The lower part of the Colorado River appeared to be little related to the location of the horsts of the Basin and Range province (Lee 1906) and it was not easy to interpret the geomorphology of this arid province of fragmented pediments, bajadas, playas, structural plains and floodplains in cyclic terms, although Blackwelder (1928, 1931) suggested that there was some evidence for the ranges being in a second cycle of erosion and some of the faults having been exhumed.

Moving westward, the Sierra Nevadas are considerably older than the Rockies, dating back to a Jurassic orogeny, and the central granite batholith was early subjected to a polycyclic interpretation. The gently sloping (100 feet/mile) western side was proposed by Lindgren and Knowlton (1896) as a tilted subdued peneplain (the Sierra surface) 1,000–3,000 feet above which rise accordant monadnocks, remnants of an older surface. The highest part of the range is made up of the upper Kern watershed where Lawson (1904) recognized three peneplains: the extensive Subsummit Plateau (11,200–11,500 feet) correlated with the Sierra Peneplain and later with the Yosemite surface (Matthes 1930), believed to be of Miocene age; above the Subsummit Plateau rise the monadnock peaks of an older peneplain; about 2,500 feet below are the valley flats of a partial later cycle. Lawson (1904) believed the Subsummit surfaces to have been produced in the Late Tertiary into which the Kern Valley was cut in the Quaternary. Matthes (1930) pointed to the valley-in-valley forms of the Yosemite as indicative of three erosional stages (broad valley, mountain valley and canyon stages) subsequently glaciated, and three surfaces in the Mount Whitney region. Lindgren (1911) and Jenkins (1935) also gave a chronology for Tertiary valley cutting on the western slope of the Sierras, later filled with placer gravels. On the steep eastern slope Knopf (1918) recognized two surfaces supposedly correlated with those of Lawson. Further north in the Cascade Range of Oregon and Washington the Cascade (Russell 1901) or Methow (Willis 1903) erosion surface was identified, but its existence was later doubted by Waters (1939).

In contrast to the Sierras, the Coast Ranges are Pliocene or Recent in age, fragmented by faulting and subject to current orogenic movements. In such an unstable and idiosyncratic environment it became difficult to conceive of sufficient lengths of time or sufficient stability to develop a typically Davisian polycyclic landscape. Nevertheless, Dudley (1936) proposed erosion surfaces at 1,700 feet and 2,100 feet in the San Bernardino Mountains; Willis (1925: 677) postulated evidence of two cycles of erosion in the Gabilan Range since the end of the Pliocene; Trask (1926: 293) that the Santa Lucia Coast Range, during the Pleistocene and subsequently, had been almost completely peneplained, then uplifted about 4,500 feet and eroded to form the present mountains; and Putnam (1942: 749) in the Ventura region was forced to compress into the same short timespan the deposition of 4,000–6,000 feet of sediments, their subsequent deformation, their erosion to late maturity, vertical uplift of about 1,000 feet, and erosion to produce the present relief. On the other hand, Sauer (1929: 212–3) saw that Penck's *Grossfalt* model (see chapter 10) could circumvent the time–compression problem by allowing simultaneous planation at different elevations, rather than the successive production of erosion surfaces, and he applied this theory to an interpretation of the Peninsular Range of California. This was naturally criticized along Davisian lines by Bryan and Wickson (1931) and by Miller (1935). Even Davis himself began to recognize the problems presented by the young mountain ranges of the Pacific coast when he moved to California where he spent his last years, for he wrote at the age of 80:

> the scale on which deposition, deformation and denudation have gone on by thousands and thousands of feet in this new-made country is ten- or twenty-fold greater than that of corresponding processes in my old tramping ground. On shifting residence from one side of the continent to the other, a geologist must learn his alphabet over again in an order appropriate to his new surroundings.
>
> (Davis 1930F: 404)

Note 1 The preparation of this chapter owes a considerable debt to the important regional works by Fenneman (1931, 1938) and Thornbury (1965).

References

Adams, G. F. (1945) 'Upland terraces in southern New England', *Journal of Geology* 53: 289–312.

Adams, G. I. (1928) 'The course of the Tennessee river and the physiography of the southern Appalachian region', *Journal of Geology* 36: 481–93.

Alden, W. C. (1924) 'Physiographic development of the northern Great Plains', *Bulletin of the Geological Society of America* 35: 385–424.

—— (1932) 'Physiography and glacial geology of eastern Montana and adjacent areas', *US Geological Survey Professional Paper* 174.

Anderson, A. L. (1929) 'Cretaceous and Tertiary planation in northern Idaho', *Journal of Geology* 37: 747–64.

—— (1947) 'Drainage diversion in the northern Rocky Mountains of east-central Idaho', *Journal of Geology* 55: 61–75.

Ashley, G. H. (1931) 'Our youthful scenery', *Bulletin of the Geological Society of America* 42: 537–46.

—— (1933) 'The scenery of Pennsylvania, its origin and development, based on recent studies of physiographic and glacial geology', *Pennsylvania Geological Survey, 4th Series, Bulletin* G6.

—— (1935) 'Studies in Appalachian mountain structure', *Bulletin of the Geological Society of America* 46:1395–436.

Atwood, W. W. (1909) 'Glaciation of the Uinta and Wasatch Mountains', *US Geological Survey Professional Paper* 61.

—— (1911) 'Physiographic studies in the San Juan district', *Journal of Geology* 19: 449–53.

—— (1916) 'The physiographic conditions at Butte, Montana, and Bingham Canyon, Utah, when the copper ores in these districts were enriched', *Economic Geology* 11: 697–740.

Atwood, W. W. (1940) *The Physiographic Provinces of North America*, New York: Ginn.

Atwood, W. W. and Atwood, W. W., Jr (1938) 'Working hypothesis for the physiographic history of the Rocky Mountain Region', *Bulletin of the Geological Society of America* 49: 957–80.

Atwood, W. W. and Mather, K. (1915) 'The Grand Canyon of the Gunnison River (Abstract)', *Annals of the Association of American Geographers* 5: 138–9.

——————— (1932) 'Physiography and Quaternary geology of the San Juan Mountains, Colorado', *US Geological Survey Professional Paper* 166.

Baker, C. L. (1912) 'Notes on the Cenozoic history of central Wyoming (Abstract)', *Bulletin of the Geological Society of America* 23: 73–4.

—— (1946) 'Geology of the northwestern Wind River Mountains, Wyoming', *Bulletin of the Geological Society of America* 57: 565–96.

Barrell, J. (1913) 'Piedmont terraces of the northern Appalachians and their mode of origin (Abstract)', *Bulletin of the Geological Society of America* 24: 688–90.

—— (1920a) 'Post-Jurassic history of the northern Appalachians (Abstract)', *Bulletin of the Geological Society of America* 24: 690–1.

—— (1920b) 'The piedmont terraces of the northern Appalachians', *American Journal of Science*, 4th series, 49: No. 294, 227–58, 327–61 and 407–28.

Bascom, F. (1921) 'Cycles of erosion in the Piedmont province of Pennsylvania', *Journal of Geology* 29: 540–59.

—— (1931) 'Geomorphic nomenclature', *Science* 74: 172–3.

Bates, R. E. (1939) 'Geomorphic history of the Kickapoo region, Wisconsin', *Bulletin of the Geological Society of America* 50: 814–80.

Behre, C. H. (1933) 'Physiographic history of the upper Arkansas and Eagle Rivers, Colorado', *Journal of Geology* 41: 785–814.

Bethune, P. de (1948) 'Geomorphic studies in the Appalachians of Pennsylvania', *American Journal of Science* 246: 1–22.

Bevan, A. (1925) 'Rocky Mountain peneplains northeast of Yellowstone Park', *Journal of Geology* 33: 563–87.

Blackwelder, E. (1909) 'Cenozoic history of the Laramie region, Wyoming', *Journal of Geology* 17: 429–44.

—— (1912) 'The old erosion surface in Idaho: A criticism', *Journal of Geology* 20: 410–4.

—— (1914) 'Post-Cretaceous history of the mountains of west-central Wyoming', *Journal of the Washington Academy of Sciences* 4: 445–6.

—— (1915) 'Post-Cretaceous history of the mountains of central Wyoming', *Journal of Geology* 23:97–117, 193–217, 307–40.

—— (1928) 'The recognition of fault scarps', *Journal of Geology* 36: 289–311.

—— (1931) 'Desert plains', *Journal of Geology* 39:133–40.

—— (1934) 'Origin of the Colorado River', *Bulletin of the Geological Society of America* 45: 551–66.

Bradley, W. H. (1936) 'Geomorphology of the north flank of the Uinta Mountains', *US Geological Survey Professional Paper* 185I, pp. 163–204.

Bryan, K. (1946) 'Cryopedology – the study of frozen ground and intensive frost-action with suggestions of nomenclature', *American Journal of Science* 244: 622–42.

Bryan, K. and Wickson, G. G. (1931) 'The W. Penck method of analysis in Southern California', *Zeitschrift für Geomorphologie* 6: 287–91.

Bucher, W. H. (1932) '"Strath" as a geomorphic term', *Science* 75: 130–1.

—— (1946) 'Biographical Memoir of Douglas Wilson Johnson (1878–1944)', *National Academy of Sciences, Biographical Memoirs*, vol. XXIV, no. 5, pp. 197–230.

Butts, C. (1904) 'Description of Kittatiny quadrangle', *US Geological Survey Folio* 115, pp. 2–3.

Campbell, M. R. (1896) 'Drainage modifications: their interpretation', *Journal of Geology* 4: 567–81, 657–78.

—— (1898) 'Description of the Richmond quadrangle', *US Geological Survey Folio* 46.

—— (1903) 'Geographic development of northern Pennsylvania and southern New York', *Bulletin of the Geological Society of America* 14: 277–96.

—— (1931) 'The alluvial fan of Potomac River', *Bulletin of the Geological Society of America* 42: 182, 825–52.

—— (1933) 'Chambersburg (Harrisburg) peneplain in the piedmont of Maryland and Pennsylvania', *Bulletin of the Geological Society of America* 44: 553–73.

Chamberlin, R. T. (1910) 'The Appalachian folds of central Pennsylvania', *Journal of Geology* 18: 228–51.

—— (1930) 'The level of baselevel', *Journal of Geology* 38:166–73.

Chamberlin, T. C. (1909) 'Diastrophism as the ultimate basis of correlation', *Journal of Geology* 17: 685–93.

Chorley, R. J. (1973) 'Douglas Wilson Johnson', *Dictionary of Scientific Biography*, vol. VII, New York: Scribners, pp. 143–5.

Clark, W. B. *et al.* (1906) 'The physical features of Maryland', *Maryland Geological Survey* 6: 27–92.

Cleland, H. F. (1928) 'A Pleistocene peneplain in the Coastal Plain', *Journal of Geology* 28: 702–6.

Cole, W. S. (1930) 'The interpretation of intrenched meanders', *Journal of Geology* 38: 423–36.

—— (1935) 'Rock resistance and peneplain expression', *Journal of Geology* 43: 1049–62.

—— (1941) 'Nomenclature and correlation of Appalachian erosion surfaces', *Journal of Geology* 49: 129–48.

Collier, A. J. and Thom, W. T. Jr (1918) 'The Flaxville Gravel and its relation to other terrace gravels of the northern Great Plains', *US Geological Survey Professional Paper* 108, pp. 179–84.

Cooke, C. W. (1931) 'Seven coastal terraces in the Southeastern States', *Journal of the Washington Academy of Sciences* 21: 503–13.

—— (1935) 'Tentative ages of Pleistocene shorelines', *Journal of the Washington Academy of Sciences* 25: 331–3.

—— (1936) 'Geology of the coastal plain of South Carolina', *US Geological Survey Bulletin* 867.

—— (1943) 'Geology of the Coastal Plain of Georgia', *US Geological Survey Bulletin* 941.

—— (1945) 'The Geology of Florida', *Florida Geological Survey Bulletin* 29.

Cotton, C. A. (1948) *Landscape: as developed by the processes of normal erosion*, 2nd edn, London: Cambridge University Press.

Crickmay, C. H. (1933) 'The later stages of the cycle of erosion', *Geological Magazine* 70: 337–47.

Cushing, H. P. (1905) 'Geology of the northern Adirondack Region', *Bulletin of the New York State Museum* 95.

Daly, R. A. (1905) 'Accordance of summit levels among Alpine mountains', *Journal of Geology* 13: 105–25.

Darton N. H. (1894) 'Outline of Cenozoic history of the Middle Atlantic Slope', *Journal of Geology* 2, 568–87.

Dudley, P. H. (1936) 'Physiographic history of a portion of the Perris block, southern California', *Journal of Geology* 44: 358–78.

Eakin, H. M. (1916) 'The Yukon-Koyukuk region, Alaska', *U. S. Geological Survey Bulletin* 631, 1-88.

Eardley, A. J. (1944) 'Geology of the north-central Wasatch Mountains, Utah', *Bulletin of the Geological Society of America* 55: 819–94.

Fairchild, H. L. (1925) 'The Susquehanna river in New York and evolution of western New York drainage', *Bulletin of the New York State Museum* 256.

Fenneman, N. M. (1908) 'Some features of erosion by unconcentrated wash', *Journal of Geology* 16: 746–54.

—— (1931) *Physiography of Western United States* New York: McGraw-Hill.

—— (1936) 'Cyclic and non-cyclic aspects of erosion', *Bulletin of the Geological Society of America* 47: 173–86.

—— (1938) *Physiography of Eastern United States*, New York: McGraw-Hill.

Fillman, L. (1929) 'Cenozoic history of the northern Black Hills', *University of Iowa Studies in Natural History* 13: (1).

Flint, R. F. (1940) 'Pleistocene features of the Atlantic Coastal Plain', *American Journal of Science* 238: 757–87.

—— (1941) 'Pleistocene strand lines: a rejoinder', *American Journal of Science* 239: 459–62.

Fridley, H. M. (1929) 'Identification of erosion surfaces in south-central New York', *Journal of Geology* 37: 113–34.

Fridley, H. M. and Nölting, J. P. (1931) 'Peneplains of the Appalachian Plateau', *Journal of Geology* 39: 749–55.

Fryxell, F. M., Horberg, L. and Edmund, R. (1941) 'Geomorphology of the Teton Range and adjacent basins, Wyoming–Idaho (Abstract)', *Bulletin of the Geological Society of America* 52: 1903.

Galloway, J. J. (1919) 'Geology and natural resources of Rutherford County, Tennessee', *Tennessee Division of Geology, Bulletin* 22.

Gilbert, G. K. (1883) 'Review of Whitney's "Climatic Changes" III', *Science* 1: 192–5.

Glock, W. S. (1932) 'Available relief as a factor of control in the profile of a land form', *Journal of Geology* 40: 74–83.

Goldthwait, J. W. (1914) 'Remnants of an old graded upland on the Presidential Range of the White Mountains', *American Journal of Science* 4th series, 37: 451–63.

Goldthwait, R. P. (1940) 'Geology of the Presidential Range, New Hampshire', *New Hampshire Academy of Science Bulletin* 1: 1–43.

Hatch, L. (1917) 'Marine terraces in southeastern Connecticut', *American Journal of Science* 4th series, 44: 319–30.

Hayes, C. W. (1899) 'Physiography of the Chattanooga district in Tennessee,

Georgia, and Alabama', *US Geological Survey, 19th Annual Report, Part 2* pp. 1–58.

Hayes, C. W. and Campbell, M. R. (1894) 'Geomorphology of the southern Appalachians', *National Geographic Magazine* 6: 63–126.

Hershey, O. H. (1896) 'Preglacial erosion cycles in northwestern Illinois', *American Geologist* 18: 72–100.

—— (1902a) 'Peneplains of the Ozark Highland', *American Geologist* 27: 25–41.

—— (1902b) 'Boston Mountain physiography', *Journal of Geology* 10:160–5.

Hickok, W. O. (1933) 'Erosion surfaces in south central Pennsylvania', *American Journal of Science* 5th series, 25: 101–22.

Hobbs, W. H. (1904) 'Lineaments of the Atlantic border regions', *Bulletin of the Geological Society of America* 15: 483–506.

Howard, A. D. (1941) 'Rocky Mountain peneplanes or pediments', *Journal of Geomorphology* 4: 138–41.

Hughes, R. V. (1933) 'The geology of the Beartooth mountain front in Park County, Wyoming', *Proceedings of the National Academy of Sciences* 19: 239–53.

Jenkins, O. P. (1935) 'New technique applicable to the study of placers', *California Journal of Mines and Geology* 31: 143–200.

Jillson, W. R. (1928) 'Peneplains in Kentucky', *Pan-American Geologist* 50: 333–8.

Johnson, D. W. (1905) 'The Tertiary history of the Tennessee river', *Journal of Geology* 13: 194–231.

—— (1916) 'Plains, planes and peneplanes', *Geographical Review* 1: 443–7.

—— (1929) 'Baselevel', *Journal of Geology* 37: 775–82.

—— (1931a) 'Evolution of the drainage system of eastern North America', *Comptes Rendus, International Geographical Congress*, tome II, first part, section II, pp. 600–6.

—— (1931b) *Stream Sculpture on the Atlantic Slope*, New York: Columbia University Press.

—— (1938–42) 'Studies in scientific method', *Journal of Geomorphology* vols. 1–5.

Johnson, W. D. (1900) 'The High Plains and their utilization', *US Geological Survey, 21st Annual Report, Part IV*.

Keith, A. (1894) 'Geology of the Catoctin belt', *US Geological Survey, 14th Annual Report, Part 2*, pp. 285–395.

—— (1896) 'Some stages of Appalachian erosion', *Bulletin of the Geological Society of America* 7: 519–29.

—— (1916) 'Topography of Massachusetts', *US Geological Survey, Water Supply Paper 415*, pp. 8–23.

Kirkham, V. R. D. (1930) 'Old erosion surfaces in southwestern Idaho', *Journal of Geology* 38: 652–63.

Knopf, A. (1918) 'A geologic reconnaissance of the Inyo Range and the eastern slope of the Sierra Nevada, California', *US Geological Survey Professional Paper* 110.

Knopf, E. B. (1924) 'Correlation of residual erosion surfaces in the eastern Appalachian highlands', *Bulletin of the Geological Society of America* 35: 633–68.

Lane, A. C. (1921) 'White Mountain physiography', *American Journal of Science* 5th series, 1: 349–54.

Lawson, A. C. (1894)'The geomorphogeny of the coast of northern California', *University of California Publications in Geology* 1: 241–71.

—— (1904) 'Geomorphogeny of the upper Kern basin', *University of California Publications in Geology* 3: 291–376.

Lee, W. T. (1906) 'Geology of the lower Colorado River', *Bulletin of the Geological Society of America* 17: 275–84.

—— (1922) 'Peneplains of the Front Range at Rocky Mountain National Park, Colorado', *US Geological Survey Bulletin* 730, pp. 1–17.

Lindgren, W. (1911) 'The Tertiary gravels of the Sierra Nevada', *US Geological Survey Professional Paper* 73, pp. 9–81.

Lindgren, W. and Knowlton, F. H. (1896) 'The age of the auriferous gravels of the Sierra Nevada', *Journal of Geology* 4: 881–906.

Lobeck, A. K. (1917) 'Position of the New England peneplane in the White Mountains region', *Geographical Review* 3: 53–60.

Longwell, C. R. (1946) 'How old is the Colorado River?' *American Journal of Science* 244: 817–35.

Love, J. D. (1939) 'Geology along the margin of the Absaroka Range, Wyoming', *Geological Society of America Special Paper* 20.

Lovering, T. S. (1929) 'Geologic history of the Front Range', *Proceedings of the Colorado Scientific Society* 12: (4).

Löwl, F. (1882) 'Die Entstehung der Durchbruchstäler', *Petermanns Geographische Mitteilungen* 28: 405–16.

Lusk, R. G. (1928) 'Gravel on the Highland Rim Plateau and terraces in the valley of the Cumberland River', *Journal of Geology* 36: 164–70.

McGee, WJ (1888) 'Three formations of the middle Atlantic slope', *American Journal of Science* 3rd series, 35: 120–43, 328–30, 367–88, 448–66.

—— (1893) 'The Lafayette formation', *US Geological Survey 12th Annual Report*, 1.

Mackin, J. H. (1938) 'The origin of Appalachian drainage – a reply', *American Journal of Science* 236: 27–53.

—— (1947) 'Altitude and local relief in the Bighorn area during the Cenozoic', *Wyoming Geological Association Field Conference in the Bighorn Basin, Guidebook*, pp. 103–20.

Mahard, R. H. (1942) 'The origin and significance of intrenched meanders', *Journal of Geomorphology* 5: 32–44.

Malott, C. A. (1920) 'Static rejuvenation', *Science* 52: 182–3.

—— (1922) 'The physiography of Indiana', in *Handbook of Indiana Geology*, part 2, pp. 59–256.

—— (1928) 'Base-level and its varieties', *Indiana University Studies* 82: 37–59.

Mansfield, G. R. (1924) 'Tertiary planation in Idaho', *Journal of Geology* 32: 472–87.

—— (1927) 'Geography, geology and mineral resources of part of southeastern Idaho', *US Geological Survey Professional Paper* 152.

Marbut, C. F. (1896) 'Physical features of Missouri', *Bulletin of the Missouri Geological Survey* 10: 11–109.

Martin, G. C. (1908) 'Accident – Grantsville Quadrangle' *US Geological Survey Folio* no. 160.

Matthes, F. E. (1930) 'Geologic history of the Yosemite Valley', *US Geological Survey Professional Paper* 160.

Melton, F. A. and McGuigan, F. H. (1928) 'The depth of the base of the Trinity Sandstone and the present altitude of the Jurassic peneplain in southern Oklahoma and south-western Arkansas', *Bulletin of the American Association of Petroleum Geologists* 12: 1005–14.

Meyerhoff, H. A. (1938) 'Tertiary marine planation in the Piedmont and southern New England (Abstract)', *Bulletin of the Geological Society of America* 49: 1954–5.

Meyerhoff, H. A. and Hubbell, M. (1928) 'The erosional landforms of eastern and central Vermont', *Report of the Vermont State Geologist* no. 16, pp. 315–81.

Meyerhoff, H. A. and Olmsted, E. W. (1934) 'Windgaps and water gaps in Pennsylvania', *American Journal of Science* 5th series, 27: 410–16.

—— (1936) 'The origins of Appalachian drainage', *American Journal of Science* 5th series, 32: 21–42.

Miller, W. J. (1917) 'The Adirondack Mountains', *Bulletin of the New York State Museum* 193.

—— (1935) 'Geomorphology of the southern Peninsular Range of California', *Bulletin of the Geological Society of America* 46: 1535–62.

Miser, H. D. and Purdue, A. H. (1929) 'Geology of the Dequeen and Caddo Gap Quadrangle, Arkansas', *US Geological Survey Bulletin* 808.

Moore, R. C. (1926) 'Significance of enclosed meanders in the physiographic history of the Colorado Plateau Country', *Journal of Geology* 34: 29–99.

Morgan, A. M. and Sayer, A. N. (1942) 'Geology and groundwater', in *The Pecos River Joint Investigation*, National Resources Planning Board, pp. 28–38.

Olmsted, E. W. and Little, L. S. (1946) 'Marine planation in southern New England (Abstract)', *Bulletin of the Geological Society of America* 57: 1271.

Peschel, O. F. (1880) *Physische Erdkunde*, vol. II, Leipzig: Duncker and Humblot.

Pond, A. M. (1929) 'Preliminary report on the peneplanes of the Taconic Mountains of Vermont', *Report of the Vermont State Geologist* no. 16, pp. 292–314.

Powers, W. E. (1935) 'Physiographic history of the upper Arkansas Valley and the Royal Gorge, Colorado', *Journal of Geology* 43: 184–99.

Price, W. A. (1944) 'Greater American deserts', *Proceedings and Transactions of the Texas Academy of Sciences* 27: 163–70.

Purdue, A. H. (1901) 'Physiography of the Boston Mountains', *Journal of Geology* 9: 694–701.

Purdue, A. H. and Miser, H. D. (1923) 'Description of the Hot Springs District', *US Geological Survey Folio* 215.

Putnam, W. C. (1942) 'Geomorphology of the Ventura Region, California', *Bulletin of the Geological Society of America* 53: 691–754.

Ray, L. L. and Smith, J. F., Jr (1941) 'Geology of the Moreno Valley, New Mexico', *Bulletin of the Geological Society of America* 52: 177–210.

Renner, G. T. (1927) 'The physiographic interpretation of the fall line', *Geographical Review* 17: 276–86.

Rich, J. L. (1910) 'The physiography of the Bishop conglomerate, southwestern Wyoming', *Journal of Geology* 18: 601–32.

—— (1914) 'Certain types of streams and their meaning', *Journal of Geology* 22: 469–97.

—— (1918) 'An old erosion surface in Idaho: Is it Eocene?', *Economic Geology* 13: 120–36.

—— (1938) 'Recognition and significance of multiple erosion surfaces', *Bulletin of the Geological Society of America* 49: 1695–722.

Robinson, H. H. (1910) 'A new erosion cycle in the Grand Canyon District, Arizona', *Journal of Geology* 18: 742–63.

Ruedemann, R. (1931) 'The tangential master-streams of the Adirondack drainage', *American Journal of Science* 222: 431–40.

—— (1932) 'Development of drainage of Catskills', *American Journal of Science* 223: 337–49.

Russell, I. C. (1901) 'Geology and water resources of Nez Perce Co., Idaho, Part I', *US Geological Survey, Water Supply Paper* 53.

Russell, W. L. (1929) 'Drainage alignment in the western Great Plains', *Journal of Geology* 37: 249–55.

Salisbury, R. D. (1895) 'Physical geography', *New Jersey Geological Survey* 4: 63.

Sauer, C. O. (1929) 'Landforms in the Peninsular Range of California as developed about Warner's Hot Springs', *University of California Publications in Geography* 3: (4): 199–290.

Shaler, N. S. (1899) 'The spacing of rivers with reference to baseleveling', *Bulletin of the Geological Society of America* 10: 263–76.

Sharp, H. S. (1929) 'The physical history of the Connecticut shoreline', *Connecticut Geological and Natural History Bulletin* 46.

—— (1932) 'The geomorphic development of central Ohio (Part 1)', *Bulletin of Denison University* 27: 1–46.

—— (1938) 'The upland of the Beartooth Mountains, Montana (Abstract)', *Proceedings of the Geological Society of America for 1937*, p. 113.

Shaw, E. W. (1918a) 'Pliocene history of northern and central Mississippi', *US Geological Survey Professional Paper* 108H, pp. 125–63.

—— (1918b) 'Ages of peneplains of the Appalachian Province', *Bulletin of the Geological Society of America* 29: 575–86.

Simpson, C. T. (1900) 'The evidence of the *Unionidae* regarding the former courses of the Tennessee river and other southern rivers', *Science* new series, 12: 133–6.

Smith, W. R. T. (1899) 'Some aspects of erosion in relation to the theory of peneplains', *University of California Department of Geology, Bulletin* 2, pp. 155–78.

Stephenson, L. W. (1926) 'Major features of the geology of the Atlantic and Gulf Coastal Plains', *Journal of the Washington Academy of Sciences* 16: 460–80.

—— (1928) 'Major marine transgressions and regressions and structural features of the Gulf Coastal Plain', *American Journal of Science* 5th series, 16: 281–98.

Stone, R. W. (1908) 'Physiography of southwestern Pennsylvania', *Pennsylvania Geological Survey, Report for 1906–08*, pp. 120–7.

Stose, G. W. (1928) 'High gravels of the Susquehanna river', *Bulletin of the Geological Society of America* 39: 1073–85.

—— (1930) 'Is the Bryn Mawr peneplain a warped surface?', *American Journal of Science* 5th series, 19: 178–84.

—— (1940) 'Age of the Schooley peneplain', *American Journal of Science* 238: 461–76.

Strahler, A. N. (1944) 'Valleys and parks of the Kaibab and Coconino Plateaus', *Journal of Geology* 52: 361–87.

—— (1945) 'Hypotheses of stream development in the folded Appalachians of Pennsylvania', *Bulletin of the Geological Society of America* 56: 45–88.

Tarr, R. S. (1898) 'The peneplain', *American Geology 1st* 21: 351–70.

—— (1902) *The Physical Geography of New York State*, New York: Macmillan.

Tarr, R. S. and Martin, L. (1914) *College Physiography Geologist*, New York: Macmillan.

Thompson, H. D. (1939) 'Drainage evolution in the southern Appalachians', *Bulletin of the Geological Society of America* 50: 1323–56.

—— (1941) 'Topographic analysis of the Monterey, Staunton, and Harrisonburg quadrangles', *Journal of Geology* 49: 521–49.

Thornbury, W. D. (1954) *Principles of Geomorphology*, New York: Wiley.

—— (1965) *Regional Geomorphology of the United States*, New York: Wiley.

Tietze, E. (1878) 'Einige Beonerkungen über die Bildung von Querthälern', *Jahrbuch Geologischen Reichsanstalt* 28: 581–610.

Trask, P. D. (1926) 'Geomorphogeny of the northern part of the Santa Lucia Coast Range, California, *American Journal of Science* 5th series, 12: 293–300.

Trowbridge, A. C. (1921) 'The erosional history of the Driftless Area', *University of Iowa Studies in Natural History* 9: (3).

Umpleby, J. B. (1912) 'An old erosion surface in Idaho: Its age and value as a datum plane', *Journal of Geology* 20: 139–47.

Ver Steeg, K. (1930) 'Wind gaps and water gaps of the northern Appalachians, their characteristics and significance', *Annals of the New York Academy of Sciences* 32: 87–220.

—— (1931a) 'Erosion surfaces of the Appalachians', *Pan-American Geologist* 56: 267-84.

——(1931b) 'Erosion surfaces of eastern Ohio', *Pan-American Geologist* 55: 93–102, 181–92.

—— (1936) 'The buried topography of western Ohio', *Journal of Geology* 44: 918–39.

—— (1940) 'Correlation of Appalachian peneplains', *Pan-American Geologist* 73: 203–10.

Von Engeln, O. D. (1942) *Geomorphology*, New York: Macmillan.

Ward, F. (1930) 'The role of solution in peneplanation', *Journal of Geology* 38: 262–70.

Waters, A. C. (1939) 'Resurrected erosion surface in central Washington', *Bulletin of the Geological Society of America* 50: 638–59.

Westgate, L. G. and Branson, E. B. (1913) 'The later Cenozoic history of the Wind River mountains', *Journal of Geology* 21: 141–59.

White, G. W. (1939) 'End moraines of north-central Ohio', *Journal of Geology* 47: 277–89.

Willis, B. (1889) 'Round about Asheville', *National Geographic Magazine* 1: 291–300.

—— (1895) 'The northern Appalachians', *National Geographical Society, Monograph* 1, pp. 169–202.

—— (1902) 'Stratigraphy and structure, Lewis and Livingstone Ranges, Montana', *Bulletin of the Geological Society of America* 13: 305–52.

—— (1903) 'Physiography and deformation of the Wenatchee–Chelan district, Cascade Range', *US Geological Survey Professional Paper* 19, pp. 49–97.

—— (1925) 'Physiography of the California Coast Ranges', *Bulletin of the Geological Society of America* 36: 641–78.

Wilson, C. W., Jr (1935) 'The pre-Chattanooga development of the Nashville dome', *Journal of Geology* 43: 449–81.

Wright, F. J. (1925) 'The physiography of the upper James River basin in Virginia', *Virginia Geological Survey Bulletin* 11, pp. 11–55.

—— (1928) 'The erosional history of the Blue Ridge', *Denison University Journal of the Scientific Laboratories* 23: 321–44.

—— (1931) 'The older Appalachians of the South', *Denison University Journal of the Scientific Laboratories* 26: 143–269.

—— (1936) 'The newer Appalachians of the South (Part 2)', *Denison University Journal of the Scientific Laboratories* 31: 93–142.

—— (1942) 'Erosional history of the southern Appalachians', *Journal of Geomorphology* 5: 151–61.

(All Davis references are given in Volume 2, pp. 793–825).

CHAPTER EIGHT

Henri Baulig and Eustatic Planation

Henri Baulig (1877–1962), a boy of humble origins, studied history and geography at the Sorbonne between 1896 and 1903 under Paul Vidal de la Blache. Encouraged by the latter, Baulig left France in 1904 to spend 6 years working under W. M. Davis at Harvard (see chapter 5). Returning to the Sorbonne in 1911, he began his monumental researches into the denudation chronology of the Central Plateau, although these were soon interrupted by a period at Rennes and service as a private soldier in a non-combatant territorial unit during the four years of the First World War. In 1919 he took up an appointment at the University of Strasbourg and returned to his researches (Juillard and Klein, 1980).

Baulig's approach to geomorphology was dominated by the fusion of two great models: the eustatic model of Eduard Suess, which has been documented in our chapter 3, and the cyclic model of W. M. Davis, with which he was inculcated at Harvard. It is significant that many of his earlier publications were on the subject of the geography and geomorphology of North America, and that his first major geomorphic contribution was on the Davisian theme of the profile of equilibrium (Baulig 1925). Fourteen years later Baulig stated his view of the 'usual cyclic theory of geomorphological evolution' wherein

> successive generations of erosional forms develop during periods of complete or nearly complete stability, separated by rapid shifts in the relative position of the landmass and of the baselevel: a conclusion fully supported ... by the obvious discontinuity of valley forms, both in longitudinal profiles of river beds and in cross-sections of valley sides. For various reasons each cycle is normally more advanced in the lower than in the upper parts of river courses. As each cyclic elementary valley becomes more and more mature as one proceeds downstream, it is comprehensible that its sides, gradually flattening, should finally coalesce with the similarly evolved sides of adjacent valleys into a sort of local peneplain. This, when bordering on higher, mountainous ground, would assume the appearance of a peripheral bench. A later cycle will in the same manner induce the development of another bench, lower than the first and bordering on it farther out. As each cyclic system of forms cannot develop except by encroaching on the preceding system, while it is losing ground to the next succeeding system, the different cycles represented in the same region, although successive in their origin, are simultaneous in their development. This simple and apparently legitimate extension of

Plate 8.1 Henri Baulig in 1929

Source: Courtesy of Professor Etienne Juillard

the cyclic theory ... underlies the many studies devoted to step-like erosion surfaces (*niveaux d'érosion étagés*).

(Baulig 1939: 298–9)

For Baulig, unlike Davis and Martonne, the shifts in the relative position of the landmass and of the baselevel were due to eustatic changes in the Suessian model.

Le Plateau Central de la France

An almost exact contemporary of Martonne (1873–1955), Henri Baulig (plate 8.1) lived virtually all his long life in the shadow of the former's achievements. Martonne developed precociously, was immensely productive, and had established himself by the First World War as France's leading physical geographer as the result of his championing of the ideas of W. M. Davis and his publication of the *Traité de Géographie Physique* (see chapter 5). In contrast, Baulig had to wait for success until he was past 50, but in 1928 he issued a clarion call for eustatism with his substantial volume *Le Plateau Central de la France et sa Bordure Méditerranéenne* (Baulig 1928a). The book is dedicated to the memory of two great sources of inspiration, Eduard Suess (1831–1914) and Paul Vidal de la Blache (1845–1918). It is Vidallian only in its regionalistic approach and Suessian in its determination to be geomorphological and to uphold whenever possible the theme of the eustatic master. His special acknowledgements include Ch. Lallemand, director of the *Nivellement Général* (General Levelling) *de la France* and Emmanuel de Martonne who vetted the manuscript and later (1929) wrote a long review of it.

We are told that the Central Plateau consists of four main structural elements (figure 8.1):

1 The so-called Hercynian block which has a complex structure dating back to Palaeozoic times and with some Carboniferous and Permian strata.
2 A sedimentary covering of Mesozoic age resting flatly on the planated Hercynian block most of which it formerly covered and still surrounds.
3 Tertiary sediments, accumulated and preserved mainly in downfaulted troughs, known as *limagnes*.
4 Volcanic structures superimposed on elements 1–3.

The spatial distribution of these elements is expressed geographically in four main regions. The west (Limousin) which is essentially crystalline, free of great Tertiary dislocations, of sedimentation and of recent volcanism; appearing as vast regular plateaus surmounted by strong monadnocks. The centre (Auvergne) which has a crystalline and metamorphic base and which experienced tremendous fracturing in Tertiary times, leading to a very rugged relief with uplifted horsts, downfaulted infilled troughs and intense,

Figure 8.1 The geology and drainage of the Central Plateau: CF, Clermont Ferrand; L, Limoges; Ly, Lyons; N, Nîmes; P, Poitiers; SE, St Etienne

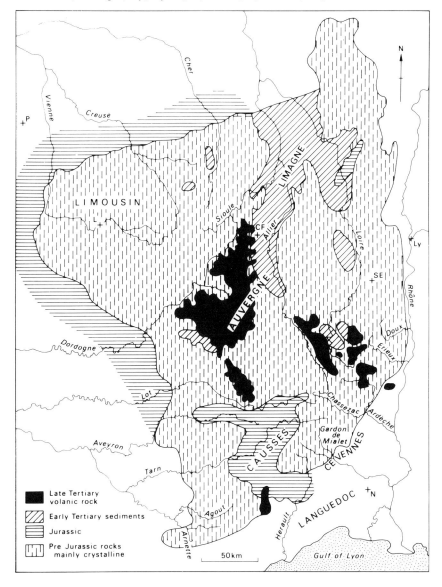

polymorphic volcanism. The south which is a clearly Hercynian structure, characterized by a relatively great extension of Permian basins and a thick Jurassic deposition in the Causses, and by east-west Tertiary fractures. The Rhône-Mediterranean border, including Languedoc, where strong earth movements have lifted up the Massif in relation to bordering regions along fractures but without volcanism.

These regions were shown by Baulig (1928a: 13–33) to have experienced a complex geological history. In pre-Carboniferous times the Massif was composed of a complex nappe structure which was later covered by extensive Carboniferous and Permian deposits. Before the tranquil sedimentation of the Mesozoic a very perfect surface of planation was produced:

> Littoral erosion seems to be the only action capable of such perfect levelling. But, nevertheless, the lower part of the Secondary deposits do not contain basal conglomerates; fossils are absent apart from a few plant imprints: when by chance we find marine terrains, their calcareous facies indicate that the sea had merely re-covered an already levelled surface.
>
> (Baulig 1928a: 18)

This post-Hercynian levelled surface appears to have been produced by a long period of semi-arid erosion (i.e. pediplanation), evidenced by residual weathered deposits, and was then overlapped by Jurassic rocks, dominantly carbonates. These limestones reached their thickest (1,400 m) in the basin forming the present Causses. Later erosion has stripped the post-Hercynian surface in parts of the north where its regularity is still a striking feature of the Central Plateau. Towards the end of the Jurassic a general uplift of the Plateau (except in the south-east) ushered in the prolonged Eogene cycle of erosion which persisted until the great mid-Tertiary dislocations associated with the Alpine orogeny. This cycle, less complete than its predecessor but also semi-arid in character, produced another extensive (Eogene) erosion surface of Early Tertiary age. There is evidence of this surface bevelling both the pre-Jurassic crystalline rocks of Limousin and the flanking Jurassic rocks. However, the fracturing and volcanism which affected the central and eastern parts of the Plateau during the Oligocene and Miocene have resulted in the Eogene surface being predominantly preserved beneath Tertiary sediments in fault troughs (*limagnes*), such as the Allier valley, or beneath Tertiary and later volcanic rocks in Auvergne. Erosion of these Tertiary deposits has exposed the Eogene surface in places so that both the post-Hercynian and Eogene surfaces of the Central Plateau are exhumed ones. The mid-Tertiary uplift of the Plateau was greatest in the south-east, producing the present Cevennes escarpment, but in the north-west there was little uplift or deformation and the landscape dips gently in that direction.

Prior to dealing with this Pliocene and later erosional history in the south-eastern region of the Mediterranean Border, Baulig (1928a: 34–59) classifies the topographic forms produced by the mid-Tertiary and earlier events into constructional, dislocated, structural and cyclic erosional.

Constructed forms consist mainly of volcanic accumulations and the terminal surfaces of alluvial terraces or infillings. The latter should be identified on at least three characteristics: a considerable thickness of alluvium situated about maximum flood-level; patches or remnants of alluviation at a height in agreement with the general slope upstream and downstream; and extensions of the terrace, of the same nature, in the lower part of the valleys of adjacent tributaries. With regard to the height concordance, Baulig affirms that the terminal surface of fluvial upbuilding (*remblaiement*) is normally an equilibrium surface constructed by the watercourse in order to adjust its slope to the characteristics of its regime. But, he adds, the establishment of a profile of equilibrium does not stop the river from modifying its profile during the progress of the cycle.

Dislocated forms include, among others, escarpments caused by flexures and faults. It is difficult to distinguish between fault scarps due directly to earth movements and fault scarps rejuvenated by erosion.

Structural forms depend partly on the nature of the contact of the beds which may be intrusive, tectonic or purely sedimentary as where a hard lower layer underlies a soft upper layer (called *surfaces*, or *platformes structurales* by Nöe and Margerie (1888: 113, 117)). Much space is given to discussing fossil erosion surfaces and the author considers that strong indifference of the hydrography to structure and the presence of river meanders incised into hard rocks (as elaborated by Vacher (1909)) favour the former existence of a relatively soft surface cover.

Forms of cyclic erosion are fluvial erosion forms which have experienced lowerings of baselevel that set in train the long-recognized cycles associated with the upstream migration of nickpoints and valley-in-valley features:

> The only way to date cyclic forms with any certainty is to attribute them to the age of their origin. Thus ... each cycle has the age of its base level.
>
> (Baulig 1928a: 47)

The first clear description of the succession and inward incision (*emboîtement*) of cyclic forms in valleys is said to be that by Albert Heim (1878: 298–9). Baulig here quotes from Martonne (1911: 4):

> One must choose between two approaches: that of A. Heim and his pupils who classify terraces according to their absolute height, and suppose that the relative altitude of old valleys increases upstream; and that of A. Penck and Ed. Brückner who suppose, on the contrary, that old valleys tend to wear down their beds in relation to the same baselevel.

In other words, one must assume either stability of baselevel and repeated uplifts of the continental mass or stability of the landmass and displacements of sea level. Each hypothesis could hold at different epochs in the same region.

Baulig goes on to consider the long profile of streams and the use of them and of transverse sections in the analysis of landforms. They may be obtained from direct levelling or constructed from existing maps (figure 8.2). In France the long profiles of many rivers had been surveyed by Departmental surveyors since 1856 – a year of great floods – and especially since 1914 in connection with hydraulic power schemes. Baulig makes elaborate use of these findings which are shown graphically in impressive diagrams, made all the more impressive by the relatively insignificant need to bolster them with extra readings from the aneroid barometer.

There follows a long critique of the previous geomorphological investigations of the Central Plateau and detailed descriptions of the 'distinctive' regions composing it and its borders (Baulig 1928a: 81–512).

The first complete geomorphological study of part of the Plateau was produced on *Limousin* by a geographer, A. Demangeon in 1910. In this large homogeneous region he distinguished landforms related to three erosional cycles: the mountain with a higher rim and isolated summits which are the remnants of a topography developed in a first cycle; the wide platform or plateau which is surmounted by the mountains and which corresponds to a second cycle during which the first was dissected; and the valleys of a third cycle which in the crystalline areas have not passed youth but on the softer beds have reached full maturity:

> As to the age of the landforms, Demangeon recognised that the surface of the plateau passes under the oldest lava flows of Mont Dore: hence it was formed between the end of the Oligocene and the Miocene. The succession of later cycles results entirely from mass movements of the region [*Limousin*] without dislocations in the real sense of that word.
>
> (Baulig 1928a: 75–6)

One physical problem concerns the origin of the Plateau. Is it a product of marine or of continental erosion? In any case, the present drainage pattern on its ancient rocks is indifferent to underlying structure, and abounds in incised meanders and other features which are most satisfactorily explained by assuming a former, almost continuous cover of softer rocks. Another problem concerns the valleys in which Demangeon found evidence of three stages or minor cycles and Baulig finds one or two more:

> If . . . the high thalwegs [valley-floor long profiles] are prolonged downstream with a gently-flattning slope, they pass well above the surface of the plateau. Two explanations of this come to mind: either, as Demangeon admits, the thalwegs are older than the Eogene plateaux, *or*, more probably, downstream they were developed on a sedimentary cover which has since disappeared with them.
>
> (Baulig 1928a: 132)

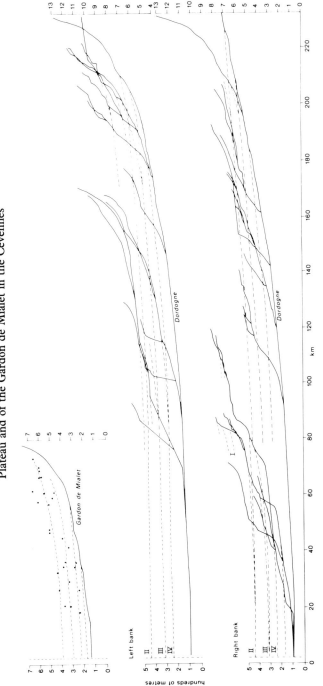

Figure 8.2 Long profiles and extrapolated baselevels of the left bank and right bank tributaries of the River Dordogne in the Central Plateau and of the Gardon de Mialet in the Cevennes

Source: After Baulig (1928a), Plate I. Courtesy Armand Colin, Paris

There follows a detailed account of the physiography of the *Auvergne* (Baulig 1928a: 185–235) but we must, neglecting volcanism and fracturing, content ourselves here with the brief remark that the Oligocene beds are best represented in the *limagne* of the Allier River and that

> in spite of inadequacy of data and difficulties of interpreting them, the incision and in-filling during the Pliocene in the valleys of the Loire and Allier is better explained not by local movements but by general movements of the ground or, preferably, by variations in the height of sea-level.
>
> (Baulig 1928a: 235)

Before discussing in detail the remaining regions of the Central Plateau and its borders (1928a: 236–450) Baulig summarizes the second important memoir on part of it, namely A. Briquet's *Sur la morphologie de la partie médiane et orientale du Massif Central* (1911). This survey covers the most complex areas of the Massif since it includes the *limagne* and the marginal horsts as well as extensive volcanic terrains (in Auvergne), the south-eastern edge of the main Plateau and the adjacent Rhône corridor.

Briquet recognized traces of four cycles of erosion, irrespective of the scanty remnants of a later cycle. The oldest cycle is expressed in a vast 'peneplain' or platform which truncated most of the Hercynian massif older than itself. Earth movements caused this peneplain to be deformed and initiated a new cycle, which on the Mediterranean drainage worked in conjunction with a rapidly lowered sea level. Repeated earth movements initiated third and fourth cycles, each of which led to the incision of rejuvenated valleys and the creation of valley-in-valley forms. The first three cycles were thought by Briquet to correspond to the three recognized by Demangeon in Limousin.

Baulig expressed misgivings on some of the geomorphological methods used:

> In analysing relief forms, Briquet employs, alternatively, two methods, which might be called morphological and stratigraphical. The first is based on the determination of topographic continuities and discontinuities, on terracing and valley-in-valley features.
>
> The second uses the obviously correct principle that an erosional surface is younger than the terrain it truncates and older than any beds which cover it. But Briquet sometimes misapplies this principle when he assumes that an erosional surface has the age of the deposits covering it. . . .
>
> He also abuses the stratigraphical principle when he assumes [wrongly] that two erosional surfaces covered by deposits of the same age are related to the same cycle.
>
> (Baulig 1928a: 78)

In addition, there are errors of detail, particularly in underestimating the number of cycles experienced in the valleys, especially in the Cevennes:

The reduction in the number of cycles in the valleys leads to the fatal error of making the slope of the upper more mature thalwegs steeper than that of the lower less mature thalwegs. Hence the inevitable conclusion that the region has been deformed during the course of the cyclic series whereas a more exact analysis shows that the thalwegs are not appreciably deformed.

(Baulig 1928a 79)

Because of these misinterpretations and inexactitudes Baulig finds that Briquet's work, in spite of its excellent grasp of the constituent elements, is not definitive.

The long section in Baulig's main volume on the *Mediterranean Border* (pp. 410–544) is of particular interest to geomorphologists as it contains chapters on Pliocene and present landforms and his mature thoughts on eustatism. The summary of Pliocene sedimentation leans heavily on M. Gignoux's *Géologie Stratigraphique* (1926) and is followed by an attempt to decide on the probable maximum altitude of Pliocene sea level, in which we read:

preserved marine deposits give a minimum value only for the height of the Pliocene Mediterranean; but, on the other hand, the presence of extensive undeformed erosion surfaces up to nearly 400 m implies base levels of about that altitude; that one of these levels, that at 280 m which exists in various regions, is dated in the Sahel of Algiers by the fact that its horizontal surface cuts across the older inclined Pliocene, intimating that it (and levels which succeed it) is later than the old Pliocene; that because the level of the base of the oldest Quaternary (Sicilian), did not apparently exceed 100 m above the present zero, levels between 280 m and 140 m are Upper Pliocene; that the level at 380 m is concordant with the lower surfaces even in the Sahel where it has been worn down and that it probably corresponds to the maximum altitude of the Pliocene Mediterranean; erosional levels below it mark standstill periods during the regression of the sea at the end of the Pliocene and in the Quaternary.

(Baulig 1928a: 489)

from the upper Pliocene onward, the horizontality of the erosional surfaces preserved on the right bank of the Rhône show that *everything had happened as if only sea-level had varied and the landmass had remained strictly immobile.*

(Baulig 1928a: 513)

Thus, during Pliocene and Quaternary times a series of relatively flat erosional platforms and benches were formed extensively in Languedoc and in the adjacent valleys of the Cevennes predominantly at 380, 280 and 180 m, but additionally at 250 m and 150 m. These benches are linked to graded valley floors (e.g. the Gardon de Mialet; figure 8.2). This fact, together with some irregularity of the bench surfaces and the fluvial character of the associated valleys, led Baulig to postulate a subaerial origin for both, produced under the influence of an intermittent fall of baselevel (Sparks 1972).

The problem, Baulig asserts, is to determine what mechanism caused new cycles related to relative displacement of sea level. Are we to follow an epeirogenic or eustatic or isostatic hypothesis?

The epeirogenic hypothesis (see chapter 1) is based on movements of the whole of the earth's crust (*écorce*). It was almost exclusively accepted among scholars in the United States (see chapter 7) and dominates the thoughts of geologists and geomorphologists in Europe.

The eustatic hypothesis (see chapter 3) is based on variations in the volume or the shape of ocean basins. It was inspired by Suess and in France was defended chiefly by Lamothe and Depéret:

> More recently in Scandinavian countries and in France, a close relationship between certain oscillations of sea-level and in the advance and retreat of great Quaternary glaciers has been established.
>
> (Baulig 1928a: 515)

The isostatic hypothesis rests essentially on the well-established fact that the oceanic parts of the earth's crust have a greater density than that of the continental parts and a hydrostatic equilibrium prevails in a more or less fluid environment. Its verity had been strengthened greatly by the probable isostatic uplift of some glaciated regions after the disappearance of their glaciers. Baulig devotes a whole appendix (pp. 559–62) to the differences between eustasy and isostasy. He mentions the difficulty of deciding whether isostasy works as a quick, continuous reaction to degradation or as an accumulated summary reaction to periods of peneplanation or planation. According to Wegener's theory of sialic continents floating in sima of one-tenth greater density, at least nine-tenths of the continents should surmount the flotation (equilibrium) line. If 100 m were eroded off a continental block, its surface – allowing for isostatic uplift – would be at about 90 m relative to former baselevel:

> In these conditions all planation and even all peneplanation becomes impossible.
>
> (Baulig 1928a: 560)

On the other aspects of this erosion, Baulig borrows from Suess (*La Face de la Terre*, 1900: II: 849), who stated that sedimentation would gradually raise sea level. But the French scholar considered that this infilling would probably not compensate for the isostatic effects already mentioned, because as the oceans cover three times the surface area of the continents, the removal of a continental layer 100 m thick would raise sea level by about 30 m only.

Moreover, there is the very important fact that continental erosion operates at different rates in different regions, so it is highly improbable that isostatic uplift consequent upon the same cycle of erosion would have the same amplitude in different regions. Therefore, 'we are neither for nor against the isostatic hypothesis in itself' (Baulig 1928a: 561–2); isostasy will simulate eustasy only exceptionally and on a restricted scale.

Baulig now returns to the eustatic hypothesis which seems to him simpler than isostasy and to be supported by the fluvial terraces and exposed shorelines demonstrated by Lamothe and Depéret in the Mediterranean and Atlantic with reasonable certainty up to about 60–65 m and decreasing certainty with increasing altitude above that height. However, if the higher erosional levels can be established as eustatic, the same explanation may be applied to lower levels. When a series of levels is clear cut and occurs in various regions, agreement is probably not fortuitous. However, it is necessary to establish, if possible, the synchronization of levels of the same altitude (Baulig 1928a: 517).

Eustatism only demands that some regions remain strictly stable for appreciable periods. It presupposes, rather than opposes, deformations of the lithosphere in so far as it depends on deformations to cause variations in the volume of the oceans:

> The eustatic development of stable continental regions is controlled essentially by the intermittent deformations of unstable regions continental or oceanic.
>
> (Baulig 1928a: 521)

The major movements of the earth's crust capable of modifying sea level may well be intermittent (in the same way as eustatic changes are intermittent). However, there is no reason why local tectonic movements of a restricted height and contrary nature should not occur.

Baulig now proceeds to press his case in other parts of France and in doing so reveals the pit-falls of eustatism and his skill as a geomorphologist and as an advocate. Roussillon has levellings at 150 m up to 225 m but the most obvious old shorelines are at 100 m and 280 m. Elsewhere, in the Sahel of Algiers, General Lamothe's work, which we have already described (see chapter 3), comes under severe scrutiny. The terraced steps of a constant height:

> carry conglomerates with marine fossils, gravels with very round quartz pebbles and consolidated dunes. The author [Lamothe] considers the steps to be old littoral platforms . . . and assigns them to 325, 265, 204, 148, 103, 60, 31, and 18 m above present sea level. Because these shorelines (*rivages*) are horizontal and occur at the same heights elsewhere on the Algerian coast, he concludes that they were formed by a purely eustatic process. All the levels, even the highest which cuts into the inclined old Pliocene . . . is 'post-Pliocene' (Upper Pliocene and Quaternary).
>
> (Baulig 1928a: 523)

For Baulig these findings are of greater interest for the problem of eustatism generally and for the identification of erosional levels on the other side of the Mediterranean in the Rhône valley area in Bas Languedoc, where, as he has already argued, there no longer exist any surfaces of general levelling or littoral erosion above the 180 m cyclic level (pp. 443, 489). He considers that whereas the lower lines of levels identified by Lamothe in Algeria agree with the traditionally accepted alluvial terraces, the breaks of slope above 148 m agree only approximately with erosional levels in Bas Languedoc and the correlation decreases rapidly with increasing height. The shorelines of the Sahel lie at a greater height than the plains or peneplains of *fluvial* erosion which seem to correspond to them on the northern side of the Mediterranean. Perhaps Lamothe's eustatic assumption for the higher levels is weaker than it appears at first sight? Certainly the old shorelines he has traced on the map need revision in places and the talus accumulations and upstanding knolls found on levellings above 148 m indicate that they are not solely platforms of littoral erosion. Moreover, he presumed the presence of the sea and the position of old shorelines are based largely on the existence of a beach, gravel, sands and sandstones, especially aeolian, and shelly conglomerates formed in shallow water offshore. But in fact there is nothing in the composition of the gravel masses that indisputably indicates a marine origin and the fossiliferous conglomerates are often too fragmentary and too sparse to be useful or decisive at some of the highest levels. So Baulig comes to the conclusion that the steps in the Sahel at least above 140 m height correspond to the erosional levels in Bas Languedoc but that these higher levels have nothing to indicate that they might not have been exclusively fashioned by fluvial erosion (p. 534).

However, the eustatic importance is reinstated in the Armorican Massif (pp. 537–42) and in the Paris Basin (pp. 542–4). The *recent* evolution of the relief of Armorica is said to have been fashioned in a purely eustatic way, as is shown by the use of a graphical method based on the frequency of spot or *summit* heights (see Baulig 1922, 1926a, b). In the Breton peninsula this method reveals more frequent summit maxima at levels a few metres either side of 290 m, 190 m, 165 m, 125 m, 104 m, 90 m (very clearly), 65 m (very clearly), 45 m, 25 m (doubtful) and 10 m (doubtful).

With regard to the Paris Basin, Baulig (1928b) has already discussed the heights of its eustatic erosion levels. Its relief is characterized by a great development of structural surfaces or of fossil surfaces more or less completely exhumed (for example, the surface of the clay-with-flints). Cyclic surfaces occupy a restricted area only. Where obvious flattenings, continuous and practically level, do occur they are always at critical altitudes – particularly at 180 m and 280 m – and are independent of structural or fossil surfaces.

Baulig, apart from his many appendices, numerous photographs and wealth of profiles and structural maps, ends his text with a plea for eustatism:

> regions, widely-spaced and totally independent from a structural viewpoint, show perfectly clearly an exactly similar geomorphological development since the Upper Pliocene. This similarity, in the present state of ideas and knowledge, admits of only one explanation: that is eustasy pure and simple.
>
> (Baulig 1928a: 543)

Baulig's Cross-channel Excursion

In October (9th–16th) 1933 Baulig repeated his theme of the importance of variations of sea level in a series of lectures at the University of London. These were published as *The Changing Sea Level* in 1935 and re-issued in 1956. Anyone familiar with *Le Plateau Central de la France* will find this summary rather repetitive and certainly more tentative but, although a few scholars found the talks not to their liking, the great majority enjoyed hearing the 56-year-old professor at the height of his powers, arguing incisively on a subject attuned to predominant British geomorphological thought. We have already dealt with eustasy in chapter 3 and in the pages immediately above, and Baulig's first lecture, entitled *Critical Retrospect*, will serve us well as a recapitulation. After some elementary observations we plunge (pp. 2–4) into glacio-eustatism which originated with an American geologist, Charles Maclaren, who in 1842 suggested that sea level would be lowered or raised between 350 and 700 feet (100 and 200 m) by the accumulation or melting of Pleistocene ice-sheets. In 1869, an English geologist, Alfred Tylor, estimated the glacial shift of sea level to be about 600 feet (180 m) and in the early twentieth century many relevant studies were made, for example, by Georges Dubois (1925, 1931) and by Ernst Antevs (1928) who estimated a glacial-controlled fall of sea level by nearly 100 m during the last maximum glaciation, provided isostatic influences are ignored.

Baulig now turns to his friend and inspirer, Eduard Suess, who attributed eustatic movements to deformations of ocean basins and who believed that Alpine mountains were generated by tangential or compressional thrusts, which led to folding and eventually to the broad uplift of folded belts (Suess 1875). Once the lateral push had ceased, the affected mass would begin to sink under its own weight and during geological time, broadly speaking, the oceans have tended to deepen and sea level to drop repeatedly. These Suessian views on orogenesis failed to be widely accepted in Europe and were strongly opposed in the United States:

> Thus Suess's eustatism – the belated echo of bygone controversies – remained a purely theoretic conception, nay, a personal opinion, until Depéret and de Lamothe revived it in a more precise and limited form.
>
> (Baulig 1935: 5)

These French geologists, with the aid of Gignoux's palaeontological work (1913, 1922) and Chaput's study of the alluvial terraces in the lower Loire basin (1917, 1919), were thought by Baulig to have demonstrated that the evolution of the western Mediterranean in post-Pliocene times fitted in well with the eustatic theory. However, he admits that these eustatic findings cannot be applied to certain very unstable areas and were, moreover, largely opposed by Davis (1905K) and others in the United States. There Douglas Johnson (1919, 1925, 1931a, b), after treating in detail the recognition and restoration of former shorelines, had come to the conclusion that existing evidence is inadequate to establish the validity of any single theory on sea-level changes (see chapter 3). With typical foresight, or perhaps because of criticism of his book on the Central Plateau of France, Baulig adds that glacio-eustasy is so complex that without further detailed examination it must be assumed that shoreline features invariably succeed one another in strict descending order of both age and altitude.

He warns also against the overfacile interpretation of river terraces. Alluvial terraces cannot be related directly to the present river unless it too has a continuous floodplain composed of its own alluvium, or in other terms has a profile graded to equilibrium. A watercourse that formerly experienced a 'glacial' regime will develop a different profile in post-glacial times when its upper course will tend to dissect the 'glacial' floodplain into terraces while its lower course will be buried beneath late glacial or post-glacial infill (figure 8.3). This glacial influence was neglected by Depéret and expressly denied by Lamothe. In any event there can be no doubt that the most reliable basis for the reconstruction of former river profiles and of coastal profiles or changes of sea level are terraces or benches cut in hard rock.

Figure 8.3 The long profile of a glacial river: profiles before (R–W), during (W) and after (p–W) the last (Würm) glaciation

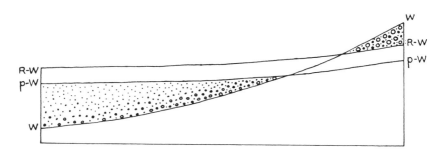

Source: From Baulig (1935), Figure 1, p. 9. Courtesy Institute of British Geographers.

In his second lecture Baulig (1935: 13–24) deals with new lines of research on high levels of erosion:

> The valley *sides* themselves, *when graded*, afford another basis for the restoration of former river profiles.
>
> (Baulig 1935:13)

The existence of valley-in-valley forms carries distinctly cyclic suggestions. Each nickpoint works its way gradually headward until eventually almost all traces of the valley incision associated with it disappear from the drainage system. But, irrespective of the number and duration of past cycles, the only cycles represented in a *limited section* of a valley are those '*which have reached a decidedly more advanced stage of development than any subsequent cycle*' (p. 14). This idea provides a means of reconstructing former river profiles from graded portions of the valley sides, that is those which show a regularly decreasing curvature. If the lower slopes of each cyclic valley are extended, with decreasing curvature, to the approximate axis of the valley, the altitude of the river bed towards the close of the corresponding cycle may be determined. Obviously this method is liable to error and normally shows only the main levels distinctly.

Longitudinal profiles offer hopes of finding evidence of nickpoints, some of which may be pre-Pliocene. Ideally the restored long profiles should harmonize with the restored cross-sections, just as the breaks of slope in the profiles of tributaries should, within reason, harmonize with those in the profiles of the main rivers. The final analysis will entail 'a certain amount of personal interpretation' (p. 16) which will be limited because:

1 All profiles concerning the same cycle must be harmonized to form a 'harmonic whole';
2 In any section of the same valley, the successive profiles are mutually dependent and, if distortion of the land is absent, the upper, more mature profiles should be less steep in each section than the lower;
3 Where distortion has occurred, it seems reasonable to assume that former levels, in the same section of the course, will have experienced at least the same amount of distortion.

The profile analysis method depends for its value partly on the character of the topography, the accuracy of the available maps, and the conscientiousness of the workers. Martonne (1911) suggested that the very high levels in the French Alps were tilted downstream. Baulig in the Central Plateau found a purely eustatic interpretation adequate in the suggested restoration of ancient profiles, although his opinion could not be considered final because of lack of adequate cross-sections and of precise profiles for most of the tributaries. On the eastern side of the Vosges, the upper levels may be tilted slightly downstream towards the Rhine rift valley but below about 500 m the restored profiles showed no signs of disturbance in spite of their proximity to the edge of the Rhine *graben* (Despois 1926, Sittig 1933).

Elsewhere, as Baulig has already explained, the belt of sedimentary rocks bordering the Massif Central, west of the Rhône and north of the Gulf of Lyons, has wide plateaus at 200 m to over 400 m which, however, show clear flattenings within a few metres of 180, 280 and 380 m respectively. Are these surfaces peneplains (the product of subaerial agencies) or platforms of marine abrasion? Since marine deposits and possible old raised shorelines have not been found Baulig assumes that:

the platforms of Lower Languedoc, in their present state, are surfaces of fluvial erosion.

(Baulig 1935: 20)

But we have yet to take account of events in the region lying east of the Central Plateau, a borderland which fronts southward to the sea and eastward to the Rhône valley in whose drainage it lies. Here the platforms are at an almost constant altitude whereas, if they had developed in relation to the Rhône, they should rise upstream at about the same gradient as the river itself. The deduction is that, when the flat surfaces in question were formed, the Rhône was not there, having been replaced by the sea. A hypothetical chronology would be: at the start of the Pliocene 'for some reason, either rapid rise of the land or sudden sinking of the sea level, but more probably the latter' (p. 20), the Rhône quickly incised its main valley and the lower courses of its tributaries, leaving their upper courses hanging above the rejuvenated stretches. Then a rapid rise of sea level drowned the Rhône valley almost up to Lyons and turned the tributary valleys into rias, in some of which existing deposits of pure marine clay indicate fairly deep water. A maximum rise of sea level of 400 m does not seem improbable:

> Thus we are led to the provisional conclusion that the platforms of Languedoc were developed in relation to successively lower levels of the Pliocene sea.
>
> (Baulig 1935: 21)

Similar levellings have been found at about 380, 280 and 180 m and attributed to the Upper Pliocene in the Sahel of Algiers, and in Dobruja, a plateau district south of the Danube delta (Nordon 1930).

Baulig now considers some of the results of his cartographic analyses of the distribution of spotheights. On the Armorican plateau flattenings were frequent especially at 180 m and 280 m; in the Paris Basin the chief levellings – of the hard rocks on the higher dip slopes of the scarplands – occur almost invariably at about 180, 280 and 380 m. The 'startling and confessedly heretical proposition' that the Pliocene sea had once occupied much of the Paris Basin was put forward in spite of the absence of known marine deposits. In Brittany the presence of tiny grains of glauconite (a mineral of marine formation) in Pliocene sands suggests the presence of the sea up to over 100 m above present sea level.

Baulig gives other examples of his levels but wisely adds that all these findings together are far from definitely establishing the eustatic hypothesis, although the observations on which it is based have not yet been disproved. It seems best to investigate whether features characteristic of eustasy could be created by movements of the earth's crust, epirogenic or isostatic (Baulig 1935: 25–32).

A problem arises here. American geologists prefer epeirogeny to be distinguished from orogeny, whereas most of their European counterparts consider that tiltings and bendings of sedimentary strata and warpings of peneplains are the result of orogenic disturbances. However, according to Baulig, both schools invariably associate differential vertical movement with both phenomena. In this respect, in spite of his familiarity with American geomorphological thinking due to long residence in the United States, we must differ from him. G. K. Gilbert (1890) coined the term epeirogeny to express the process of continental formation and to distinguish it from orogeny or mountain formation (see chapter 1). He and his compatriots were not concerned with absolute differential uplift but with the uniformity, gentleness and widespread nature of the vertical movement. For example, Douglas Johnson (1931a: 51) considered that 'ancient (marine) levels result from uniform uplift of a large block of the earth's crust'. In fact, Baulig takes a regional viewpoint, an outlook typical of the inhabitants of an untypically fragmented, peninsular-indented, west European landmass. He talks of 'epeirogenic uplifts' of small portions of the earth's crust (p. 26) which may partake of the nature of folds and he seeks the location of the outer belts of displacement at the edges of the Pliocene platforms of Languedoc or the Paris Basin. He suggests that wholesale uplift of wide areas of land 'thousands of kilometres across' would result in definite fold lines and become orogenic and that uparching due to tangential pressure would eventually entail the transference of subcrustal material to beneath the uplifted block. If uplift affected all the continents there would be a general transference of material from the oceans to beneath the continents and the ocean floor would sink, causing, because the oceanic area is roughly three times that of the land-masses, an absolute fall of sea level of about one-third of the amount of continental elevation.

In fact since Late Tertiary times the continents may have become unusually elevated above sea level (Barrell 1917). Their present altitude averages about 875 m and much of this seems attributable to the Late Tertiary and Quaternary. If allowance is made for surface erosion of the land, it must be assumed that the present altitude of the land incorporated 'a total drop of sea level amounting to hundreds of metres' (p. 27). But this idea may be taken a step farther. The existence of three main levels of erosion – as in Languedoc – implies three periods of practically stationary sea level and seems to indicate that shifts of marine baselevel in recent geological times have been few, rapid and large:

This apparently signifies simultaneous re-arrangement of large segments of the earth's crust, as if in response to disturbed equilibrium, and leads us to the discussion of isostasy and some of its morphological consequences.

(Baulig 1935: 27)

Here we may interpolate into Baulig's argument the fact that many geologists long after G. K. Gilbert felt a need for epeirogeny of some kind. For instance

Otto Jessen (1943) wrote of *Randschwellen* and Lester King (1962) coined the term *cymatogeny* to express a condition intermediate between epeirogeny – the broad relatively uniform uplift of large areas such as the Canadian Shield – and orogeny, the creation of mountains, especially in linear belts of deformed rocks. Cymatogeny is the heaving – arching and occasional doming – of large undulations, 'scores and sometimes hundreds of miles' across, in the earth's crust, in which there may be considerable uplift but rock deformation is at a minimum (King 1968: 240–2).

However, to return to isostasy, Baulig outlines the accepted quality of the concept, a task we have already attempted (see chapter 1). He then quantifies a simplified and idealistic example applied to two columns, one of ocean (B), the other of land (A), but each of density 0.9 and 'floating' freely on a medium of density unity. Suppose a rock layer 100 m thick is eroded from A and deposited on B. When isostatic effects have been completed a fixed mark on the land (A) will appear to have risen by 100×0.9 m (90 m) and the sea level by only 3 m, being about 33 m rise for deposition on a sea floor three times the area of the land, and about 30 m (33×0.9 m) drop due to isostatic depression of the sea floor by weight of sediment. Consequently the rise of sea level which actually affected the landmass would seem to be 87 m or, allowing for better estimations of the densities involved, at least 80 m. But, in reality, it seems highly improbable that isostasy works perfectly and continuously. Rather it may await critical pressure stress upon its elastic resistance and, Baulig surmises, might vary in potency as, for example, through variations in radioactive emission whereby the inner earth may undergo alternate but trifling phases of contraction (favourable to crustal stability) and of expansion (favourable to crustal instability). Such variations might to some extent simulate negative eustatic shifts of sea level.

He then confronts the problem of whether isostasy and erosion combined would suffice to cause a baselevel variation as large as the 380 m, which presumably has occurred since Late Pliocene times. If the estimated rate of denudation of the United States, 1 inch in 760 years (Dole and Stabler 1909, Baulig 1910), is accepted as a rough measure of continental erosion, about 33 m of surface would have been removed in the 1 million years attributed to the Pleistocene. During that period, the valleys of Lower Languedoc were deepened about 100 m but were not widened substantially in hard rocks. Yet as each of the 380, 280 and 180 m cycles caused distinct peneplanation of the same hard rocks:

> we are perhaps not far amiss in assigning a duration of at least five million years to each of these cycles. Adding one million years for each 100-metre drop from one base level to the next ... we may conclude that that part of Pliocene times beginning with the 380-metre cycle can hardly have lasted less than eighteen million years.

> (Baulig 1935: 30)

During this time, erosion at the present rate would have removed about 594 m of the land surface, resulting isostatically in a relative rise of 594 × 0.8 m (475 m) or more. But the apparent drop of sea level from the 380 m to the 100m level, which is considered by Gignoux to be the highest level of the Pleistocene sea, includes some 50 m attributable to the formation of small Pliocene polar ice-caps, leaving about 230 m to be accounted for otherwise. So the statistics given for Pliocene denudation could be doubled and would still provide enough scope for isostatic adjustment in conjunction with erosion to produce, *on an average*, the required effect.

But because erosion works at very different rates in different regions is it likely that the isostatic adjustment it stimulates will produce uniform land uplifts which simulate eustatic lowerings of sea level? Much will depend on the investigator's concept of isostasy, on whether it is considered local or regional or continental.

If at this stage of the argument we should feel inclined 'to throw overboard either isostasy or, more probably eustatism' (p. 31), Baulig suggests that we should first consider the following observed facts and reasonable inferences:

1 During the Tertiary and Quaternary periods great changes in the relative level of the land and the sea have resulted in abnormally high landmasses.
2 Epeirogenic and isostatic movements appear correlative.
3 'During the period in question', major changes in the relative position of land and sea must have been approximately synchronous globally. But the topography of Lower Languedoc indicates that most of the Pliocene was taken up in the modelling of three main platforms of erosion appertaining to three prolonged phases of stable sea level. Presumably these phases prevailed globally irrespective of how *the landmasses behaved locally* (p. 32):

> The major changes of level, affecting both land and sea, must be placed in the intervening periods.

4 Because these intervening periods were relatively short they offered, other things being equal (p. 32),

> no opportunity for the development *anywhere on the globe* of peneplains comparable in extent and perfection to those of Lower Languedoc.

Elsewhere the record will vary according to local conditions, such as rock resistance, erosional efficiency, proximity to the sea or baselevel, and even instability of landmass.

These considerations may encourage the required attempts to obtain world-wide correlations from stable regions not far from the sea. In any event (p. 32):

> the search for eustatic – or pseudo-eustatic – levels of erosion appears promising; it has already brought to light some new and unexpected facts and will probably reveal more.

Figure 8.4 Complex polygenetic surfaces developed on a granitic massif. A: Intersecting buried fossil peneplains forming a faceted surface. An initial horizontal erosion surface (A) was covered by Cretaceous rocks. This was tilted, and a second horizontal erosion surface (B) formed which was then covered by Eocene rocks. Tilting was resumed and a third horizontal erosion surface (C) formed which was then covered by Oligocene rocks. After further tilting, a Miocene erosion surface was cut intersecting both the Oligocene and the oldest granitic rocks. Removal of the remaining Cretaceous-Oligocene strata would reveal exhumed (resurrected) erosional facets. B: A landscape composed of a dome dissected by three cycles of valley cutting controlled by an intermittent eustatic drop in baselevel. An erosion surface was formed on the granite, which was buried by sedimentary strata (stippled) and then domed up. At this time sea level stood close to the top of the dome and as it fell by stages (I', II', III') most of the sedimentary cover was eroded away, together with a staircase of valleys cut into the granite, and parts of the original erosion surface were exhumed. The present valley bottoms and interfluves are shown in solid lines. The three interfluve areas of exhumed surface (E), the polycyclic river profiles (I–II, II–III, III–III'), the corresponding marginal granitic benches (I'', II'', III'') and the flanking sedimentary platform (P) are shown

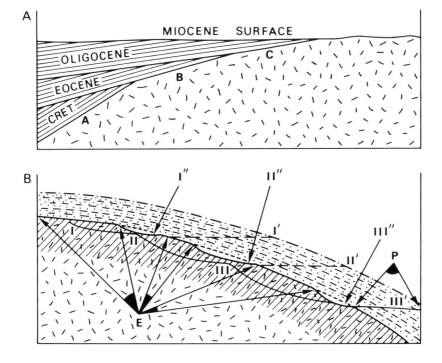

Source: After Baulig, (1928a), Figures 9 and 10, pp. 39 and 40. Courtesy Armand Colin, Paris

In his fourth lecture (pp. 33–46) Baulig concentrated on 'the search for more facts' and discussed difficulties of geomorphological and geological interpretation. In searching for secure evidence of eustasy, avoid seismic areas, glaciated isostatic regions and recently folded and faulted tracts; do not confuse true fault scarps with 'erosional fault-scarps' (as defined by Spurr, (1901)), nor consider that marine deposits are necessarily proof of a marine origin for the bench they rest on; make use of accurate river profiles, when in equilibrium, to indicate stability or no differential movement and of 'platforms of erosion', if not deformed, to indicate former baselevels; face up to the difficulty of distinguishing between structural platforms and cyclic platforms and to the still greater difficulties of intersecting peneplains and resurrected surfaces (figure 8.4).

In difficult cases we are advised to use the statistical method of spot-height frequencies for compiling frequency curves but there is always some vagueness in geological stratigraphy and

Truth in geomorphology, indeed, is seldom more than increasing probability.

(Baulig 1935: 44)

Correlations of ancient levels on the basis of continuity and of similar altitude are always meeting with gaps and deformities over long distances.

Baulig concludes his analysis with a study of four imaginary 'extreme possibilities and intermediate cases', and ends with:

Whatever the final outcome, if only not entirely negative, . . . research . . . will lead to important conclusions, the bearing of which on geology, geomorphology, and geophysics we can only surmise.

(Baulig 1935: 46)

Thus, to us as critics, Baulig's efforts on behalf of eustasy ended not with a bang but a whimper. Some people found his approach cautious and scholarly, while others found it insubstantial and dithering.

A true patriot, he served his country in the First World War, and when transferred with Strasbourg University to Clermont Ferrand during the Second World War he was arrested by the Gestapo early in 1944 for his work for distressed students and for the 'frank expression of his views'. With peace, he returned to Strasbourg in 1945 and 2 years later retired to enjoy a most productive old age. In later life he expressed frustration at the persistent disagreements, national bias and growing confusion which he saw as hampering the progress of eustatic studies. Despite his early use of statistical methods in terrain analysis, he opposed the development of the 'new' quantitive geomorphology in France and it must be admitted that his critical and opinionated approach in later years inhibited landform studies in that country. Baulig survived until 1962 when, aged 85, in the manner of old soldiers, he simply faded away.

Baulig's Critics in France

Although Tricart, Baulig's successor at Strasbourg, characterized the period 1910–50 in France as one of blind belief in eustatism, Baulig's *magnum opus* soon evoked a review by Martonne (1929) which was not entirely laudatory. This extensive review praised the precision, vigour and authority of Baulig's denudation chronology and it is clear that Martonne was inclined to accept much of it, providing that its implications could be disentangled from a reliance on the more excessive eustatic movements. Although Martonne could countenance an early Quaternary sea level some 100 m above the present, he baulked at Baulig's belief that the Pontian (Mio-Pliocene) sea level stood 400 m above the present, that it then fell suddenly to 100 m below the present (i.e. involving a drop of 500 m) at the start of the Pliocene and then rose rapidly to 380 m, subsequently to drop in jerks to 280, 180 and finally to 100 m at the start of the Quaternary. Martonne's difficulties in accepting this radical eustatism included:

1 He could envisage no possible mechanism, either glacial or tectonic, to explain 500 m of sea-level change (p. 128).
2 The errors inherent in any reconstruction of ancient thalwegs and, especially, of transverse valley profiles. Martonne believed that even small errors could introduce potentially great discrepancies in projected elevations.
3 The complete absence of high *shoreline* features (i.e. cliffs, littoral deposits or deltas) in the Rhône valley (p. 130).
4 The apparent immunity of the Central Plateau to local Alpine tectonic movements during the supposed eustatic phases of the Pliocene, when it had been so susceptible to them up to and including the Miocene (p. 131).

Martonne (1929: 131–2) concluded:

> One would like to be wrong in disagreeing with such an author, so much does one admire his combination of ingenuity, conscientiousness, commitment and subtlety, because one finds so many facts and interesting and novel ideas on every page of his book.
>
> In conclusion , we are presented with a radical eustatic theory, from which some useful generalizations remain! ...
>
> But it is clear that the author attaches supreme importance to his eustatic theory. It is this theory which has sustained him during years of work, which has led him to advance his researches step by step in order to produce new arguments. A working hypothesis can be a springboard: and the eustatic theory is such in the hands of Baulig. The future can confirm it, by sweeping away the criticisms, which seem to us to be substantial, or in disproving the theory altogether. The eustatic hypothesis will always inspire demanding and productive research.

In reality, the research foreseen by Martonne tended to concentrate on the geomorphic effects of glacial eustatism, rather than on the pre-Quaternary variety which formed the heart of Baulig's theory. As we shall see in chapter 9, it was only in Britain that the latter was enthusiastically taken up. After the Second World War the rise of a new generation of French geomorphologists, together with the general decline in belief in non-glacial eustatism (see chapter 3), led to increased criticism of Baulig's ideas. This was particularly the case in respect of his statistical use of altitudes to identify high-level erosion surfaces and of his attempts to link surfaces above 100 m with purely eustatic mechanisms. An example of such criticism was that by Baulig's successor at Strasbourg, Jean Tricart, whose strong opposition to the cyclic ideas of W. M. Davis embraced any work having a similar flavour. In his *Géomorphologie Structurale* (1968) he devotes a long section to variations in sea level and the position of shorelines, discussing the mechanisms of eustasy and outlining the generally held view of the close relationship between eustatic equilibrium, which acts rapidly, and isostatic equilibrium, which acts slowly. He shows (p. 57) that when the eustatic theory was formulated, it was influenced by the prevailing catastrophic conception of tectonics, such as formed the basis for Davis' simplified cycle of erosion, namely that periods of intense tectonic activity or mountain building were separated from each other by periods of calm. Today, Tricart says, this theory has been gradually abandoned and replaced by one incorporating a much more continuous evolution, 'paroxysms' being separated or followed by periods of lesser tectonic activity. What is more, these paroxysms are now considered to have been regional and not universal and so it is difficult to see how they could have had a great effect on general sea level. Also, improved knowledge of stratigraphy seems to undermine the former belief in worldwide transgressions and regressions. For example, the great transgression during the Middle and Upper Cretaceous, which nineteenth-century scholars considered universal, appears essentially to be restricted to the Hercynian regions of Europe.

The clearest examples of strong marine advances and retreats occurred in the Mediterranean basin during the Neogene, with the Pontian regression and the Plaisancian transgression which followed it. During the regression, erosion had time to cut, in many regions, deep valleys which the transgression converted into rias. But this rather too elegant example is an exceptional case lacking general application; it is purely regional, being particularly tied to the Mediterranean which was a closed sea until the opening of the Strait of Gibraltar at the very end of the Tertiary. Thus, despite the work of Henri Baulig, we can hardly condone the application of the eustatic theory to pre-Quaternary times:

> Eustatic variations of sea level capable of explaining, in conjunction with the Davisian cycle of erosion, staircases (tiers) of flattenings in regions and, worse still, on a global scale, are unrealistic. Their pretended existence is based on illusions nourished with the help of false graphical devices. Such is the case with the

imaginary levellings at 180 m and 280 m which have been attributed to the Pliocene and which in the Paris Basin would have been fashioned by a phantom sea which has not left any deposits.

(Tricart 1968: 58)

Having relegated eustatism, in its extreme interpretation, to the rank of useless theories, Tricart goes on to praise glacial eustatism which, in conjunction with isostasy, he sees as having been most effective in causing repeated changes of relative sea level throughout the Quaternary period.

Foreign Responses to Baulig's Eustasy

Apart from in Britain, the tectonic preoccupations of much of the rest of Europe were not conducive to a sympathetic reception of Baulig's eustasy. We have already noted in chapter 3 the opposition of Davis and his supporters to the eustatic theory and in chapter 7 the underlying belief in the production of baselevel changes in the Appalachians and elsewhere by epeirogenic movements. It was natural that the major American review of Baulig's work on the Central Plateau should have come from Douglas Johnson (1929), an expert on matters to do with sea level and whose important work on the denudation chronology of the Appalachians was under way. Johnson's treatment of the more extravagant aspects of Baulig's eustatism was strangely muted but, as usual, his criticisms were very much to the point:

1 Baulig had failed to correlate convincingly the reconstructed stream profiles in the Plateau with the erosional platforms on its south-eastern border.
2 There are great dangers in trying to decipher erosional cycles from stream profiles (which are also affected by such non-cyclic causes as rock variations, tectonic movements, stream capture, etc.), especially when correlations and interpretations are made by projecting the stream profiles 'through the air for great distances'.
3 Baulig's eustatic theory stands or falls on the existence of erosional platforms on the south-eastern borderland. The extent of these is not shown on any map, their exact surface form is not given, neither is the amount or direction of their slope nor the extent of their preservation or of destruction by subsequent stream erosion. The book as a whole is difficult to follow and the arguments obscure because it is so poorly illustrated.
4 It is not clear how these platforms have been defined in terms of elevation, in order that one should know within what limits possible warping of the supposedly horizontal platforms might have occurred without this warping having been detected. While the platforms are said to maintain a rigorous horizontality, 'the altitude figures given by the author vary so widely that it

is not clear how this supposed rigid horizontality was determined'. Indeed, the 'mechanical regularity' of the spacing of the three major platforms suggested to Johnson that they were approximations and raised doubts as to the accuracy of their identification on the dissected terrain (Johnson 1929: 666).

5 Alternative theories, especially the epeirogenic theory, were dismissed by Baulig without discussion. For example, the case for broad uparching of the Plateau was not considered.

6 Baulig's use of the lower eustatic levels (i.e. below 150 m), which had been identified by previous workers, was irrelevant to the discussion of the high Pliocene levels.

Despite these criticisms, which were typical of the American view of eustatic geomorphology, Johnson recognized the work as a sophisticated thesis of high scientific rank. To him, as to Martonne, it was clearly a radical work:

> If successive cycles are propagated into a still-standing continental mass as 'waves of erosion' in such a manner that today a Pliocene cycle of erosion may yet be in full swing at high levels and there attacking deposits of late glacial age, with cycles of various later dates simultaneously following it inland at lower levels, so as to give erosion planes of different altitudes and different geological dates of origin, *literally produced at the same time*, it is highly important that this history be fully established and that we reexamine Appalachian and other topography with the possibilities of this theory in mind.
>
> (Johnson 1929: 665)

One has only to consult Thornbury's standard American text on the *Principles of Geomorphology*, published a quarter of a century later, to see how little headway Baulig's thesis was to make in that country. His ideas were dealt with in passing mainly in connection with complications of the fluvial cycle, negative movements of baselevel and glacial eustasy. References to Baulig were dwarfed by those to Davis, Johnson and others.

By contrast, Cotton's *Landscape* (1948) gave more prominence to variations in sea level as a motor for the production of polycyclic landforms and referred to Baulig's concept of eustatic lowerings of sea level during which older cycles of erosion continue to operate upstream inland long after newer cycles have begun at lower levels. The effect of this, as Johnson noted, was that a later surface may be dissected by an erosional wave initiated in earlier times. Although Cotton's work contains relatively little on eustasy, it is much more reflective of the British approach to Baulig's ideas than of that of the Americans.

We shall see in chapter 9 how, mainly under the influence of Sidney Wooldridge, Baulig's eustatic ideas found their most congenial home in the British geomorphological establishment until about 1960. Wooldridge arranged Baulig's lecture series at King's College, London, in 1933 and, despite the misgivings of some regarding the 'mischief' which the lectures might cause (see chapter 5), the eustatic theory was soon an established part of British geomorphological thinking. In contrast to Thornbury's textbook, one has only to consult the excellent British text on *Geomorphology* by B. W. Sparks (1959, revised 1972) to appreciate the impact of Baulig's work at all levels of British scholarship. Sparks devoted considerable space to the geomorphic effects of movements of sea level; to the use of long profiles, valley cross-sections and other morphometric information in deducing these movements; and, particularly, to a very complete account of Baulig's work on the Central Plateau. Although Sparks placed more emphasis on the operation of processes and of climatic changes than Baulig had done, there is no concealing his admiration for the latter. Just as Baulig had done, many leading British geomorphologists were to find in aspects of the eustatic theory the promise of a great unifying generalization.

References

Antevs, E. (1928) 'The last glaciation', *American Geographical Society, Research Series* no. 1.

Barrell, J. (1917) 'Rhythms and the measurement of geologic time', *Bulletin of the Geological Society of America* 28: 745–904.

Baulig, H. (1910) 'Écoulement fluvial et denudation d'après les travaux de l'United States Geological Survey', *Annales de Géographie* 19: 385–411.

—— (1922) 'Questions de morphologie vosgienne et rhénane', *Annales de Géographie* 31: 132–54, 385–401.

—— (1925) 'La notion de profil d'équilibre: Histoire et critique', *International Geographical Congress (Cairo)*, vol. 3, pp. 51–63.

—— (1926a) 'Sur un methode altimetrique', *Bulletin de l'Association Géologique de la France* no. 10, pp. 7–9

—— (1926b) 'Le relief de la Haute Belgique', *Annales de Géographie* 35: 206–35.

—— (1928a) *Le Plateau Central de la France et sa Bordure Méditerranéenne: Etude morphologique*, Paris: Colin.

—— (1928b) 'Les hautes niveaux d'érosion eustatique dans le Bassin de Paris', *Annales de Géographie* 37: 289–304, 385–406.

—— (1935) 'The changing sea level', *Institute of British Geographers, Publication* no. 3.

—— (1939) 'Sur les gradins de piedmont', *Journal of Geomorphology* 2: 281–304.

Briquet, A. (1911) 'Sur le morphologie de la partie médiane et orientale du Massif Central', *Annales de Géographie* 20: 30–43, 122–42.

Chaput, E. (1917) 'Recherches sur les terraces alluviales de la Loire et de ses principaux affluents', *Annales de l'Université de Lyon* N. S., vol. 1, fasc. 41, Paris: Bailliére.

—— (1919) 'Les variations de niveau de la Loire', *Annales de Géographie* 28: 81–92.

Cotton, C. A. (1948) *Landscape*, 2nd edn, London: Cambridge University Press.

Demangeon, A. (1910) 'Le relief du Limousin', *Annales de Géographie* 19: 120–49.

Despois, J. (1926) 'Les formes du relief dans les vallées de la Bruche', *Bibliographie Alsacienne, (Strasburg)* 2: 312.

Dole, R. B. and Stabler, H. (1909) 'Denudation', *US Geological Survey, Water Supply Paper* no. 234, pp. 78–93.

Dubois. G. (1925) 'Sur la nature des oscillations de type atlantique des lignes de rivages quaternaires' *Bulletin de la Societé Géologique de la France* 4th ser., 25: 857–78.

—— (1931) 'Essai statistique sur les états glaciaires quaternaires et les états correspondants du niveau marin', *Annales de Géographie* 40: 655–8.

Gignoux, M. (1913) 'Les formations marines pliocènes et quaternaires de l'Italie du sud et de la Sicile', *Annales de la Université de Lyon* N. S., 1: (36).

—— (1922) 'Les rivages et les faunes des mers pliocènes et quaternaires dans la Méditerranée occidentale', *Comptes Rendus du Congrès Géologique Internationale* 3rd fasc.

—— (1926) *Géologie Stratigraphique*, Paris: Masson (4th edn 1950).

Gilbert, G. K. (1890) 'Lake Bonneville', *US Geological Survey Monograph* 1.

Heim, A. (1878) *Untersuchungen über dem Mechanismus der Gebirgsbildung*, 2 vols, Basel.

Jessen, O. (1943) *Die Randschwellen der Kontinente*, Gotha.

Johnson, D. W. (1919) *Shore Processes and Shoreline Development*, New York: Wiley.

—— (1925) *The New England-Acadian Shoreline*, New York: Wiley.

—— (1929) 'The Central Plateau of France: A review', *Geographical Review* 19: 662–7.

—— (1931a) 'The correlation of ancient marine levels' *Comptes Rendus du Congrès Internationale du Géographie (Paris)*, fasc. 1, pp. 42–54.

—— (1931b) 'The correlation of ancient marine levels', *Geographical Review* 22: 294–8.

Juillard, E. and Klein, C. (1980) 'Henri Baulig 1877–1962', *Geographers: Biobibliographical Studies*, vol. 4, London: Mansell, pp. 7–17.

King L. C. (1962) *The Morphology of the Earth*, Edinburgh: Oliver & Boyd.

—— (1968) 'Cymatogeny', in R. W. Fairbridge (ed.) *Encyclopedia of Geomorphology*, *Encyclopedia of Earth Sciences Series*, vol. 3, New York: Reinhold, pp. 240–2.

Martonne, E. de (1909) *Traité de Géographie Physique*, vol. 2, *Le Relief du Sol*, Paris: Colin, (4th edn 1926, pp. 499–1057).

—— (1911) 'L'erosion glaciaire et la formation des vallées alpines', *Annales de Géographie* 20: 1–27.

—— (1929) 'La morphologie du Plateau Central de la France et l'hypothèse eustatique', *Annales de Géographie* 38: 113–32.

Noë, G. D. de la and Margerie, E. de (1888) *Les Formes du Terrain*, Paris.

Nordon, A. (1930) 'Questions de morphologie dobrodgéenne', *Bibliotechque de l'Institut Français, Hautes Etudes Roumanie* 3: 17–32.

Sittig, C. (1933) 'Topographie préglaciaire et topographie glaciaire dans les Vosges alsaciènnes du Sud', *Annales de Géographie* 42: 248–65.

Sparks, B. W. (1972) *Geomorphology*, 2nd edn, London: Longman (1st edn 1959).

Spurr, J. E. (1901) 'Origin and structure of the Basin Ranges', *Bulletin of the Geological Society of America* 12: 217–70.

Suess, E. (1875) *Die Enstehung der Alpen*, Wien: Braumüller.

—— (1900) *La Face de la Terre*, vol. 2, trans. E. de Margerie, Paris: Colin.

Thornbury, W. D. (1954) *Principles of Geomorphology*, New York: Wiley.

Tricart, J. (1968) *Géomorphologie Structurale*, Paris: Société d'Edition d'Enseignement Supérieur (translated into English by S. H. Beaver and E. Derbyshire (1974) as *Structural Geomorphology*, London: Longman).

Vacher, A. (1909) 'Rivières à méandres et terrains à méandres', *Annales de Géographie* 18: 311–17.

CHAPTER NINE

The Subaerial–Marine Compositors

The composite approach to landform evolution, which dominated British geomorphology during the half-century preceding the Second World War, not only grew out of earlier native studies (see volume 1, pp. 391–401, 403–4, 411–17) but was fired by external influences which were characterized by two important visits. The first of these was by W. M. Davis who wrote:

> After much study of English maps and writings at home, I have been tempted to entertain certain theories regarding the development of English rivers.
>
> (Davis 1895E: 128)

This visit occurred in the summer of 1894 and Davis' publication in the following year provided the first application of fluvially dominated polycyclic ideas to landform development in the classic region of south-east England. The second visit was in October 1933 when, at the invitation of S. W. Wooldridge, Henri Baulig (see chapter 8) delivered four lectures at the University of London on the eustatic theory, entitled 'The changing sea level' (Baulig 1935).

These threads were drawn together largely in the person of Sidney William Wooldridge (1900–63) (see plate 5.7), who later paid tribute to both Davis and Baulig:

> In southern England, it has been widely assumed, following Davis, that the accordance of the higher summits marks a former peneplain which terminated an earlier cycle of erosion and of which the uplift initiated the current cycle. Since the last major folding of the area is post-middle Oligocene, and the peneplain, if such it be, is demonstrably pre-lower Pliocene, the Miocene period is, in round terms, available for the first cycle, while the evolution of the existing hills and valleys has occupied the Pliocene and Pleistocene periods.
>
> (Wooldridge 1951: 171)

It is not too much to claim that, in its contribution to earth-history, geomorphology seems within grasping distance of a great unifying generalization. Just at the point where the stratigraphic record fails or becomes incomplete, an alternative principle of inter-regional correlation is offered in the fact that old sea-levels have engraved their mark on the margins of the lands and that the same levels are recognizable inland as 'terraces' or 'platforms'. The latter-day record of the continents seems to have been one of 'uplift' starting long before Pleistocene times – or, more probably, in view of the uniformity of the record, of successive negative eustatic shifts of sea-level. Here, in one of its major fields of advance, geomorphology converges on geophysics.

(Wooldridge 1951: 173)

Wooldridge was awarded an Sc.D. at King's College, London, at the early age of 27, and later became a Fellow of the Royal Society (Kirkaldy 1963). He was appointed to a chair at Birkbeck College in 1944 and returned to King's in 1947. Between the late 1920s and the late 1950s Wooldridge occupied a position of increasing pre-eminence in British geomorphology. By those who acknowledged this and followed his lead, he was rightly regarded as a brilliant researcher, an unequalled field teacher and a jovial colleague, ever ready to appear in amateur productions of Gilbert and Sullivan. To those who followed different precepts, especially as ill health set in during his later years, Wooldridge presented a rather different image.

South-east England

Together with the Pennsylvania Appalachians and the Central Plateau of France, south-east England became one of the three major world regions whose denudation chronology is now considered classic. We have already referred to Topley's pioneer studies of the Weald (1875; see volume 1, pp. 411–17) and these formed the effective basis for more than a century of subsequent work. William Topley (1841–94) (plate 9.1) worked for the Geological Survey as a field geologist for the period 1862–80 (Kirkaldy 1975) and was an earth scientist of broad vision:

> Shingle is like running water; a landowner may use it and pass it on; but not abstract it.
>
> (W. Topley, unpublished Presidential Address to the Geologists' Association 1885)

In his classic *Geology of the Weald* (1875), Topley recognized the region as consisting of a dome with a number of minor east-west-trending anticlines and flexures having steeper northern dips and cut by east-west faults dominantly downthrown on their southern sides. These structures were seen to be truncated by a high-level surface (297–220 m), having a perceptible eastward tilt, some remnants of which formed the flattened crest of the North Downs, (overlain in places with deposits of clay-with-flints) the more undulating and deposit-free South Downs and the Forest Ridges of the central Weald. Topley believed that the present landforms evolved by differential fluvial erosion and scarp retreat acting on an uplifted plain of marine denudation (Topley 1875: 272). The South Downs were held to be in a more advanced stage of fluvial denudation than the North Downs and the water gaps guided by transverse structures. The fact that the present river gravels, and those of abandoned terraces as much as 100 m above the valley floors, are of a composition relating entirely to the present catchment areas

Plate 9.1 William Topley

Source: From Kirkaldy (1975), facing p. 373

appeared to point clearly to long-term fluvial incision of at least this magnitude (Kirkaldy 1975).

Davis's work 'On the origin of certain English rivers' (1895E) was concerned with a question which he took up more generally in a paper published in the following year (1896F; see volume 2, p. 247), namely to what extent, and by what means, is it possible to distinguish between uplifted and fluvially dissected peneplains and plains of marine denudation? Referring to Jukes-Browne's (1883) discussion of the possible arrangement of drainage on plains of marine denudation, Davis concluded that, for the 'coastal plain' drainage of eastern and southern England, the good structural adjustment of the drainage network, especially that of subsequent streams, was indicative of the mature stage of a second cycle of subaerial denudation, rather than a first-cycle superimposition from a surface of marine planation:

> in two significant respects, the features of the second (subaerial) cycle differ from those of the first. In the first place, at the beginning of the initial cycle there were no subsequent streams; all the drainage was consequent. At the beginning of the second cycle, a considerable share of the drainage may be along revived subsequent streams; and with this opportunity so early afforded, the adjustments of the second cycle may exceed those of the first. Consequently, the adjustments in the maturity of a second cycle, following the older age of a first cycle, may reach a high degree of perfection. In the second place, the crest-lines of the escarpments or ridges in the second cycle will for some time retain the evenly bevelled form to which they were reduced at the close of the first cycle. When a region presents these two special features together, it can hardly be doubted that two cycles of subaerial denudation have been more or less completely passed through in its geographical development.
>
> (Davis 1895E: 135–6)

> the rivers of eastern England are now in the mature stage of the second cycle of subaerial denudation of a great mass of gently dipping sedimentary rocks, and . . . they have in this second cycle extended the adjustments of streams to structures that were already begun in the first cycle.
>
> (Davis 1895E: 146)

> the rivers of to-day, in the mature stage of the present cycle of denudation, appear to be the revived and matured successors of a well-adjusted system of consequent and subsequent drainage inherited from an earlier and far-advanced cycle of denudation
>
> (Davis 1895E: 128)

> The rivers of Eastern England have been developed in their present courses by the spontaneous growth of drainage lines on an original gently inclined plain, composed of sedimentary strata of varying resistance. In the course of this development, the land has been at least once worn down to a lowland of faint relief, and afterwards broadly uplifted, thus opening a second cycle of denudation, and reviving the rivers to new activities; and in the second cycle of denudation, the adjustment of streams to structures has been carried to a higher degree of perfection than it could have reached in the first cycle.
>
> (Davis 1895E: 127)

At this point the simple marine versus fluvial origin of a single high-level surface in south-east England began to be replaced by the notion that the surface was of a complex origin, being a fluvial peneplain remnant in its higher parts (e.g. the central Weald) and exhibiting the structural disharmonies of drainage superimposed from a surface of marine denudation in others. As early as 1891 Andrews, together with Jukes-Browne (1895), identified a summit surface in Dorset and Wiltshire, covered in places with gravel and clay-with-flints, thought to be of marine origin and tentatively dated as Pliocene or earlier. Mackinder (1902) supported the idea of a summit surface but proposed a partial submergence of the Wealden Dome during its later erosional history. Reid (1902) believed that the mid-Tertiary folds of Hampshire had been subjected to marine planation, that the area was subject to fluvial erosion beginning in the later Pliocene, especially by the longitudinal consequent of the 'Solent River' running west-east along the axes of the Hampshire and Dieppe Basins, and that river captures had later fragmented the drainage. Stoddart (1987: 158–60) has reviewed the early British reactions to Davis' ideas.

The first major attack on Davis' two-cycle fluvial theory was delivered by Bury (1908, 1910) with particular reference to the Wealden Dome. He pointed out that the gravels lying on the crest of the North Downs chalk cuesta must have been transported across a surface which covered what is now the adjacent upstream clay vale, implying that the subsequent vale is the product of the latest fluvial erosion cycle. This pointed to fluvial erosion having been initiated by superimposition from a thin Pliocene marine cover on to a pre-existing marine surface here which bevelled the longitudinal folds and cuestas, and was possibly warped after the marine withdrawal. Bury (1910) was at pains to point out the lack of structural adjustment of the drainage of the western Weald in that:

1 Longitudinal streams do not coincide with longitudinal folds;
2 The relative erosional success of streams located over the presumed planed-off anticlinal crests;
3 Evidence exists for stream extension and capture by the originally poorly adjusted drainage

This third point had provided the focus for Bury's earlier article (1908) where, besides dealing with the possible capture of the River Blackwater by the Wey (to which Davis had previously alluded), he showed the possibility of there having been a northward extension of the Rivers Adur and Arun across the Wealden axis at the expense of the north-flowing Wey (see volume 1, Figure 77, p. 415; also Linton, 1930). The River Wey was shown to be obviously discordant over the Peasemarsh Anticline and thus probably superimposed from a high-level marine surface.

Although White (1910) suggested that a summit surface in Hampshire and

Dorset might have been subaerially produced, he noted the structural discordances of the south-flowing Rivers Itchen and Test across anticlines at Winchester and Stockbridge, implying superimposition, and identified river captures in the Hampshire Basin, further evidence of an originally poorly adjusted drainage (these captures were disputed by Bury, (1926)). In 1911 Jukes-Browne proposed a single Pliocene marine transgression in south-east England but the problem of explaining the structural disharmonies of drainage still lingered on and in 1932 Linton found it necessary to deny both that the rivers of Wessex could have maintained their courses across rising flexures and that they were consequent on Miocene folds. Although, in association with Wooldridge, he later abandoned this view, at this time he postulated superimposition of Wessex drainage from a sheet of fluvially aggraded gravel, rather than from a marine-eroded surface. As late as 1936 Bury's view that stream courses in the Isle of Purbeck (south Dorset) were independent of Miocene folding was attacked as hypothetical (Linton 1969).

Before the late 1920s considerable attention had been paid to the detailed form, origin and dating of the higher parts of the Chalk. Much of this area forms a fairly flat surface lying between 150 and 210 m on the North and South Downs, the Chilterns and the Hampshire Chalk. Associated with these elevations are certain scattered deposits, three of which are poorly fossiliferous:

1 The Lenham Beds of east Kent are thin and discontinuous sands lying on the Chalk and piped down into solution hollows. They were originally assumed to be of Pliocene or earlier marine origin (Prestwich 1858, Reid, 1890) but later correlations with the Continent placed them as Diestian (e.g. Late Pliocene, correlated with the Coralline Crag).

2 Farther west at Netley Heath, Surrey, pebbly marine sands and chatter-marked beach gravels were tentatively dated as Late Miocene or Early Pliocene (Bury 1922: 100–1), or as Pliocene or Early Pleistocene and correlated with the Red Crag of East Anglia (Chatwin 1927, Dines and Edmunds 1933).

3 Somewhat similar marine sand and gravel marine deposits occur on the Chalk of the Chiltern Hills in Hertfordshire near Berkhamstead (Gilbert 1920) and Rothamsted (Dines and Chatwin 1930), which were also dated as Plio-Pleistocene

In 1923 White made the important suggestion that the foregoing surface could be differentiated from an Early Eocene surface, overlain by Eocene outliers, forming the main dip slope of the Chalk and intersecting the later more-or-less horizontal surface (White 1923, Figure 10, p. 75; see also Linton 1931) (figure 9.1). Attention had also been drawn to the less regular, rolling highest parts of the Chalk (215–275 m), thought to be pre-Early Pliocene (Miocene?) in age. This landscape is covered in parts by a surficial

Figure 9.1 Cross-section suggesting the intersection of two erosion surfaces on the Chalk Downs near Shaftesbury

C. Chalk   E. Eocene Beds.   aa'. Bevelled surface due to early Eocene erosion lately exposed about *s*.   bb'. Bevelled surface due to Late Miocene? erosion, forming flat top of Downs about *f*.

Source: From White (1923), Figure 10, p. 75

deposit of clay-with-flints which is commonly 'a mass of rusty clay with fractured flints' (Sparks 1972: 205) but elsewhere of rather different character, sometimes occurring below 210 m. It was originally proposed by Hull and Whitaker (1861) that the clay-with-flints was the residue of the subaerial weathering of the Chalk, but Reid (1899) and Jukes-Browne (1906) showed that impossibly excessive original thicknesses of Chalk would have been required to give the observed residual thicknesses of clay-with-flints and that the deposit could not have been simply derived from Chalk weathering. Reid (1903), followed by Barrow (1919), believed that the clay-with-flints had been derived by subaerial weatherng at least partly from the Tertiary deposits above the Chalk. A weathered, decalcified glacial till origin proposed by Sherlock and Noble (1912) was rapidly dismissed and by the late 1920s the modern view was beginning to predominate, namely that the clay-with-flints is of complex origin deriving from Chalk and Tertiary subaerial weathering residues, perhaps resorted by periglacial processes.

In writing the history of any science it is always tempting to be 'over tidy' in giving the impression that the main body of scholarship was at any given time moving inexorably towards the formulation of a single all-embracing theory. However, such was very much the case with respect to the geomorphology of south-east England in the later 1920s and 1930s. In 1922–3 the Weald Research Committee was established, including A. J. Bull, F. Gossling, J. F. N. Green, J. F. Kirkaldy, and the increasingly dominant S. W. Wooldridge. The last-named published on the River Mole basin with Bull in 1925 postulating that the area had been reduced to a fluvial surface of low relief by the end of the Miocene, been subject to a rapid and short-lived marine transgression (depositing thin sand and shingle) in the Pliocene, and then uplifted some 200 feet and fluvially dissected in the late Pliocene. Wooldridge (1927) followed this with a major work on the whole London Basin showing that the Chalk cuestas were bevelled at about 500–700 feet (150–210 m) either completely or partially, the partial bevelling being limited

Figure 9.2 Generalized features of the Upper Chalk cuestas of south-east England showing the three facets of the Mio-Pliocene Peneplain, the Calabrian (Plio-Pleistocene) marine surface and the sub-Eocene surface

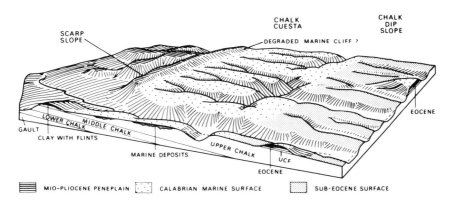

Source: After Wooldridge and Linton (1939), Figure 16, p. 60. Courtesy Institute of British Geographers.

by a low degraded bluff (old marine cliff?) behind which higher, rolling Chalkland occurs (figure 9.2). Indeed, the higher parts of the region in excess of some 700 feet appeared to consist of remnants of Davis' (1895E) high-level subaerial peneplain (this was accepted by Linton in 1932). In contrast, the bevelled surface, particularly apparent on the North Downs at about 180 m (600 feet), was covered with patches of marine sands and gravels similar to deposits at the same height resting unconformably on Eocene outcrops in the London Basin. These deposits at Netley Heath, Lenham and Berkhamstead had been dated as Pliocene and Wooldridge used morphological and heavy-mineral evidence to correlate them, to identify them as marine and to differentiate them from the older (Eocene) Reading Beds which occupy a similar stratigraphic position. It was therefore assumed that, after possibly some little uplift, the Late Miocene or Early Pliocene peneplain had been partially trimmed by a rapidly transgressive Late Pliocene sea which spread thin deposits over a wide area, and that later relative movements involving a 180 m (600 feet) fall in baselevel had occurred.

The Pleistocene denudation chronology of south-east England has naturally been evaluated with primary reference to the Lower Thames Valley (figure 9.3). It had long been apparent that this stretch of the river had had a complex recent history and in 1912 Sherlock and Noble followed Salter (1905) in proposing that the existence of abandoned river gravels in the Vale of St Albans, stretching east into Essex, pointed to a pre-glacial course of the Lower Thames (Thames I) which they thought had probably been later displaced southward by an ice advance east of the River Colne. Subsequently

Figure 9.3 Geomorphology of the Chiltern Hills and the Lower Thames Valley

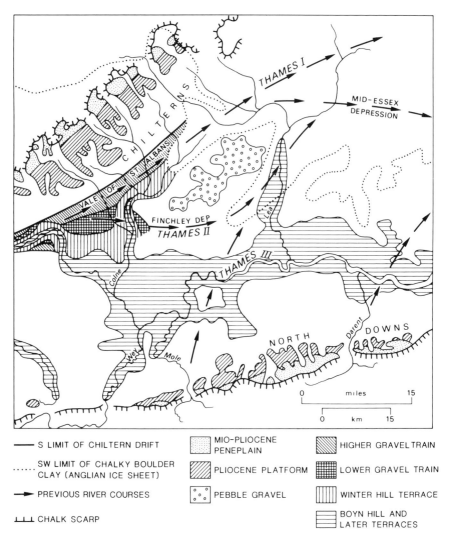

—— S LIMIT OF CHILTERN DRIFT	⬚ MIO-PLIOCENE PENEPLAIN	⬚ HIGHER GRAVEL TRAIN
······ SW LIMIT OF CHALKY BOULDER CLAY (ANGLIAN ICE SHEET)	⬚ PLIOCENE PLATFORM	⬚ LOWER GRAVEL TRAIN
➜ PREVIOUS RIVER COURSES	⬚ PEBBLE GRAVEL	⬚ WINTER HILL TERRACE
⊥⊥⊥ CHALK SCARP		⬚ BOYN HILL AND LATER TERRACES

Source: After Wooldridge and Linton (1939) Figure 26, p. 111 and Jones (1981), Figure 6.4, p. 155 and Figure 8.2, p. 200

Barrow (1919) described the thin (less than 2 m thick) Pebble Gravel of assumed greater antiquity concentrated mainly on the South Hertfordshire Plateau and the Chiltern dip slope at elevations of more than 120 m (370–500 feet). Barrow, following Whitaker (1889), assumed that these 'Plateau

Gravels' were equivalent in age to a number of other higher-level deposits in the area, including in particular the Pliocene Platform deposits, especially as they contained Lower Greensand pebbles from north-west of the Chiltern cuesta. The problem of the former courses of the Thames was further confused by Gregory (1922) who, while supporting the idea of a pre-glacial course extending into Essex, thought that there was at the same time a course farther south, and also by Hawkins (1923) who believed that the pre-glacial Thames flowed at a higher level than at present up the present course of the Lower Colne and down the Lower Lee. However, Sherlock (1924) reiterated the evidence for the existence of a Thames I flowing north-east to the Wash at the foot of the Eocene escarpment.

Despite earlier suggestions of glacial action south of the Thames in the Weald (Fisher 1866, Prestwich 1890) and the South Downs (Martin 1920), by the 1920s it was becoming apparent that the glacial evidence was confined to the north of the river. This indicated two glacial advances:

1 The Older (Chiltern) Drift, not obviously glacial in origin, consisting of scattered, weathered, pebbly clays, containing Triassic debris from central England (Barrow 1919, Hawkins 1923, Wells and Wooldridge 1923) and locally associated with gravels, lying on the Chiltern dip slope at elevations of 120–105 m. This was described as 'Western Drift' (Barrow 1919, Wells and Wooldridge 1923) and was ascribed to ice moving south or south-east from a local Chiltern ice-cap (Barrow 1918, 1919) immediately post-dating the Pebble Gravel phase (Sherlock 1924).
2 The Younger (Chalky, Eastern or Anglian) Drift, a characteristic till occuring at elevations of less than 100 m in Essex and associated with an ice advance into the London Basin from the north-east (Harmer 1902)

The advent of the researches of S. W. Wooldridge brought an overall structure to geomorphic work on south-east England. His early paper on the structural evolution of the London Basin (1926) showed that it had developed sporadically from the pre-Eocene and for a long period had been occupied by a longitudinal consequent, the proto-Thames. In a more important work (1927) he differentiated the Pebble Gravels on the South Hertfordshire Plateau from the deposits on the Pliocene Platform on mineralogical and sedimentary grounds. The former were found to include not only Lower Greensand pebbles from the north but also cherts from the Hythe Beds of the Weald to the south. Wooldridge, attempting to reconstruct a palaeogeography for the London Basin, postulated that the unfossiliferous Pebble Gravels were the fluvial equivalent of the middle glacial sands and gravels of East Anglia (Prestwich 1890, Solomon 1935). The Pebble Gravel surface (500-370 feet) appeared unwarped and probably correlated with Déperet's Sicilian terrace in the western Mediterranean. In contrast, the Pliocene Platform (>650–550 feet) was slightly warped, possibly Calabrian in age, and covered with marine deposits of local origin which possibly correlated with the Pliocene deposits of the North Downs. In this same paper and in two others (Wooldridge 1928, Saner and Wooldridge 1929) a 200 foot (60 m) surface, previously identified at the foot of the North Downs scarp in Kent (Dewey *et al.* 1925), was traced over parts of the London Basin and held to indicate prolonged baselevelling, tentatively of Milazzian (Mid-Pleistocene) age. This surface was later described in the Mole Basin (Bull *et al.* 1934), the western Weald (Green 1936) and the southern Weald (Kirkaldy and Bull 1940).

Thus, for the next 15 years or so, evidence from longitudinal profiles, terrace remnants, gravel cappings, truncated escarpment spurs, altitude frequencies and erosion gaps was marshalled to support Baulig's ideas of a eustatically controlled 'geomorphological staircase' caused by an intermittent drop of sea level from a maximum of some 200 m during the Pleistocene without appreciable warping of the land (Jones 1981: 136). So the idea was established that below the Pliocene Platform there is a flight of Pleistocene terraces, each one associated with a glacial sea-level fall, a still-stand and a subsequent aggradation due to a rise in sea level. Gossling (1936, 1940) examined the basin of the upper Darent noting gravel cappings at 520–470 feet; curves were fitted to the upper 26 miles of the River Mole and extrapolated via terrace remnants to possible old baselevels at 500–400, 230–200, 175–165 and 80 feet (Bull *et al.* 1934); Bull (1932, 1936, 1940, 1942, Kirkaldy and Bull 1940) identified surfaces in the southern Weald and South Downs at 400, 320, 220–200, 130 and 80 feet; and Green (1943) gave a general review of the age of the raised beaches of southern England.

A more detailed proposed reconstruction of Pleistocene events in the London Basin was developed by Wooldridge (1932, 1938), Wooldridge and Linton (1939), and King and Oakley (1936). The work of Wooldridge concentrated on the proposed two-phase southerly displacement of the Thames (I and II) to its present location (III) which tends to disregard some of the more obvious topographic trends. This displacement was associated with the production of four depositional surfaces and the occurrence of two glaciations, as follows (figures 9.3 and 9.4) (see fine review by Jones 1981):

1 Pebble Gravel Platform (500–370 feet). On this lie the earliest of the modern Thames (I) fluviatile deposits laid down at a high elevation as the river flowed along the Vale of St Albans
2 First glaciation (Older; Chiltern; Western).
3 Higher Gravel Train (370–340 feet). Formed by the Thames I in association with a post-glacial rise in sea level. The surface slopes eastward at about 6 feet/mile and its rapid deposition may have contributed to the southerly displacement of the Thames (to Thames II).
4 Lower Gravel Train (320–290 feet: 98–88 m). At this stage the Thames (II) flowed along the Finchley Depression with reference to a falling (Sicilian?: Early Pleistocene) sea level.

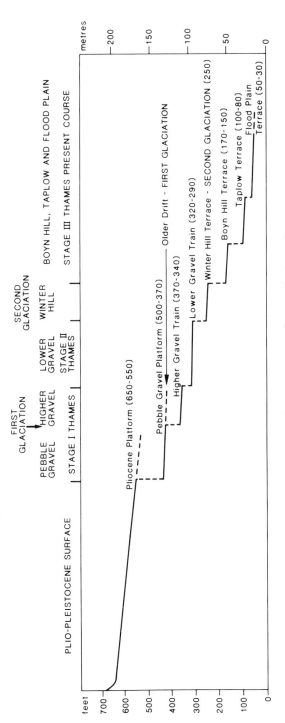

Figure 9.4 The denudation chronology of the London Basin

Source: After Wooldridge and Linton (1939) Figure 24, p. 102. Courtesy Institute of British Geographers.

5 Winter Hill Terrace (250 feet). Situated generally south-east of the Lower Gravel Train.

6 Second glaciation (Younger; Chalky; Anglian; Eastern). This ice advance deflected the Thames into its present (Thames III) course.

King and Oakley (1936) investigated the three main terraces of the present Thames (III) and gave the first complete interpretation of the Thames terraces based on a simple eustatic control theory. This interpretation accorded with Wooldridge's ideas and was adopted by him (Wooldridge 1938, Wooldridge and Linton 1939). The three most important latest terraces of the Thames III, the Boyn Hill (170–150 feet: 52–46 m), the Taplow (100–80 feet; 30–24 m) and the main Floodplain (50–30 feet: 15–9 m), were first identified and named by Whitaker (1889). The eustatic model of King and Oakley (1936) postulated phases of glacial sea-level lowering, associated with fluvial incision and planation, followed by interglacial sea-level rise and aggradation. The Boyn Hill Terrace (Milazzian?) was believed to exhibit two phases of erosion and aggradation, and these were followed by erosion below present sea level and then aggradation to give the Taplow Terrace (Tyrrhenian?: Later Pleistocene). Similar events were held to have produced the upper and lower elements of the Floodplain Terrace (Monastirian?: Late Pleistocene), the final excavation below sea level during the last ice age being followed by Holocene infilling to give the present river channel.

It remained for Wooldridge and Linton (1938a, b, 1939) to set out the broad 'classical' denudation chronology for south-east England which was to remain substantially unquestioned until the 1970s. The major features of this were:

1 Planation of the Chalk to produce the sub-Eocene unconformity and the pre-Eocene initiation of the broad flexures of the London Basin and the Weald.

2 Eocene-Middle Oligocene marine deposition which was terminated by a marine withdrawal leaving the two major synclinal consequents (the proto-Thames and the Solent River) fed by lateral consequents down the north and south flanks of the developing Wealden Dome (figures 9.5 and 9.6).

3 This stream network was disrupted by the Late Oligocene-Early Miocene (Alpine) orogeny which produced many secondary fold axes, especially on the Wealden Dome (figure 9.5).

4 The development of the Mio-Pliocene Peneplain.

5 The marine transgression into the areas of the North Downs, London Basin and Chilterns, and marine planation of the marginal parts of this peneplain during Late Pliocene-Early Pleistocene (Calabrian) times and the later superimposition of discordant drainage 'seaward' of the Wealden Island and the Chiltern Hills during a possible combination of uplift and sea-level fall.

Figure 9.5 The principal features of the denudation chronology of south-east England

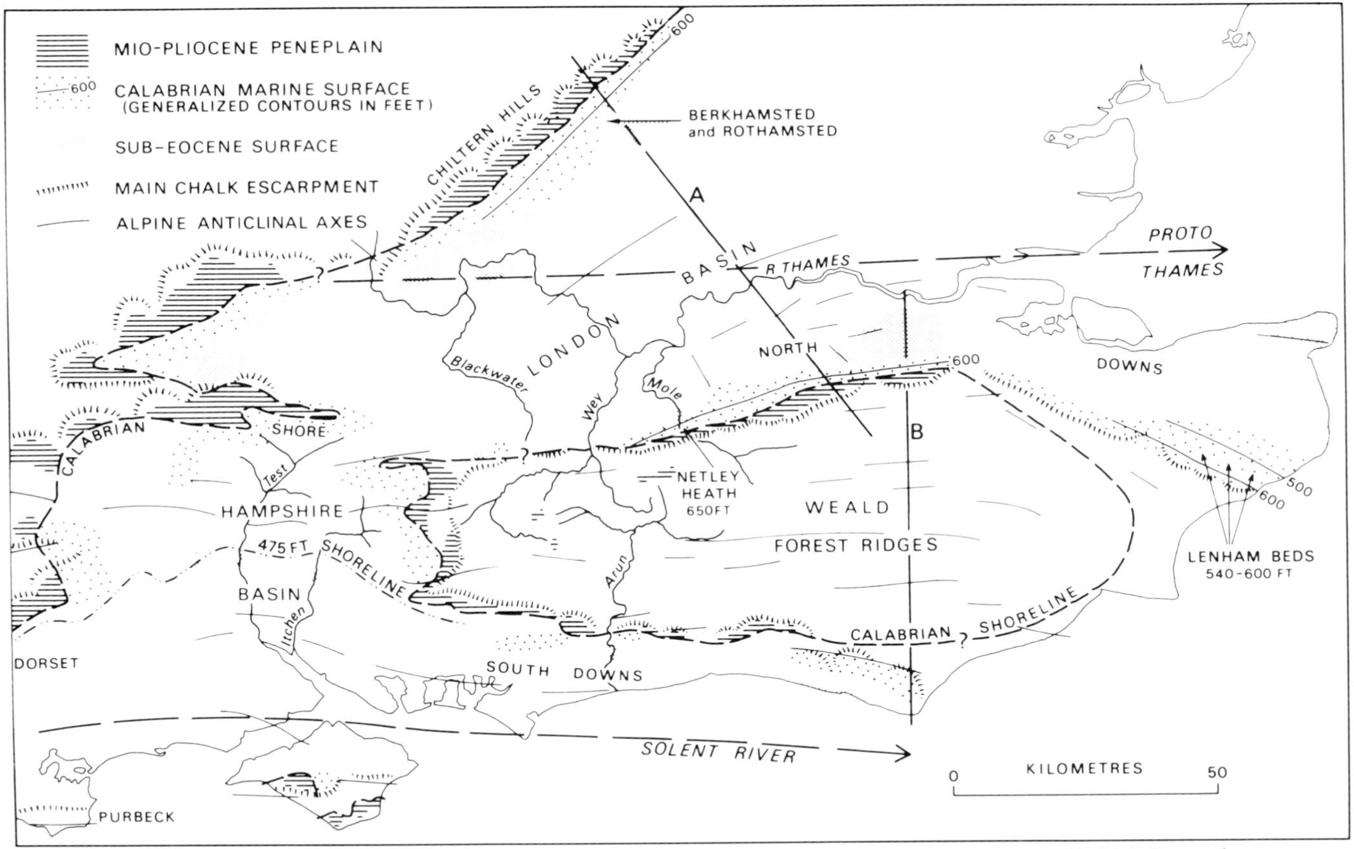

MIO-PLIOCENE PENEPLAIN

CALABRIAN MARINE SURFACE
(GENERALIZED CONTOURS IN FEET)

SUB-EOCENE SURFACE

MAIN CHALK ESCARPMENT

ALPINE ANTICLINAL AXES

600

PROTO THAMES

BERKHAMSTED
and ROTHAMSTED

CHILTERN HILLS

600

LONDON BASIN

Blackwater

Wey

Mole

600

NORTH DOWNS

A

B

Netley Heath
650FT

WEALD

FOREST RIDGES

Arun

CALABRIAN SHORELINE ?

SOUTH DOWNS

Adur

SHORELINE

475FT

HAMPSHIRE BASIN

CALABRIAN SHORE

Test

Itchen

SOLENT RIVER

DORSET

PURBECK

LENHAM BEDS
540-600 FT

500

600

0   50

KILOMETRES

Source; After Wooldridge and Linton (1939), Figure 15, p. 58 and Figure 17, p. 62. Courtesy Institute of British Geographers.

Figure 9.6 The topographical and geological cross-sections shown in figure 9.5 relating to the London Basin (A) and the Wealden Dome (B)

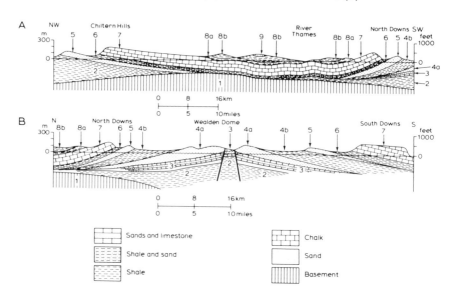

Source: After Stamp

6   A Pleistocene sequence of eustatic falls of sea level producing, with two glaciations, a complex development of surfaces associated with the Thames I-III and a flight of marine and river terraces elsewhere. Sparks (1949) identified up to eight marine terraces between 475 and 180 feet on the dip slope of the South Downs.

7   The protracted (post-mid-Pliocene) fluvial erosion of the central Weald and (post-Early Pleistocene) fluvial erosion of the Chalk cuestas of the Downs and Chilterns.

The most persistent object of interest in the latter respect had long been the Chalk dry valleys. In 1887 Reid, followed much later by Bull (1940), suggested that these valleys were cut by the considerable meltwater runoff during Pleistocene permafrost conditions on the chalk. In contrast, Chandler (1909) and more particularly Fagg (1923) believed that the retreat of the Chalk scarp slope and the decrease in elevation of the basal Chalk contact with the Gault Clay had caused a progressive drop in the water table of the Chalk scarp and dip slopes leaving valleys which previously had perennial rivers as now dry. Bull (1940) opposed Fagg's hypothesis on the grounds that the existence of scarp-slope terraces and the absence of valley-head notches at the crest of the South Downs showed that little Chalk scarp recession had in fact occurred.

Western and Northern Britain: Drainage Systems

In 1950 Wooldridge gave a comprehensive analysis of British denudation chronology beginning with the Cretaceous transgression, the assumed starting point for the development of modern landforms (Hull 1882, Jukes-Browne 1922), and drew attention to the two methods of investigating the largely unknown Tertiary history of British palaeogeography, namely:

1 The analysis and evolutionary interpretation of drainage systems, that is the spatial aspects of terrain; and
2 The recognition of past baselevels of erosion, that is the vertical aspects of terrain.

In doing so, Wooldridge pointed to the difficulties in achieving absolute, or even relative, dating which would link these evolving drainage systems and changing baselevels. It is now very obvious that one of the reasons why south-east England provides the keystone for British denudation chronology is that it does establish such a link. This link resides in the existence of a datable (Late Pliocene-Early Pleistocene) erosion surface of known (marine) origin lying in close vertical proximity (150–210 m, about 600 feet) to an older (Mio-Pliocene) surface of assumed fluvial origin (greater than about 700 feet). The geographical extent of the Plio-Pleistocene sea thus limits two distinct drainage patterns – an older, adjusted one from a younger network, less adjusted to geological structure. Elsewhere such relationships were more obscure and much more speculative geomorphic evolutionary sequences had to be adopted.

It will be recalled that Davis (1895E) proposed that the present drainage of Wales, central and south-east England, now in a second cycle of erosion, developed from an uplifted and east-south-easterly-tilted fluvial peneplain of Early or Middle Tertiary age, remnants of which can be found as elevated surfaces in central Wales, above which rise monadnocks such as Plynlimon. Davis postulated that this original drainage pattern, flowing into the Kennet–Lower Thames proto-consequent, had been fragmented, largely by the development of subsequents like the Lower Severn–Avon, leaving beheaded, misfit rivers such as can be found in the Cotswold Hills (Davis 1899M) (figure 9.7). According to this view, the present Thames and Ouse river systems are second-cycle dismembered remnants of a drainage system which originally formed on a cover of Mezozoic rocks extending westward across Wales.

Davis, of course, was not the first to postulate north-westerly links for the Thames drainage. Hull (1855) and Lucy (1872) followed Buckland in proposing the glacio-marine transportation of Severn pebbles across the Cotswold escarpment into the Cherwell and Evenlode valleys (see volume 1, pp. 116–7), although Phillips (1871) disputed this. Shrubsole (1890) described gravels in the London Basin as containing Triassic pebbles from the Midlands and Gregory (1894) pointed to gaps in the Chalk Downs

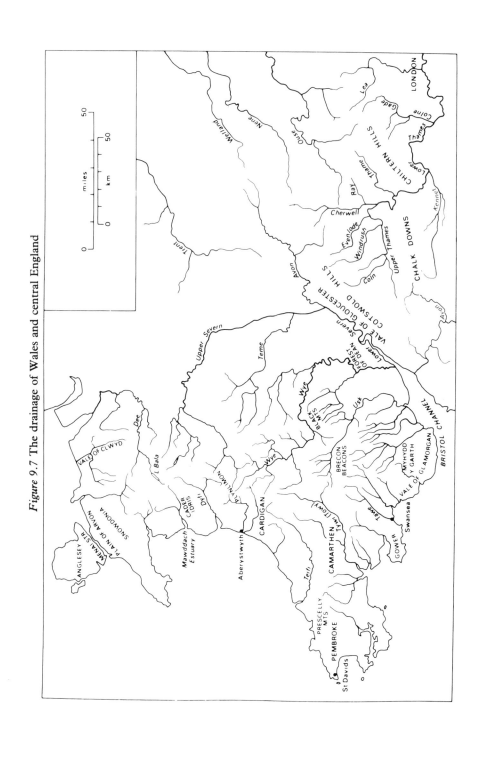

Figure 9.7 The drainage of Wales and central England

possibly corresponding to earlier south-south-easterly extensions of the Cotswold drainage. Most particularly, however, Ellis (1882) had hinted that the original chalk cover had extended westward over Wales as far as Northern Ireland and that on this had developed a superimposed consequent initial drainage towards the south-east which had been later dismembered by the north-eastward extension of the Lower Severn-Avon strike system.

In the decade or so following Davis' contributions to British geomorphology these ideas were rapidly exploited. Buckman (1899) outlined in detail an extensive proposed initial consequent system including south and central Wales, the Bristol Avon and the Trent system, as well as the Thames (figure 9.8), the beheaded parts of which comprised the underfit Cotswold tributaries. Lake (1900; 245) agreed with Buckman that the Lower Severn is a subsequent, developed along a soft strike outcrop, which captured the headwaters of rivers which originally flowed off a North Wales dome southeastward across the site of the present Vale of Gloucester to join the proto-Thames. His theory also involved the fragmentation of this original drainage in Wales by the development of secondary drainage along more-or-less concentric fault systems in Wales (figure 9.9). More than 30 years later Lake (1934) elaborated this scheme which began with a Cretaceous cover which was then updomed along a Snowdon-south-east Ireland axis giving radial drainage. The collapse of much of the dome resulted in dominant southeastward drainage off Snowdonia which was later complicated by renewed collapse. This collapse affected the central parts of the dome successively more than the outer parts, causing these radial rivers to be deflected along transverse valleys located in relation to the arcuate fracture zones (e.g. Towy Valley, Lake Bala depression). Strahan (1902: 219–21) noted in South Wales the discordance between the present drainage lines and the Palaeozoic structural trends, proposing superimposition from an Upper Cretaceous cover leading to post-Oligocene initiation of drainage on the exhumed Palaeozoic rocks. Avebury (1902: 338–62) gave detailed consideration to the complementary fragmentation of English drainage by the development of subsequents along the clay vales. He supported the view that a river extending north-east from the Bristol Channel captured successively the Usk, Wye and Upper Severn from the proto-Thames (although he could not decide whether the original south-eastward connection of the Upper Severn had been by the route of the Cherwell or Evenlode), leaving misfits, wind gaps and dry valleys in the Cotswold Hills and Chalk Downs. In a similar way the south-westerly extension of the Ouse may have beheaded the northerly tributaries of the Lower Thames (e.g. Colne and Lea) and may be on the point of doing the same to the Cherwell. However, it should not be thought that superimposition of the original Welsh drainage was universally supported at this time in that Fearnsides (1910: 821), although he recognized the possibility of initiation on a surface of marine denudation or on a surface from

Figure 9.8 Suggested reconstruction of the former consequent drainage of the English Midlands. Consequent drainage lines dotted; anticlinal axes as heavy lines

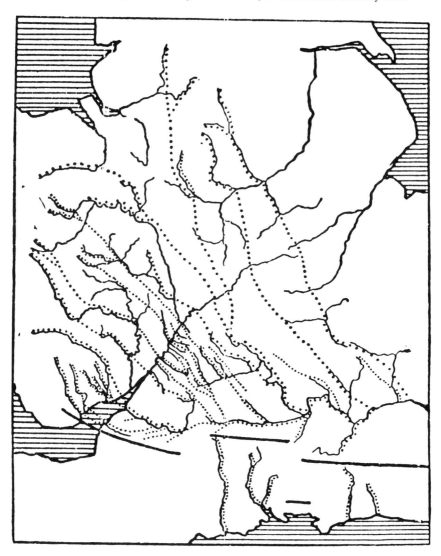

Source: From Buckman (1899) and Wooldridge and Morgan (1937)

Figure 9.9 The suggested development of Welsh drainage. Former consequent streams dotted; fracture belts along which subsequent streams have developed shown in heavy lines and numbered

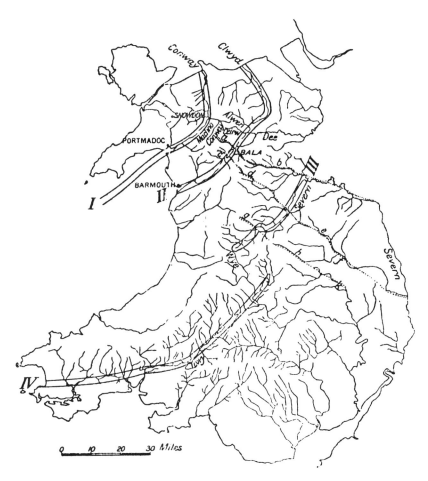

Source: From Lake (1934, p. 26) and Wooldridge and Morgan (1937)

which the Mesozoic deposits had been stripped, favoured a more Davisian notion of the development of drainage on an early peneplain later rejuvenated in Miocene times. Similarly, Jukes-Browne (1911: 334–5) followed Ramsay (see volume 1, chapter 16) in postulating the original surface as being one of Cretaceous marine erosion.

During the interwar period, work on drainage pattern evolution concentrated on glacial changes in central and eastern England and on the elaboration of earlier ideas in Wales. O. T. Jones (1930: 79–80) postulated successive regional tilts in South Wales, the first (south-easterly; Cretaceous–Eocene) mainly affecting eastern parts and the second (south or south-west) involving the warping of the high plateau surface in central and south-west Wales before much previous subaerial erosion had occurred. Later Jones (1951) thought that the latter surface may have been an uplifted and resurrected extensive Triassic plain of desert erosion from which south-east-flowing consequent drainage had stripped off the overlying Jurassic and Cretaceous rocks. Dewhirst (1930) followed Lake and Strahan in postulating the initiation of radial consequents on an updomed Cretaceous cover which were later disrupted by uparching of the Wales-south-east Ireland dome, the development of collapsed curved fractures following the pre-existing Caledonian trend along which disruptive subsequent rivers extended. Clarke (1936) also followed Strahan in assuming a set of south-east-flowing rivers superimposed over the Black Mountains of South Wales from a Cretaceous cover but, unlike Lake, he assumed that the drainage fragmentation by subsequents took place before the cover had been eroded and not after superimposition on to the underlying Palaeozoics. R. O. Jones (1931) postulated an original south-south-east Tawe drainage on a Cretaceous cover which may have persisted long after incision into the Palaeozoic basement until tectonic displacement towards the south-west. He later (1939: 563) came to doubt the existence of the Cretaceous cover and thought that the Tawe drainage may have originated on a warped and southerly-tilted erosion surface. A radically different thesis for the origin of South Wales drainage (especially the Towy) was proposed by George (1942: 120–1) who opposed the idea of superimposition in favour of the progressive south-western development of the drainage by the intermittent uplift and emergence of successive platforms of marine abrasion, particularly of a 600 foot platform in Pliocene times – a rare attempt at drainage network dating.

In other regions of Britain variants of the superimposition model of drainage evolution were prevalent. The most popular, as expressed by Wooldridge, was that the highest surface in the west was peneplained by east-flowing rivers superimposed from a Cretaceous cover (Chalk is found in Northern Ireland and western Scotland) and that the drainage of the whole country was then fragmented by adjustment to structure, probably in more than one cycle of erosion. The Bristol Avon was thought to have been superimposed across Palaeozoic ridges from a Triassic cover (Trueman 1938) and in the Lake District the radial drainage was held to have been superimposed from a dome of Mesozoic sediments at least 2,000 feet higher than the existing elevations (Goodchild 1888–9, Marr 1900, 1906). To the east of the Lake District the drainage of the Eden Valley (Hollingworth 1929) and of the Howgill Fells (McConnell 1939) was believed to have originated on a north-or north-easterly-tilted Early Tertiary surface, later disrupted by the

Figure 9.10 Drainage development on the Alston Block, northern England

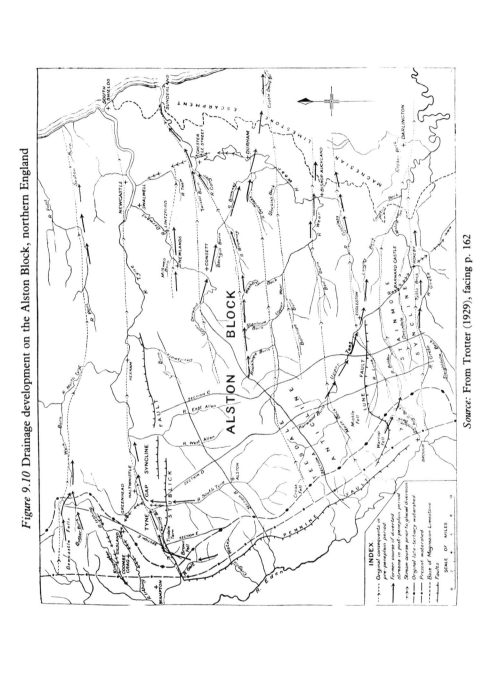

Source: From Trotter (1929), facing p. 162

Figure 9.11 Stages in the development of the Yorkshire drainage: A, initial and B, drainage in second cycle

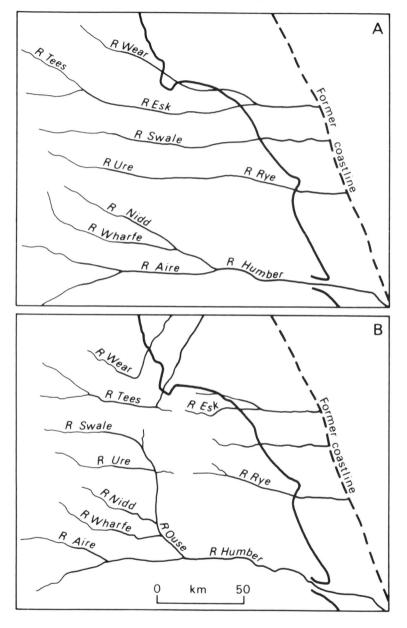

Source: After Reed (1901), Figure II, p. 59 and Figure III, p. 65. Courtesy Cambridge University Press.

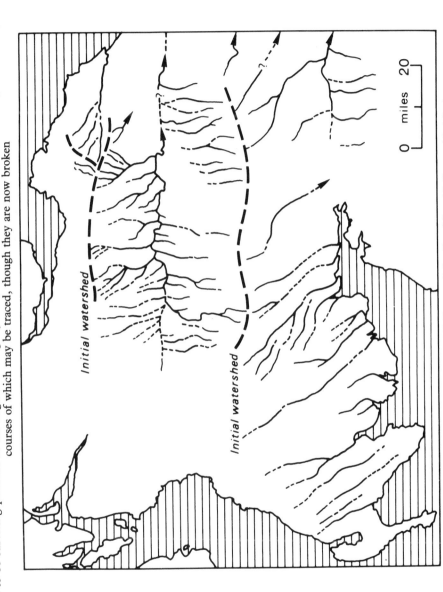

Figure 9.12 The consequent elements of the drainage of southern Scotland: The continuous lines represent flowing streams which are believed to be surviving portions of the original drainage system; the broken lines indicate other consequent streams, the former courses of which may be traced, though they are now broken

Source: After Linton (1933), Figure 1, p. 168. By permission Royal Scottish Geographical Society.

growth of an east-flowing subsequent which captured the headwaters of the River Eden due to unequal slopes, asymmetrical uplift and headward growth along lines of structural and lithological weakness (King 1976: 42).

The major work on drainage development in northern England was by Trotter (1929) who postulated the depression of the Alston Block during the Hercynian, its covering with 1,225–1,525 m of Mesozoic rocks and then Early Tertiary uplift and the development of major east-flowing consequents (figure 9.10) which were later dismembered by the growth of subsequents (e.g. the Yorkshire Ouse), possibly in association with peneplanation and rejuvenation in Late Tertiary times. Variants of such dismemberment by capture were proposed by Reed (1901) (figure 9.11) and Versey (1937, 1942), both of whom invoked peneplanation. Reed (1901) believed in drainage initiation by superimposition from a Cretaceous cover followed by the production of a peneplain below the base of the Chalk extending westward over the Pennines. This peneplain was uplifted and tilted eastward in the mid-Tertiary resulting in river extension, incision and capture.

Ideas regarding drainage development in Scotland have been grouped under three headings by Sissons (1976):

1 Derived from an original south-east-flowing system. Geikie (1865) believed that the Tay drainage had evolved from an earlier south-east-flowing system and this idea was extended by Mackinder (1902) to the initiation of the whole Scottish drainage on a south-east-tilted peneplain with some rivers (e.g. the Tweed) later being diverted eastward. This initial south-east drainage was supported by Mort (1918) in southern Scotland and by Peach and Horne (1930) who identified abandoned wind gaps in the Pentland Hills. Contrary views were expressed by Linton (1933) who did not think that Mackinder could explain the evolution of the rivers flowing north into the Tweed and by Bremner (1942) who did not believe that a drainage system could not develop *ab initio* on an uplifted, tilted peneplain.

2 Derived from an original east-flowing system. This was proposed by Ramsay (1878), Cadell (1886) and Baily *et al.* (1916). Linton (1933, 1934) transferred the 'English' model north of the border with an original east-flowing consequent drainage system superimposed from a Cretaceous cover and subsequently partly dismembered (figure 9.12), and Bremner (1942) supported the idea of eastward drainage on a Cretaceous cover with a major divide well to the west.

3 Derived from drainage extending over the discontinuously emerged marine surfaces around high ground. This view of Hollingworth (1938), similar to that of George (1942) in Wales, required large-scale subsidence but had the advantage of explaining the occurrence of drainage lines in many directions.

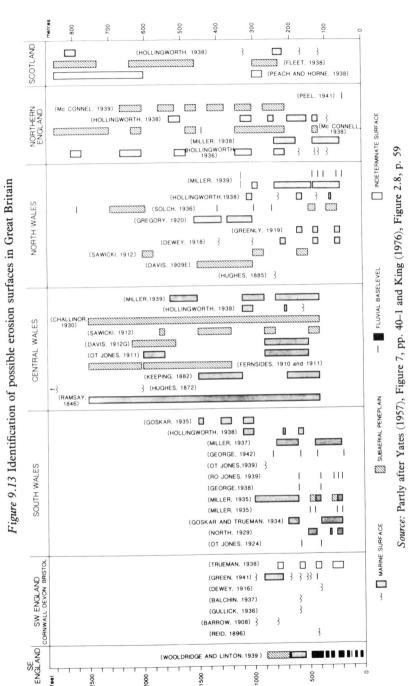

Figure 9.13 Identification of possible erosion surfaces in Great Britain

Source: Partly after Yates (1957), Figure 7, pp. 40–1 and King (1976), Figure 2.8, p. 59

Baselevels of Erosion

The recognition and dating of past baselevels of fluvial and marine erosion outside south-eastern England presented profound problems to British geomorphologists in the period before 1945. This was due to several obvious reasons: the lack of datable deposits in the predominantly erosional environments of higher relief; the interruptions provided by glacial and fluvio-glacial events: and the heavy reliance which was placed on purely morphological techniques geared to dating below 650 feet, established in south-east England, and at all levels on the assumptions underlying the simple eustatic theory. These morphometric methods, leading to the supposed identification of old baselevels, involved the projection of 'graded' river reaches by fitted mathematical approximations (e.g. Jones 1924, Miller 1935), projections using river-terrace remnants (e.g. Green 1936) or wind gaps or generalized contours to identify erosion surfaces (e.g. Miller 1937), projected profiles (Miller 1937), hypsographic curves/height frequencies (e.g. Hollingworth 1938), and clinographic curves (average contour strip slopes) (e.g. Hanson-Lowe 1935). Needless to say, the welter of speculative results produced by such methods (figure 9.13) led to increased criticism, culminating in the paper by Miller (1939a) which was critical of the ability of purely morphometric methods to identify old cliff lines satisfactorily; or to identify platforms of marine abrasion, especially inclined platforms of marine encroachment (e.g. Goskar 1935); or extrapolated graded river reaches, wherein alternative reasonable curves can differ by up to 50 feet/mile (e.g. Jones 1924); or cross-sections of rejuvenated valleys (e.g. Baulig 1935) and altitude frequencies (e.g. Hollingworth 1938). Stoddart (1987: 160–2) has highlighted British researches on erosional history during the period.

Studies of erosion surfaces and baselevels in south-west England were very sparse before the 1920s, but after the mid-1930s work was undoubtedly influenced by that in south-east England. In 1890 Reid identified a 430 foot platform, believed to be of marine origin, which he correlated with the St Erth Beds (Early Pliocene). Eighteen years later Barrow (1908) postulated two marine surfaces (800 and 1,000 feet) on Bodmin Moor, and in 1916 Dewey identified a 400 foot marine surface in Cornwall (dated as Mio-Pliocene by Wills (1929)). A marine surface at about 600 feet was described in Cornwall by Gullick (1936) and Balchin (1937), while in 1941 Green directly extended the Wooldridgean chronology into east Devon. Green's proposed eustatic baselevels included six marine surfaces above 500 feet (1,000 foot on Cretaceous rocks; a broad surface at 920–750 feet correlating with Barrow's (1908) lower surface; 690 feet; 595 feet; 530 and 505 feet), whereas valley-side flats at 440 feet in the Axe Valley were considered subaerial (Wooldridge 1950). Farther north-east and linking with the South Wales area, Trueman (1938) proposed marine erosion surfaces at 800–750, 600–550, 450–400 and 300–200 feet in the Bristol region of possibly 'late Tertiary' age.

The considerable elevation of Wales, together with the assumed previous connection of its drainage system to that of south-east England, ensured that the area would be intensively subjected to the search for ancient baselevels, particularly following the pioneer work of Ramsay (1846; see volume 1, pp. 307–13) who postulated a marine surface sloping down from 2,500 to 400 feet. As a prelude to his own analysis of the denudation chronology of Wales (see Volume 4), Brown (1960) gave an exhaustive description of previous research, which forms the basis of what follows here.

In 1872 Ramsay proposed that the present British drainage had evolved on a Cretaceous overlap which had abutted on the Welsh uplands and, after the Severn had been formed, an eastward tilting caused the Chalk scarp to retreat exposing the earlier rocks which, in turn, developed younger erosional scarps. Commenting on this paper, Hughes (in Ramsay 1872: 157–8) postulated two high marine surfaces in Wales: Ramsay's original one at a little over 2,000 feet and another at 3,000 feet. He disagreed with Ramsay's Carboniferous date for the former surface (which Ramsay reaffirmed in 1876) and suggested that it was formed during the removal of the Cretaceous cover from Wales. In 1885 Hughes (p. 6) postulated the existence of another marine surface at about 800 feet in North Wales, and a few years earlier Keeping (1882: 251) had identified two sloping planes of marine denudation east of Aberystwyth, the lower at 400–500 feet at the coast rising to 700 feet inland and the higher beginning at 1,100–1,200 feet rising gradually to 1,500 feet where it abuts on Plynlimon.

As we have seen in volume 2, W. M. Davis (1909E) was interested in his ancestral Wales and proposed the existence of a pre-glacial peneplain rising from 1,100–1,200 feet in the north to 1,400–1,500 feet in central Wales. In 1911 Davis conducted an international party to Wales (Davis 1912G; see volume 2, pp. 363–6), assisted by J. E. Marr and O. T. Jones, which led Ludomir Sawicki of Cracow to identify the remnants of three peneplains in Wales and Devon: at 400–600 feet; at 800–900 feet, rising to 1,200 feet in the valleys of Snowdonia; and at 1,800–2,000 feet (Sawicki 1912). This natural link with south-west England was followed by Dewey (1918) who saw his 400 foot surface as the equivalent of the 430 foot break of slope between Snowdonia and the Plain of Arvon and also suggested the existence of a 200–300 foot marine surface there. Dewey's work was succeeded by Greenly's (1919) who identified three marine surfaces in Anglesey: the Menaian at 200–300 feet, averaging 275 feet and sloping west at 7 feet/mile; the Tregarth at 430 feet; and the summit monadnock surface at 500–600 feet. A year later, Gregory (1920) supported Greenly's Tregarth surface, together with another marine one at 1,250–1,400 feet in Snowdonia, possibly representing the upper part of Sawicki's intermediate surface. By 1936 Sölch was speculatively describing seven sets of valley-floor or erosion surfaces in Snowdonia, namely at 2,593 feet (800 m; Miocene?), 1,969–2,297 feet

Plate 9.2 Owen Thomas Jones

Source: Courtesy *Proceedings of the Geological Society of America*, 1967, p. 219

(600–700 m), 1,312 feet (400 m), 984 feet (300 m), 820 feet (250 m; Pliocene?), 427–492 feet (130–150 m) and 230–328 feet (70–100 m; Pleistocene?). Hollingworth's (1938) statistical analysis of the frequency of spot heights from Ordnance Survey maps in Highland Britain only led him to identify three possible surfaces in North Wales, at 280 feet (Menaian?), 400 feet (Tregarth?) and, questionably, at 200 feet.

From the time of Ramsay's earliest work, central (with southern) Wales had proved an especially fruitful region for hunters of old baselevels. Following Davis, Fearnsides (1910: 820–1) identified a peneplain at 2,000 feet near Cader Idris, sloping down to some 1,200 feet on the north coast of Wales and to lower levels to the south in Cardiganshire. During the whole period prior to the Second World War the most significant worker on Welsh denudation chronology was Owen Thomas Jones (1878–1967) (plate 9.2) who studied at Cambridge under J. E. Marr, worked for the Geological Survey in Wales under Strahan and, after professorships in Aberystwyth and Manchester, was elected to the Woodwardian Chair of Geology at Cambridge in 1930 (Pugh 1967). In 1911 he observed a high plateau in north Cardiganshire sloping south at 1:250 from a high point of 2,000 feet, together with a wavecut plateau reaching up to 900 feet; although, as Brown (1960) pointed out, he did not note Keeping's two surfaces in the area. In a major paper in 1924, Jones fitted a mathematical curve of the form $y = c - k \log (x + a) + b (x + a)^3$, where y is elevation and x is horizontal distance from an arbitrary origin, to reaches of the Towy system in an attempt to reconstruct earlier baselevels (figure 9.14A). This led to the proposal for baselevels at 580 feet (Nant Stalwyn) and 400 feet (Fanog). Almost 20 years later, George (1942), working on the long profiles of the Towy and Usk, supported the Nant Stalwyn and Fanog baselevels, adding the 830 foot (Ffos Tarw) and 185 foot (Talley) stages. In 1930 Jones postulated that the supposed Welsh Summit Plain (2,100–3,500 feet) was of Triassic date and arid origin, similar to the Charnwood Forest stripped unconformity in Leicestershire (Watts 1903).

Along the south coast of Wales, North (1929) proposed the existence of a widespread 200 foot marine platform in Gower, a westward-sloping surface from 250 to 200 feet in Pembroke and three surfaces in the Vale of Glamorgan (540–440, 320–300 and 250–200 feet). Goskar and Trueman (1934) used projected profiles to identify surfaces west of Carmarthen at 200, 400 and 600 feet, comparable with those in Cornwall, and George (1938) identified a 400 foot marine platform and a 600 foot ancient shoreline near Swansea. Goskar (1935) extended this work to higher elevations in South Wales to locate marine surfaces at 1,000, 1,200–1,300 and 1,450–1,500 feet. The former was possibly associated with Miller's (1935) 900 foot marine surface in the Forest of Dean, where former baselevels of the River Wye system were identified at about 400 and 200 feet and were believed to correlate with those farther west (figure 9.14B). Further work by Miller (1937) employed

Figure 9.14 A: Long profiles of the River Towy and its tributary the Camdowr with the upper reaches projected to the Nant Stalwyn (above) and Fanog (below) baselevels by the use of a mathematical formula. B: Stream profiles of the River Wye and its tributaries (the horizontal scale of the latter exaggerated 5 times) projected to suggest two ancient baselevels

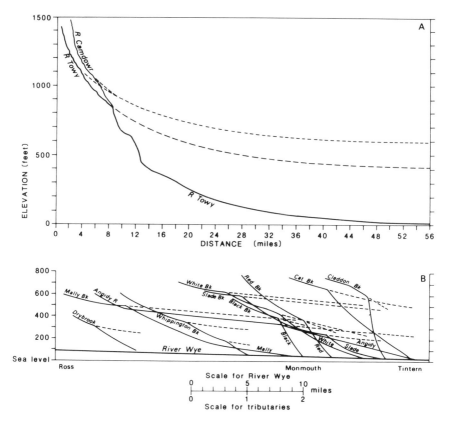

Sources: A, after Jones (1924), plate XLV. Reproduced by permission of the Geological Society from the *Quarterly Journal*; B, after Miller (1935), fig. 3, p. 168, by permission of the Royal Geographical Society from the *Geographical Journal*.

generalized contours (figure 9.15) and projected profiles (figure 9.16) in Pembroke and Carmarthen to propose a 400 foot marine surface (observed in the field to be backed by a degraded marine cliff) and a plateau at about 600 feet, probably representing part of a fluvial peneplain. Working northward to Cardigan, Miller (1938) proposed a 700–900 foot surface in Cardigan, a 900–1,100 foot surface east of Aberystwyth and a 1,500–1,700 foot high plateau in mid-Wales. He also used spur and river profiles to identify old fluvial baselevels near the Dyfi and Mawddach estuaries at about 200, 450,

700 (?) and 1000 (?) feet, and it appeared possible to him that earlier marine surfaces may have been subaerially modified and even merged. A more radical idea had previously been suggested by Challinor (1930) who thought that similar elevations could be produced by the uniform distribution of erosional forces and equal stream spacing, and that therefore they might have nothing to do with former baselevels. Hollingworth's (1938) statistical analysis of elevations in Highland Britain did not identify the 200 foot platform in South Wales, probably because the area has not been sufficiently dissected to produce the discrete hilltop spot heights on which this method depended (Brown 1960), but there were elevation frequency maxima at 400–450, 550–600, 720–740 and 1,000–1,100 feet.

The end result of all this effort directed towards the denudation chronology of Wales before the Second World War was that a number of erosion surfaces could be identified but that their correlation was questionable, their origin obscure and their dating non-existent.

To the east the lower elevations and glacial history of Midland England created a different kind of problem and Marr (1919) and Sandford (1924) showed that in such an area river terraces at about the same level can be of very different age and that the oldest terraces are not always the highest. Towards the east coast surfaces were assumed to have been obscured by the downwarping of the North Sea Basin in Pleistocene times (Wooldridge 1950). Two conventional pieces of work were published by Swinnerton (1936), who described the Mansfield Plateau (180–195 feet) north of the River Trent and proposed that it formed part of an extensive plain of marine denudation truncating parts of present Nottinghamshire and Lincolnshire, and by Versey (1937) who used elevation frequencies to suggest planation at about 800 feet in the Yorkshire Wolds. Studies of river profiles and terraces were largely bound up with reconstructions of glacial history and in 1886 Deeley identified an interglacial terrace on the Trent and Derwent near Derby, in 1896 Bemrose and Deeley described a 30 foot terrace in the valley of the Nottingham Trent, and in 1925 Tomlinson proposed a sequence of postglacial terraces in the valley of the Warwickshire Avon. O. T. Jones (1923–4) described four Pleistocene terrace sequences on the River Mersey, near Manchester, graded to below 80 feet. By far the most important regional work of this genre was that by Wills (1924, 1929, 1937, 1938) who gave a coherent account of the relations of the terraces to glacial fluctuations in the Severn Valley. Starting with a pre-glacial Severn farther to the east, and the Stratford Avon farther to the south, Wills proposed a complex history of ice-front change and drainage migration and incision with five main terraces extending progressively north along the Severn with the ice retreat, graded to the following Severn-mouth baselevels, 200, 110–75, 65–35, 35–15 and less than 25 feet, the four lowest being continuous with terraces on the Stratford Avon.

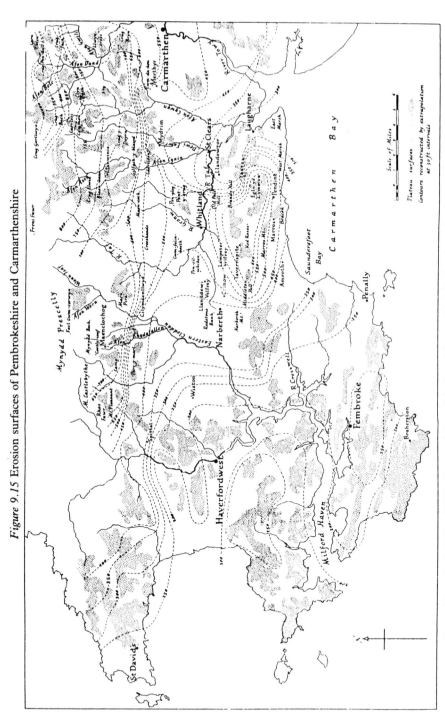

Figure 9.15 Erosion surfaces of Pembrokeshire and Carmarthenshire

Source: From A. A. Miller (1937). By permission of the Royal Geographical Society from the *Geographical Journal*.

Figure 9.16 Superimposed interfluve profiles (above) south of Mynydd Prescelly and (below) west of Afon Cynin (See Figure 9.15).

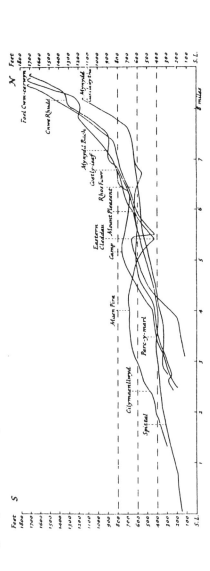

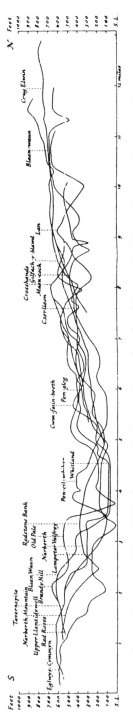

Source: From A. A. Miller (1937) facing p. 151. By permission of the Royal Geographical Society from the *Geographical Journal*.

In northern England and Scotland the recognition of former baselevels of erosion was even more speculative and rootless than in Wales. In the Lake District Hollingworth (1936) postulated the following unwarped surfaces of Mio-Pliocene age: 2,650–2,550, 2,200–2,000, 1,700–1,600, 1150–1,000*, 800–700*, 570 and 400* feet (the best developed ones are marked with asterisks and the lowest ones are probably marine). McConnell (1938) proposed a somewhat different scheme of subaerial erosion surfaces at 2,800–2,300, 2,100–2,000, 1,600–1,500, 1450, 1150–700 and 450–400 feet. Miller, working around the Irish Sea Basin (1938) and in southern Ireland (1939b), identified widespread levels at 200 feet, an incline or succession of surfaces rising to a 450 foot cliff line, and another rising to a 800 foot cliff line with flatter areas especially marked at 600–800 feet. The previously referred to statistical analysis by Hollingworth (1938) (figure 9.17) led him to propose principal planation levels in Highland Britain at 2,000, 1,070–1,000, 800–730 and 430 feet, indicating marine planations associated with successive drops of baselevel, together with secondary or less distinct surfaces at 2,600–1,800 (various), 1,170–1,130, 920–900, 570–550 and 320 feet. Moving eastward from the Lake District, McConnell (1939) described the following subaerial surfaces in the Howgill Fells at 2,000+, 1,850–1,700, 1,600–1,500, 1,400–1,250, 1,150–1,000 and 900–700 feet, providing good elevation correlations with his Lake District surfaces except that the two upper surfaces are higher in the Lake District, possibly due to differential uplift. In the Pennines, Trotter (1929) identified a 'Pennine' surface forming part of a subaerial peneplain on the Alston Block and near the east coast Versey (1937) postulated the existence of a summit peneplain on the North York Moors. At lower elevations Peel (1941) mathematically extrapolated the upper section of the River Tyne along terrace remnants to recognize an earlier baselevel at about 150 feet.

In Scotland, Geikie (1901) ascribed summit accordance in the Southern Uplands and Highlands to Tertiary baselevelling and Barrow et al. (1912) believed that the Grampian Plateau was the result of exhumation by the stripping off of the Old Red Sandstone (Sissons 1976). On lower elevations (400–350 feet) in the Buchan area of north-east Scotland Read (1923) found two rock platforms covered with rounded gravel and flints with interbedded sands and clays which he assumed were of Pliocene marine origin. In the Central Lowlands two surfaces, the higher at 750–500 feet, were believed to be marine. As part of their classic tectonic study of the structure of the north-west Highlands, Peach and Horne (1930) identified two surfaces at 2,950–1,950 feet (High Plateau) and at less than 1,000 feet (Intermediate Plateau). Farther east in the Grampians Fleet (1938) made a cartographic study and named the Grampian Main Surface (3,000–2,300 feet) above which the Cairngorm monadnocks rise, the Grampian Lower Surface (2,000–1,400 feet) and the Grampian Valley Benches (1,000–750 feet), the latter being

Figure 9.17 Hollingworth's (1938, Plate 10) altimetric curves generalized for six regions of Britain (redrawn). These were constructed on the basis of the frequency of summits, grouped at 20 foot intervals

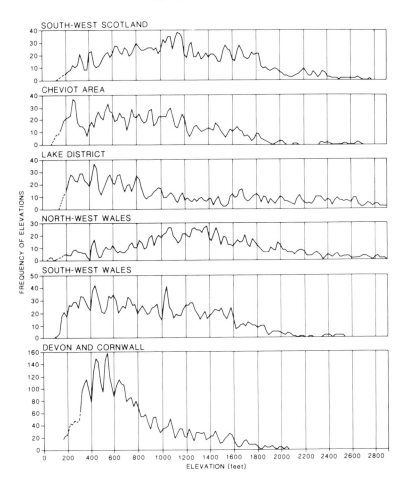

Source: Reproduced by permission of the Geological Society from the *Quarterly Journal.*

of relatively small extent. Hollingworth's (1938) statistical study inferred marine surfaces in south-west Scotland at about 2,700–2,600, 1050, 800–720, 550 and 400 feet, together with those at 1,070, 920, 760 and 560 feet in the Cheviot Hills.

What can be made of this welter of observation and inference directed towards the identification of former baselevels of marine and subaerial erosion? For his part Wooldridge (1950) was convinced that Britain as a whole provides ample evidence of Tertiary eustatic shifts of sea level, clearly supporting the work of Baulig (1935) in respect of his 180 and 280 m surfaces. A more realistic view might have been that, in a maritime region affected by Alpine tectonics and glacial isostasy and eustasy, it is not possible to deduce a single and widespread clear denudation chronology for much of the 60 million years of Tertiary history, particularly in the general absence of deposits of known origin and date.

References

Andrews, W. R. (1891) 'The Origin and mode of formation of the Vale of Wardour', *Wiltshire Archaeological and Natural History Magazine* 26: 258–69.

Avebury, Lord (1902) *The Scenery of England: and the causes to which it is due*, London: Macmillan.

Bailey, E. B. *et al.* (1916) *The Geology of Ben Nevis and Glen Coe*, Memoirs of the Geological Survey of Scotland.

Balchin, W. G. V. (1937) 'The erosion surfaces of North Cornwall', *Geographical Journal* 90: 52–63.

Barrow, G. (1908) 'The high level platforms of Bodmin Moor', *Quarterly Journal of the Geological Society of London* 64: 384–400.

—— (1918) 'Excursion to Chorley Wood', *Proceedings of the Geologists' Association* 29: 140–8.

—— (1919) 'Some future work for the Geologists' Association', *Proceedings of the Geologists' Association* 30: 36–48.

Barrow, G. *et al.* (1912) *The Geology of the Districts of Braemar, Ballater and Glen Clova*, Memoirs of the Geological Survey of Scotland.

Baulig, H, (1935) 'The changing sea level', *Institute of British Geographers, Publication* no. 3.

Bemrose, H. H. and Deeley, R. M. (1896) 'Discovery of mammalian remains in the old river gravels of the Derwent near Derby', *Quarterly Journal of the Geological Society of London* 52: 497–510.

Bremner, A. (1942) 'The origin of the Scottish river system', *Scottish Geographical Magazine* 58: 15–20, 54–9, 99–103.

Brown, E. H. (1960) *The Relief and Drainage of Wales*, Cardiff: University of Wales Press.

Buckman, S. S. (1899) 'The development of rivers, and particularly the genesis of the Severn', *Natural Science* 14: 273–89.

Bull, A. J. (1932) 'Note on the geomorphology of the Arun Gap', *Proceedings of the Geologists' Association* 43: 274–60.

—— (1936) 'Studies in the geomorphology of the South Downs', *Proceedings of the Geologists' Association* 47: 99–129.

—— (1940) 'Cold conditions and landforms in the South Downs', *Proceedings of the Geologists' Association* 51: 63–71.

—— (1942) 'Pleistocene chronology', *Proceedings of the Geologists' Association* 53: 1–45.

Bull, A. J., Gossling, F., Green, J. F. N., Hayward, H. A., Turner, E. A. and Wooldridge, S. W. (1934) 'The River Mole: Its physiography and superficial deposits', *Proceedings of the Geologists' Association* 45: 35–69.

Bury, H. (1908) 'Notes on the River Wey', *Quarterly Journal of the Geological Society of London* 64: 318–34.

—— (1910) 'On the denudation of the western end of the Weald', *Quarterly Journal of the Geological Society of London* 66: 640–92.

—— (1922) 'Some high level gravels in north-east Hampshire', *Proceedings of the Geologists' Association* 33: 81–103.

—— (1926) 'The rivers of the Hampshire Basin', *Proceedings of the Hampshire Field Club* 10: 1–12.

—— (1936) 'Some anomalous river features in the Isle of Purbeck', *Proceedings of the Geologists' Association* 47: 1–10.

Cadell, H. M. (1886) 'The Dumbartonshire highlands', *Scottish Geographical Magazine* 2: 337–47.

Challinor, J. (1930) 'The hill-top surface of North Cardiganshire' *Geography* 15: 651–6.

Chandler, R. H. (1909) 'On some dry Chalk valley features', *Geological Magazine* 6: 538–9.

Chatwin, C. P. (1927) 'Fossils from the ironsands on Netley Heath (Surrey)', *Memoirs of the Geological Survey Summer Programme 1926* pp. 154–7.

Clarke, B. B. (1936) 'The post-Cretaceous geomorphology of the Black Mountains', *Proceedings of the Birmingham Natural History and Philosophical Society* 16: 154–72.

Deeley, R. M. (1886) 'Pleistocene succession in the Trent basin', *Quarterly Journal of the Geological Society of London* 42: 437–80.

Dewey, H. (1916) 'On the origin of river gorges in Cornwall and Devon', *Quarterly Journal of the Geological Society of London* 72: 63–76.

—— (1918) 'On the origin of some landforms in Caernarvonshire', *Geological Magazine* 55: 145–57.

Dewey, H., Wooldridge, S. W., Cornes, H. W. and Brown, E. E. S. (1925) 'The geology of the Canterbury district', *Proceedings of the Geologists' Association* 36: 257–84.

Dewhirst, M. (1930) 'The rivers of Wales in relation to structure lines', *Geography* 15: 374–83.

Dines, H. G. and Chatwin, C. P. (1930) 'Pliocene sandstone from Rothamsted (Hertfordshire)', *Memoirs of the Geological Survey Summer Programme 1929* pp. 1–7.

Dines, H. G. and Edmunds, F. H. (1933) *The Geology of the Country around Reigate and Dorking*, Memoirs of the Geological Survey.

Ellis, T. S. (1882) 'On some features in the formation of the Severn Valley as seen near Gloucester', *Transactions of the School of Science and Philosophical Society of Gloucester* pp. 2–15.

Fagg, C. C. (1923) 'The recession of the Chalk escarpment and the development of Chalk valleys in the regional survey area', *Proceedings and Transactions of the Croydon Natural History Society* 9: 93–112.

Fearnsides, W. G. (1910) 'North and Central Wales', in *Geology in the Field*, Geologists' Association, pp. 786–825.

Fisher, O. (1866) 'On the probable glacial origin of certain phenomena of denudation', *Geological Magazine* 3: 483–7.

Fleet, H. (1938) 'Erosion surfaces in the Grampian Highlands of Scotland', *Rapp. Comm. Cartog. des Surfaces d'App. Tertiaire, International Geographical Union* pp. 91–4.

Geikie, A. (1865) *The Scenery of Scotland*, 1st edn, London: Macmillan (3rd edn 1901).

George, T. N. (1938) 'Shoreline evolution in the Swansea District', *Proceedings of the Swansea Scientific and Field Naturalists' Society* 2: 23–48.

—— (1942) 'The development of the Towy and Upper Usk drainage pattern', *Quarterly Journal of the Geological Society of London* 98: 89–137.

Gilbert, C. J. (1920) 'On the occurrence of extensive deposits of high level sands and gravels resting upon Chalk at Little Heath near Berkhamsted', *Quarterly Journal of the Geological Society of London* 75: 32–43.

Goodchild, J. G. (1888–9) 'The history of the River Eden', *Transactions of the Cumberland and Westmorland Association*.

Goskar, K. L. (1935) 'The form of the high plateau in South Wales', *Proceedings of the Swansea Scientific and Field Naturalists' Society* 1: 305–12.

Goskar, K. L. and Trueman, A. E. (1934) 'The coastal plateaux of South Wales', *Geological Magazine* 71: 468–77.

Gossling, F. (1936) 'Note on a former high level erosion surface about Oxted', *Proceedings of the Geologists' Association* 47: 316–21.

—— (1940) 'A contribution to the Pleistocene history of the upper Darent Valley', *Proceedings of the Geologists' Association* 51: 311–40.

Green, J. F. N. (1936) 'The terraces of southernmost England', *Quarterly Journal of the Geological Society of London* 92: 58–88.

—— (1941) 'The high platforms of east Devon', *Proceedings of the Geologists' Association* 52: 36–52.

—— (1943) 'The age of the raised beaches of Southern Britain', *Proceedings of the Geologists' Association* 54: 129–40.

Greenly, E. (1919) *The Geology of Anglesey*, Memoirs of the Geological Survey no. 2, 2 vols.

Gregory, J. W. (1894) 'The evolution of the Thames', *Natural Science* 5: 97–108.

—— (1920) 'The pre-glacial valleys of Arran and Snowdon', *Geological Magazine* 57: 148–64.

—— (1922) *The Evolution of the Essex Rivers and the Lower Thames*, Colchester: Benham.

Gullick, C. F. W. R. (1936) 'A physiographical survey of west Cornwall', *Transactions of the Royal Geological Society of Cornwall* 14: 380–99.

Hanson-Lowe, J. (1935) 'The clinographic curve', *Geological Magazine* 72: 180-4.

Harmer, F. W. (1902) 'A sketch of the later Tertiary history of East Anglia', *Proceedings of the Geologists' Association* 7: 416–79.

Hawkins, H. L. (1923) 'The relation of the River Thames to the London Basin', *Report of the British Association for the Advancement of Science 1922* pp. 365–6.

Hollingworth, S. E. (1929) 'The evolution of the Eden drainage in the south and west', *Proceedings of the Geologists' Association* 40: 115–38.

—— (1936) 'High level erosional platforms in Cumberland and Furness', *Proceedings of the Yorkshire Geological Society* 23: 159–77.

—— (1938) 'The recognition and correlation of high-level erosion surfaces in Britain: A statistical study', *Quarterly Journal of the Geological Society of London* 94: 55–84.

Hughes, T. M. (1885) 'Notes on the geology of the Vale of Clywd', *Proceedings of the Chester Society of Natural Science and Literature* 3: 5–37.

Hull, E. (1855) 'Physical geography and Pleistocene phenomena of the Cotteswold Hills', *Quarterly Journal of the Geological Society of London* 11: 477–94.

—— (1882) *Contributions to the Physical History of the British Isles*, London: Stanford.

Hull, E. and Whitaker, W. (1861) *The Geology of Parts of Oxfordshire and Berkshire*, Memoirs of the Geological Survey.

Jones, D. K. C. (1981) *The Geomorphology of the British Isles: Southeast and Southern England*, London: Methuen.

Jones, O. T. (1911) 'The physical features and geology of Central Wales', *Aberystwyth and District National Union of Teachers Souvenir* p. 25.

—— (1923–4) 'The origin of the Manchester Plain', *Journal of the Manchester Geological Society* 39–40: 89–123.

—— (1924) 'The upper Towy drainage system', *Quarterly Journal of the Geological Society of London* 80: 568–609.

—— (1930) 'Some episodes in the geological history of the Bristol Channel region', *Report of the British Association for the Advancement of Science* pp. 57–82.

—— (1951) 'The drainage systems of Wales and the adjacent regions', *Quarterly Journal of the Geological Society of London* 107: 201–25.

Jones, R. O. (1931) 'The development of the Tawe drainage', *Proceedings of the Geologists' Association* 42: 305–21.

—— (1939) 'The evolution of the Neath-Tawe drainage', *Proceedings of the Geologists' Association* 50: 530–66.

Jukes-Browne, A. J. (1883) 'On the relative ages of certain river valleys in Lincolnshire', *Quarterly Journal of the Geological Society of London* 39: 596–610.

—— (1895) 'The origin of valleys of the Chalk Downs of North Dorset', *Proceedings of the Dorset Natural History and Antiquarian Field Club* 16: 8.

—— (1906) 'The Clay with Flints: Its origin and distribution', *Quarterly Journal of the Geological Society of London* 62: 132–64.

—— (1911) *The Building of the British Isles*, 3rd edn, London: Stanford (4th edn 1922).

Keeping, W. (1882) 'The glacial geology of Central Wales', *Geological Magazine* 19: 251–7.

King, C. A. M. (1976) *The Geomorphology of the British Isles: Northern England*, London: Methuen.

King, W. B. R. and Oakley, K. P. (1936) 'The Pleistocene succession in the lower parts of the Thames Valley', *Proceedings of the Prehistoric Society* 2(1): 52–76.

Kirkaldy, J. F. (1963) 'Sidney William Wooldridge', *Proceedings of the Geologists' Association* 74: 122–6.

—— (1975) 'William Topley and "The Geology of the Weald"' *Proceedings of the Geologists' Association* 86: (4): 373–88.

Kirkaldy, J. F. and Bull, A. J. (1940) 'The geomorphology of the rivers of the southern Weald', *Proceedings of the Geologists' Association* 51: 115–50.

Lake, P, (1900) 'Bala Lake and the river systems of North Wales', *Geological Magazine* decade IV, 7: 141, 204–15, 241–5.

—— (1934) 'The rivers of Wales and their connection with the Thames', *Science Progress* 113: 23–39.

Linton, D. L. (1930) 'Notes on the development of the western parts of the Wey drainage system', *Proceedings of the Geologists' Association* 41: 160–74.

—— (1931) 'On the relations between the early and mid-Tertiary planation surfaces of south-east England', *Comptes Rendus, International Geographical Congress (Paris)* 2: 1–6.

—— (1932) 'The origin of the Wessex rivers', *Scottish Geographical Magazine* 48: 149–66.

—— (1933) 'The origin of the Tweed drainage system', *Scottish Geographical Magazine* 49: 162–75.

—— (1934) 'On the former connection between the Clyde and the Tweed', *Scottish Geographical Magazine* 50: 82–92.

—— (1969) 'The formative years in geomorphological research in south-east England', *Area* no. 2, pp. 1–8.

Lucy, W. C. (1872) 'On the gravels of the Severn . . .', *Proceedings of the Cotteswold Naturalists' Field Club* 5: 71–142.

McConnell, R. B. (1938) 'Residual erosion surfaces in mountain ranges', *Proceedings of the Yorkshire Geological Society* 24: 76–98.

—— (1939) 'The relic surfaces of the Howgill Fells', *Proceedings of the Yorkshire Geological Society* 24: 152–64.

Mackinder, H. J. (1902) *Britain and the British Seas*, Oxford (2nd edn 1907).

Marr, J. E. (1900) *The Scientific Study of Scenery*, London: Methuen (9th edn 1943).

—— (1906) 'The influence of the geological structure of the English Lakeland upon its present features', *Quarterly Journal of the Geological Society of London* 62: 64–128.

—— (1919) 'Submergence and glacial climates during the accumulation of the Cambridgeshire Pleistocene deposits', *Quarterly Journal of the Geological Society of London* 75: 204–29.

Martin, E. A. (1920) 'The glaciation of the South Downs', *Transactions of the South-Eastern Naturalists' Union* 25: 13–30.

Miller, A. A. (1935) 'The entrenched meanders of the Herefordshire Wye', *Geographical Journal* 85: 160–78.

—— (1937) 'The 600 ft. platform in Carmarthenshire and Pembrokeshire', *Geographical Journal* 90: 148–59.

—— (1938) 'Pre-Glacial surfaces around the Irish Sea basin', *Proceedings of the Yorkshire Geological Society* 24: 31–59.

—— (1939a) 'Attainable standards of accuracy in the determination of preglacial sea levels by physiographic methods', *Journal of Geomorphology* 2: (2): 95–115.

—— (1939b) 'River development in southern Ireland', Proceedings of the Royal Irish Academy 45: section B, 14: 321–54.

Mort, F. W. (1918) 'The rivers of southwest Scotland', *Scottish Geographical Magazine* 34: 361–8.

North, F. J. (1929) *The Evolution of the Bristol Channel*, Cardiff: National Museum of Wales.

Peach, B. N. and Horne, J. (1930) *Chapters in the Geology of Scotland*, Oxford.

Peel, R. F. (1941) 'The North Tyne Valley', *Geographical Journal* 28: 5–19.

Phillips, J. (1871) *The Geology of Oxford and the Valley of the Thames*, Oxford.

Prestwich, J. (1858) 'On the age of some sands and iron-stones on the North Downs', *Quarterly Journal of the Geological Society of London* 14: 322–35.

—— (1890) 'On the relation of the Westleton Beds, or Pebbly Sands of Suffolk, to those of Norfolk and on their extension inland', *Quarterly Journal of the Geological Society of London* 46: 84–154.

Pugh, W. J. (1967) 'Owen Thomas Jones 1878–1967', *Biographical Memoirs of Fellows of the Royal Society* 13: 223–43.

Ramsay, A. C. (1846) *The Denudation of South Wales and the Adjacent English Counties*, Memoirs of the Geological Survey, no. 1.

—— (1872) 'On the river courses of England and Wales', *Quarterly Journal of the Geological Society of London* 28: 148–60.

—— (1876) 'On the physical history of the Dee, Wales', *Quarterly Journal of the Geological Society of London* 32: 219–29.

—— (1878) *The Physical Geography and Geology of Great Britain*, London: Stanford.

Read, H. H. (1923) *The Geology of the Country Around Banff, Huntly and Turriff*, Memoirs of the Geological Survey of Scotland.

Reed, F. R. C. (1901) *The Geological History of the Rivers of East Yorkshire*, Cambridge.

Reid, C. (1887) 'On the origin of dry Chalk valleys and of Coombe rock', *Quarterly Journal of the Geological Society of London* 43: 364–73.

—— (1890) *The Pliocene Deposits of Britain*, Memoirs of the Geological Survey.

—— (1899) *The Geology of the Country Around Dorchester*, Memoirs of the Geological Survey.

—— (1902) *The Geology of the Country Around Ringwood*, Memoirs of the Geological Survey.

—— (1903) *The Geology of the Country Around Chichester*, Memoirs of the Geological Survey.

Salter, A. E. (1905) 'On the superficial deposits of central and parts of southern England', *Proceedings of the Geologists' Association* 19: 1–56.

Sandford, K. S. (1924) 'River gravels of the Oxford district', *Quarterly Journal of the Geological Society of London* 80: 113–79.

Saner, B. R. and Wooldridge, S. W. (1929) 'River-development in Essex', *Essex Naturalist* 22: (5): 244–50.

Sawicki, L. (1912) 'Die Einebnungsflächen in Wales und Devon', *Comptes Rendus des Séances de la Société Scientifique de Varsovie* 5th année, fasc. 2, pp. 123–34.

Sherlock, R. L. (1924) 'The superficial deposits of south Buckinghamshire and south Hertfordshire, and the old course of the Thames', *Proceedings of the Geologists' Association* 35: 1–28.

Sherlock, R. L. and Noble, A. H. (1912) 'On the glacial origin of the Clay-with-Flints and on the former course of the Thames', *Quarterly Journal of the Geological Society of London* 68: 199–212.

Shrubsole, O. A. (1890) 'Valley-gravels about Reading', *Quarterly Journal of the Geological Society of London* 46: 582–94.

Sissons, J. B. (1976) *The Geomorphology of the British Isles: Scotland*, London: Methuen.

Sölch, J. (1936) 'Alte Flächensysteme und pleistozäne Talformung im Snowdongebiet', *Sitzungsberichte der Heidelberger Akademie der Wissenschaften, Mathematisch-naturwissenschaftliche Klasse* 5: 3–31.

Solomon, J. D. (1935) 'The Westleton Series of East Anglia: Its age, distribution and relations', *Quarterly Journal of the Geological Society of London* 91: 216–38.

Sparks, B. W. (1949)'The denudation chronology of the dip-slope of the South Downs', *Proceedings of the Geologists' Association* 60: 165–215.

—— (1972) *Geomorphology*, 2nd edn, London: Longman.

Stoddart, D. R. (1987) 'Geographers and geomorphology in Britain between the wars', in R. W. Steel (ed.) *British Geography 1918–1945*, Cambridge, pp. 156–76.

Strahan, A. (1902) 'On the origins of the river system of South Wales and its connection with the Thames', *Quarterly Journal of the Geological Society of London* 58: 207–22.

Swinnerton, H. H. (1936) 'The physical history of east Lincolnshire', *Transactions of the Lincolnshire Naturalists' Union*.

Tomlinson, M. E. (1925) 'River terraces of the lower valley of the Warwickshire Avon', *Quarterly Journal of the Geological Society of London* 81: 137–68.

Topley, W. (1875) *The Geology of the Weald*, Memoirs of the Geological Survey.

Trotter, F. M. (1929) 'The Tertiary uplift and resultant drainage of the Alston Block and adjacent areas', *Proceedings of the Yorkshire Geological Society* 21: 161–80.

Trueman, A. E. (1938) 'Erosion levels in the Bristol district and their relation to the development of the scenery', *Proceedings of the Bristol Naturalists' Society* 4th series, 8: 402–28.

Versey, H. C. (1937) 'The Tertiary history of east Yorkshire' *Proceedings of the Yorkshire Geological Society* 23: 302–314.

—— (1942) 'The build of Yorkshire', *The Naturalist* pp. 27–37.

Watts, W. W. (1903) 'The Charnwood Forest: A buried Triassic landscape', *Geographical Journal* 21: 623–33.

Wells, A. K. and Wooldridge, S. W. (1923) 'Notes on the geology of Epping Forest', *Proceedings of the Geologists' Association* 34: 244–52.

Whitaker, W. (1889) *The Geology of London and Part of the Thames Basin*, Memoirs of the Geological Survey.

White, H. J. O. (1910) *The Geology of the Country Around Alresford*, Memoirs of the Geological Survey.

—— (1923) *The Geology of the Country South and West of Shaftesbury*, Memoirs of the Geological Survey.

Wills, L. J. (1924) 'The development of the Severn Valley in the neighbourhood of Iron-Bridge and Bridgnorth', *Quarterly Journal of the Geological Society of London* 80: 274–314.

—— (1929) *The Physiographical Evolution of Britain*, London: Arnold.

—— (1937) 'The Pleistocene history of the West Midlands', *British Association for the Advancement of Science* Nottingham, pp. 71–94.

—— (1938)'The Pleistocene development of the Severn from Bridgnorth to the sea', *Quarterly Journal of the Geological Society of London* 94: 161–242.

Wooldridge, S. W. (1926) 'The structural evolution of the London Basin', *Proceedings of the Geologists' Association* 37: 162–96.

—— (1927) 'The Pliocene history of the London Basin', *Proceedings of the Geologists' Association* 38: 49–132.

—— (1928) 'The 200-foot platform in the London Basin', *Proceedings of the Geologists' Association* 39: 1–26.

—— (1932) 'The physiographic evolution of the London Basin', *Geography* 17: 99–116.

—— (1936) 'River profiles and denudation chronology in southern England', *Geological Magazine* 73: 1–16.

—— (1938) 'The glaciation of the London Basin and the evolution of the Lower Thames drainage system', *Quarterly Journal of the Geological Society of London* 94: 627–67.

—— (1950) 'The upland plains of Britain: Their origin and geographical significance', *Report of the British Association for the Advancement of Science, Section E*, pp. 162–75.

—— (1951) 'The progress of geomorphology', in G. Taylor (ed.) *Geography in the Twentieth Century*, London: Methuen, pp. 165–77.

Wooldridge, S. W. and Bull, A. J. (1925) 'The geomorphology of the Mole Gap', *Proceedings of the Geologists' Association* 36: 1–10.

Wooldridge, S. W. and Linton, D. L. (1938a) 'Some episodes in the structural evolution of south-east England', *Proceedings of the Geologists' Association* 49: 264–91.

—— (1938b) 'Influence of the Pliocene transgression on the geomorphology of south-east England', *Journal of Geomorphology* 1: 40–54.

—— (1939) 'Structure, surface and drainage in south-east England', *Institute of British Geographers, Publication no. 10*.

Yates, R. A. (1957) 'Physiographical evolution', in E. G. Bowen (ed.) *Wales*, London: Methuen, pp. 19–52.

CHAPTER TEN

German Tectonic Geomorphology

Albrecht Penck

We have already devoted considerable space in volume 2 and in chapter 1 of the present volume to the oversimplified Davisian assumption of a common tectonic pattern of sudden, short uplift followed by prolonged downwearing, as well as to its critics. Of the latter, there were none more powerful than Albrecht Penck and his son, Walther. In 1894 Albrecht Penck provided geomorphologists with their first unified text, *Morphologie der Erdoberfläche*, which has been referred to several times previously in this volume and later will furnish certain morphometric aspects for discussion in chapter 11. Similar in historical position, age and status, Albrecht Penck (1858–1945) constantly invites comparison with W. M. Davis (1850–1934). Their careers, however, were strikingly different. Davis began with strong geographical leanings and only in later years did he produce his best geomorphology; Penck was a superb young geomorphologist but became increasingly enmeshed in a racially dominated human geography which had disastrous consequences.

Penck (plate 10.1) was born on 25 September 1858 at Reuditz near Leipzig in Germany which happened to be close to the southern limit of the maximum advance of the Scandinavian Quaternary ice-sheet (Louis 1958, Beckinsale 1974). In 1875 he went to study natural sciences – chemistry, geology, botany, mineralogy and petrography – at the University of Leipzig and in that year Otto Torell of Stockholm delivered a forceful lecture at Berlin which persuaded listeners that the glacial clay of the north European plain was carried by a continental ice-sheet and not by floating icebergs. Within 2 years Penck had written about a northern 'basalt' erratic embedded in the diluvium near Leipzig. In 1879 he used detailed analyses of the sequence of glacial sedimentation to postulate at least three main ice advances interspersed with two interglacial phases during which rivers laid down normal bedded deposits. In the following year he worked under Karl von Zittel at Munich and so found himself near the outer or northern limit of Alpine ice-sheet advances. Thereafter he wrote extensively on Alpine glaciation and eventually compiled, in collaboration with Eduard Brückner, the three-volume *Die Alpen im Eizeitalter* (1901–9) (see chapter 2). Penck's climatic geomorphology will be treated in chapter 12.

Plate 10.1 Albrecht Penck

Source: Courtesy G. J. Martin

In the meanwhile in 1885, when 27 years old, Penck was elected professor of physical geography at the University of Vienna where his distinguished students and foreign visitors during the next two decades included E. Brückner, J. Cvijić, F. Machatschek, A. Gründ, O. Lehmann, E. de Martonne and L. von Sawicki. He travelled and lectured in many parts of the world and continued to do so after 1906 when he accepted the chair of geography in succession to Ferdinand von Richthofen at the Geographisches Institut, Berlin.

Volume 2 (see chapter 13) traced Davis' initial friendship with Albrecht Penck at around the turn of the century. In the summer of 1897 they travelled to the Rockies and across Canada with a party and in 1905 they both took part in the British Association field trip to South Africa. Davis was involved in an academic exchange with Penck during the year 1908–9 (see volume 2, chapter 15) and even went so far as to write a poem for Penck to insert in his copy of *Geographical Essays*. However, the publication of Davis' two important works in German (1911M, 1912J) began a cooling in their relationship, which became more frigid during the First World War (see volume 2, chapter 21). After the war the situation was made worse by Davis' (1920G) critical reviews of the Penck *Festband* (1920G) and of Penck's (1919) *Gipfelflur* paper (see volume 2, chapter 22).

After the First World War Penck concentrated on Quaternary and Alpine problems but became increasingly engrossed with administration and with cultural themes. He retired from the chair of geography at Berlin in 1926 but his influence persisted through his writings, his editorial work and his notable students. In 1927 the wartime breach of friendship with W. M. Davis was largely healed after the untimely death of Walther Penck whom Davis admired (see volume 2, p. 583). In 1928, in his seventieth year, Albrecht presided with distinction over the centennial celebrations of the Berlin *Gesellschaft für Erdkunde* and the combined meeting of Oceanic conferences. During the Second World War his house in Berlin was damaged by bombs and he moved to Prague where he died in 1945 at the age of 87.

Any assessor of Albrecht Penck's contributions to landform studies faces the complication that he was never reluctant to accept new theories nor to change his own. For example, his early descriptions were free of Davisian ideas, then the cycle concept was introduced for a few years, only to be sharply refuted after 1918. At first Penck agreed with Suess that secular variation in the relative altitude of land and sea was eustatic rather than diastrophic but by 1900 he had accepted that independent crustal movement was a factor in the raising or lowering of coastlines. All told he published over 400 books and articles and from a scientific viewpoint his chief contributions were to Quaternary geology, geomorphology, hydrology and cartography. To geomorphology his main contributions involved the general classification of landforms, the knowledge and description of individual landforms, especially glacial, and the significance of climatic change in landform analysis. We will discuss his notable work on Quaternary chronology and on glacial topics in volume 4, but here we return to his influential ideas on Alpine summits (see volume 2, pp. 528–36).

The Level of Mountain Summits

According to Machatschek (1969: 83):

> In the gradual development of a mountain landscape, one phenomenon appears common to almost all mountains: neighbouring peaks are about the same height and the highest peaks lie in almost one and the same plane.

Albrecht Penck and others recognized this accordance of summit levels quite early and in 1879 E. von Mojsisovics had attempted to explain it by a certain degree of uplift of individual parts of the range which maintained the balance of the whole mountain range by ensuring that downward destruction was offset by uplift from below. In 1887 Penck expressed the opinion that the tangential maximum summit plane was caused by levelling brought about by the presence of an upper limit above which uplift cannot elevate the mountains because maximum uplift rates do not exceed the high denudation rates encountered at these elevations. In contrast, W. M. Davis used earlier ideas to interpret the high-level accordances of various mountain ranges as raised peneplains, so assuming that denudation had preceded uplift. The raised peneplain summit model was first used to explain the constancy of summit heights in ranges of plateau-like character but some American scientists began to apply it also to Alpine-type ranges in the west of North America. However, in 1905 R. A. Daly in his discussion of the accordance of summit levels among Alpine mountains clearly grasped the difference between the two main theories relating to the phenomenon. He considered that the almost-level summit planes had been either inherited from earlier levelled areas or were upper denudation levels which appeared only during the creation of a mountain range. Thus, the 'inherited' accordances were on the way to extinction, whereas the 'newly created' accordances were in process of formation and maintenance. Although the peneplain model was shortly afterwards applied to the Alps (Staff and Rasmuss 1912), at about this time both Davis (1912J: 275, 286) and Hettner (1913) accepted the possibility of an upper limit imposed by denudation.

At this point Albrecht Penck, under the influence of his son Walther, re-enters the argument with his *Die Gipfelflur der Alpen* or, in English, *The Summit Plane* (or peak-level) *of the Alps* (1919). He describes the typical Alpine summits, where ridges are sharp and exposed to rapid destruction because of the great intensity of mechanical weathering and the steepness of the slopes. This weathering increases with altitude and, other conditions being equal, will tend in the long term to lessen the difference in height between the summits and bases of ridges. After a certain time, however, the deposition of debris will protect surfaces, slowing up this process, unless the debris is carried away and the steepness of the slopes maintained.

Penck now embarks on an account of the action of snow and ice on mountain landforms, including the formation of sharply pointed ridges and also of steep-walled, flat-floored cirques which we will deal with in volume 4. But, he continues, the ridges of the Alps are all sharp and do not appear to indicate the presence of an earlier peneplain (*Rumpffläche*) from which the mountains might have been carved. The sharpened ridges between very steep valley sides appear in fact to be largely independent of glacial action, although it has effected some alterations such as increased sharpening higher up and truncation lower down. The landforms are those which must necessarily develop between deeply cut valleys when the valley depth becomes greater than one-quarter of the intervalley spacing. In the highest parts of the Alps, as in every mountain range with sharp ridges, the height of peaks and ridges is influenced by the depth of the valleys. As adjacent valley floors are usually on the same level they conform to a similar pattern and, because in those regions of the Alps the deepening of valleys still proceeds, the ridges are sharpened repeatedly. It seems pointless to consider these sharp ridges as shapes inherited from an earlier peneplain; rather the summits should be looked upon as the result of the elevation of the mountain range.

Apart from sharpened ridges there occur, especially in limestone regions in the north and south of the Alps, rounded crests and extensive plateaus. Von Staff thought that they were the remnants of former *Rumpfflächen*, but Penck points out that their altitudes vary widely and that the extensive coarse deposits north of the Alps indicate the presence there of a great Tertiary Alpine mountain system. The rounded crests, Penck suggests, might well be the sloping part of destroyed rounded mountains and it is probable that rounded ridges and sharp ridges have existed for a long time. It appears that in different regions both rounded and sharp-edged shapes can develop from a common base and that rounded shapes can develop from sharp-edged shapes and vice versa. For Penck the downwearing process may follow one of three possible sequences of transformations, differentiated largely by differences in the intensity and duration of the uplift (see volume 2, pp. 528–31).

The first transformation sequence is characterized by strong and prolonged uplift. Valley incision is rapid, but at a lesser rate than uplift, so that the valley floors become elevated above the original plain. Between the valleys, sections of the uplifted plain rise up as *Reidels* (upland strips) which are gradually eaten away by the backwearing of the valley sides, and finally vanish when the slopes of adjacent valleys intersect in a sharp edge. If strong uplift continues, the sharp ridges do not keep pace with the general rise of the land surface but rise only in proportion to the rise of the valley floors. Thus the height difference between the ridges and the valley floors remains almost constant. If the time comes when river erosion becomes strong enough to counteract the land upthrust, this erosion and destruction on the slopes will create a balance between uplift and levelling. The upper uplift level has now been reached, but as long as uplift continues peaks and ridges that have developed in the mountain range will maintain a constant height. Only when uplift slackens is it possible for rivers to cut down the valley floors in absolute terms and so bring about the lowering of the sharp ridges between them. Eventually vertical erosion also slackens, valley floors broaden, and sharp-crested ridges become blunted and rounded. Finally, vertical erosion virtually ceases, valley floors become wide and shallow, intervalley ridges become flattened and the land is nearly flat. In this development sequence the most noteworthy stage is that during which the sharp edges remain for a long period at a constant height. Then their level expresses the upper limit of elevation, a plane of peak limit (*Grenz-gipfel-flur*) beyond which the land surface cannot rise under the circumstances. Obviously, this plane does not persist as long as the sharp ridges which occur when the landscape is developing both upward towards, and downward from, this upper limit.

The second transformation sequence is characterized by strong uplift of limited duration. As during the first sequence, *Reidels* are formed quickly but uplift ceases before they can be destroyed by the development of very steep slopes. Sharp ridges are not formed and the mountain range fails to grow as high as the upper limit of elevation but remains of moderate height, as also do internal height differences within it. Its later variety of forms comes from the transformation of the *Reidels* by rounding and levelling.

The third transformation sequence is associated with, and lasts only a little longer than, a period of very slow uplift. There is naturally little river incision, valleys are wide and flat, and the interfluves (all of a small height) are eventually flattened to form a plain (*Rumpf*). The different contents or aspects of the three transformation series are summarized in a diagrammatic table.

Penck then goes on to discuss his general theme in which the three sequences might be associated with one another because separate parts of a mountain range could be uplifted at different rates, and he gives many details of the probable correlation of Alpine morphometry with the rate of uplift. In the eastern Alps, for example, the lines of the great longitudinal valleys are interpreted as strips that indicate a slackening of uplift and are not directly related to geological structure. In the Alps Penck would like to recognize signs of a *Grossfaltung* (i.e. broad folding) (see figures 10.5 and 10.10) affecting an older range in the process of destruction. Individual parts of that range were raised more than others and the folds elevated to the upper limit of elevation (levelling), leaving the flat trough floors below that limit. However, because these troughs were raised well above the lower levelling limit, they continued to be destroyed. There is geological proof for such a broad folding in that many indications of post-Pliocene uplift can be found in the outer parts of the Alps especially in the south where the marine Pliocene deposits in the valleys rise notably in height toward the main Alps:

> Whereas we earlier concluded (*Alpen im Eiszeitalter*, pp. 743, 771, 910) from this that the Alps, as a whole, arched upward after folding, we now tend to believe that we are not dealing here with the formation of one single uparching but with the creation of broad folds which became decisive in the development of the [summit] heights of the mountain range.
>
> (Penck 1919: 268)

Walther Penck in his posthumous *Die morphologische Analyse* (1924; trans. 1953: 231–2) followed his father in suggesting that the uniformity of Alpine summit levels is independent of glacial reshaping, of ancient structural folding and, within wide limits, also of lithology or natural rock resistance to weathering. The upper summit level is probably neither inherited nor carved from a hypothetical peneplain which stretched over the whole Alps, but it does show a strong connection with the distribution of slope form. It seems, in fact, to resemble an undulating arch having ridge zones and downwarped longitudinal valley zones. In the latter, very flat slope units are to some extent preserved on the summits of the intervalley divides; on the ridge zones the *Gipfelflur* coincides with equally elongated, narrow landforms, which, however, lack flattish features because these have been removed by the backwearing of very steep slopes. That the ridge zones once had some summit flattenings is indicated by the convexity which still persists in their slope profiles. Albert Heim (1927) also supported Albrecht Penck's suggestion that an upper denudation level existed due to the balance of strong uplift and denudation. Many other European geologists considered that the peak summit level had developed from a flat relief, commonly as a former peneplain of variously dated Tertiary age, that had survived recent elevation. Others postulated that the *Gipfelflur* revealed appreciable differences in relief (Aigner 1925–6, Krebs 1928; Leutelt 1929). At the same time, the idea that the accordance of summit levels might be associated with snow lines and tree lines on mountains was further developed by Richter (1928) and was demeaned by Henri Baulig (1952) who, in his protracted review of flattened surfaces, invoked great variations of those levels, especially during the Pleistocene.

Fritz Machatschek in his long-lived *Geomorphologie* (1919–69) has much to say on *Gipfelflur*. He thinks that regional variations in its level, where not assignable to lithologic and orographic influences, may be imputed to different amounts of uplift caused by broad-scale warping. A mountain area composed of uniform rock and with equal spacing and depth of valleys would develop highly accordant peak and ridge heights, but he recognized that such areas which have existed undisturbed by tectonic movement long enough to develop *Gipfelflur* must indeed be rare (1969: 83–4). He also mentions some European geologists who, like Richter (1928) and Annaheim (1936), associated the peak summit level with the terraced structure of the Alps. They invoked intermittent uplift or 'stages' in the creation of the mountain range which led to the formation of a tier of peak levels, or peak-level stairway (*Gipfelflurtreppe*), a subject to which we will return later.

In France where Davisian methods were popular, *Gipfelflur* found a place in Martonne's masterly *Traité de Géographie Physique* (e.g. eighth edition 1948: 612–4) but, not surprisingly, it was given a very cool reception. There could be, the author says, other explanations, notably the position of certain intrusive masses (as suggested by R. A. Daly in 1905) or more simply the exhumation of a buried erosion surface. The development of summits at a uniform level simply by the backwearing of slopes seems to presuppose for Martonne two conditions that rarely exist; namely, a geological structure that varies very little from one valley to another in its resistance to erosion, and a uniform drainage density.

The criticism by French scholars paled beside the forthright opposition aroused in leading physiographers in the United States. Davis (1920G) took upon himself the task of answering the condemnations of his cyclic theme expressed in Penck's *Festband* (1918). Penck was greatly disturbed by Davis' unfeeling review and replied at length by letter on 4 December 1920. Its contents may be safely inferred from Davis' reply of 3 April 1921 which we have printed elsewhere in full (see volume 2, pp. 520–7).

This splendid letter was written about 2 years after the publication of Albrecht Penck's article on the *Gipfelflur* concept which was reviewed by Davis (1923J) as 'The cycle of erosion and the summit level of the Alps' (see volume 2, pp. 528–36). The review repeated most of the arguments already made in the private correspondence and showed clearly that Davis could only consider Penck's ideas as elaborations of the cycle of erosion:

> [Penck's article] proposes several refinements of the cycle of erosion; and it employs deduction to an extent that ... has been decried by certain other German geographers ... .
>
> (Davis 1923J: 2)

However, Davis does reserve some praise for Penck:

> Nevertheless, the interaction of upheaval and erosion has never been presented by myself or by anyone else in the beautiful manner deduced by Penck in his first ideal cycle of the 'Gipfelflur' essay; and for that reason his essay should be regarded as marking an extension of the previous treatment of the cycle scheme.
>
> (Davis 1923J: 30)

Plate 10.2 Walther Penck in the Puna de Atacama, 1913

Source: Courtesy Helmut Penck

The American, now over 70 years old, was doomed to face many repetitions of the Penckian attack on the cyclic concept but these were now always incorporated in the writings of Walther Penck. Albrecht survived to the ripe old age of 87, having far outlived his own son and Davis, his former 'dear friend'.

Walther Penck and the Puna de Atacama

Walther Penck (plate 10.2) was born in Vienna on 3 August 1888 and early in life showed a skill in mountaineering and a keen interest in the natural sciences. When his father moved in 1906 to the chair of geography in the University of Berlin, Walther became a student there. In 1908–9 he accompanied his father to the United States where they travelled widely, meeting many famous American geologists, including G. K. Gilbert. They returned home across the Pacific, visiting Hawaii, Japan and China and traversing Siberia. Soon afterwards Walther enrolled and later graduated at the University of Heidelberg (Chorley 1974, Piotrovsky 1984).

Here we must re-introduce Isaiah Bowman (1878–1950), a very competent physiographer who, on the direct advice of W. M. Davis, trained himself to be a complete geographer (see chapter 4). He moved east from Michigan to study under Davis in September 1902, the move being facilitated by the provision of a scholarship at Harvard. In addition, every morning that he lectured Davis provided the impoverished young student with a large breakfast. In the autumn of 1904 when, at Davis' instigation, Albrecht Penck came to Harvard to lecture on the theme of his *Morphologie der Erdoberfläche*, Bowman was thrilled:

> One of my ambitions has been to see and hear Penck – the real *Penck*! I can scarcely wait to see him. I suppose you know of the monograph he's getting out, in sections, on the glacial features in the Alps.
>
> (Quoted in Martin 1980: 12)

Bowman was loaned to Penck to help him with the composition and pronunciation of the lectures he was about to give and he was rewarded with a tactful, generous gift. Later, in 1908–9, when Albrecht Penck was appointed Kaiser Wilhelm Professor of Political Science at Columbia University, he visited Yale to give the Silliman lectures and again met Bowman who had been on the staff of the geology department there since 1905. Young Walther Penck was then temporarily studying at Yale, so he and Bowman became well acquainted. The American had joined an expedition to the Andes in 1907 and went on other surveys there in 1911 and 1913, so accumulating a considerable knowledge which was expressed in many articles and several notable books, including *The Andes of Southern Peru* (1916) and *Desert Trails of Atacama* (1924) (plate 10.3). In the former Bowman adopted a regional diagram of land type and land use which became very popular and also made important contributions to the origin of coastal terraces, the formation of sand dunes and the development of cirques. On his first visit to the central Andes in 1907 he had expected from the textbooks to see a wild, picturesque relief but found instead, on travelling inland from Iquique in Chile, a view to the east of a 40 mile stretch of level Cordilleran skyline:

Plate 10.3 Isaiah Bowman in Andean South America

Source: Courtesy of G. J. Martin

Riding up over the piedmont and the western flank of the Andes I found bevelled surfaces in cross-section on the high quebrada walls as well as at the top of the country. The Central Andes had been peneplaned and the lofty mountains whose altitude seemed so attractive on the map were but moderate sized cones perched on top of extensive lava flows (Western Cordillera) or residual remnants (Eastern Cordillera) rising above old erosion surfaces now in process of destruction.

(Quoted in Martin 1980: 38)

Bowman saw here:

A rare opportunity of making the first scientific interpretation of Andean physiography based on modern analysis of land forms in the field . . .

(Quoted in Martin 1980: 37–8)

This shows, however, that he was basing his analysis on Davisian methods.

It seems more than chance that in 1912 Walther Penck was appointed geologist to the mining department of Buenos Aires, Argentina, and that he should concentrate on surveying and geologically mapping the southern edge of the Puna de Atacama (figure 10.1) (Penck 1933), close to the area described by Bowman (1909). When the First World War broke out in Europe, Walther happened to be on leave in Germany and he served for a while in the German army. At the end of 1915 he became professor of mineralogy and geology at the University of Constantinople and spent the next 30 months mainly on tectonic and stratigraphic studies in Anatolia and around the Sea of Marmora. During the summer of 1918, repeated bouts of malaria forced him to return to Germany and he began he publish substantial accounts of his Turkish fieldwork. Unable to resume his studies in Turkey after the end of the war, he acted as an unsalaried professor at the University of Leipzig and, despite straitened financial circumstances and rampant inflation, continued to investigate the German highlands and to see his Argentine studies through the press. Davis in his belated answer on 3 April 1921 (see volume 2, p. 520) to Albrecht Penck's letter of 4 December 1920 wrote:

It is pleasant news that your son, Walther, is established as professor in Leipzig where his father long ago studied. As he may have told you, I have enjoyed reading parts of his Argentine monograph, an able piece of work, and I have written asking him to specify the difficulties he finds in accepting the cycle theory.

It was during the early 1920s that Walther Penck compiled his most influential comments on the intepretation of landforms through an analysis of the relations between endogenetic (diastrophic) and exogenetic (erosional) processes. His *Wesen und Grundlagen der morphologischen Analyse* (Nature and basis of morphological analysis) (Penck 1920b) was the shortest and least significant of his three important treatises on the subject. He died of cancer of the mouth on 29 September 1923, leaving a devoted wife and two small sons. His saddened father did his utmost to publish what he could of Walther's

Figure 10.1 The southern part of the Puna de Atacama where Walther Penck conducted much of his work in Argentina during the period 1912–14. Contours are at 500, 2,000 and 4,000 m. The area around Fiambalá (F) geologically mapped by Penck is outlined

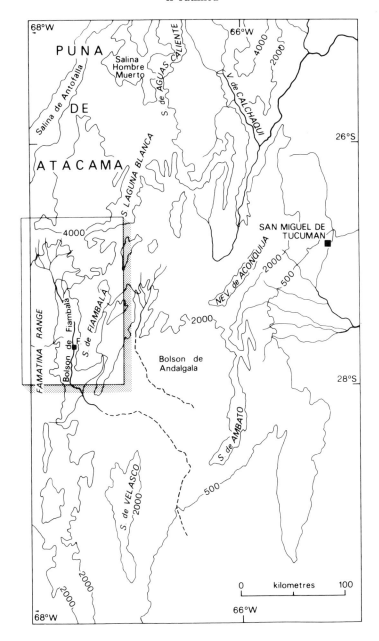

literary remains. Albrecht produced, without any serious editing, the hurriedly written and obscure notes of *Die morphologische Analyse* (Penck 1924) which were, in fact, only part of a contemplated larger opus. Soon afterwards he combined and published two lectures given by Walther at Leipzig in December 1921 on the piedmont flats of the southern Black Forest (Penck 1925). The latter was 'translated' and commented on by Davis (1932G) in a most biased manner (see volume 2, pp. 699–714). The former had to wait for full translation until 1953, during which interval most non-linguistic geomorphologists had to make do with interpretations almost as unclear as Walther Penck's often telegrammatic phrases.

The *Südrand der Puna de Atacama* (Penck 1920a) was completed in Constantinople in the summer of 1918 before Albrecht had finished his *Gipfelflur* paper (1919), and it contains the first statements of many ideas which were refined and elaborated later (e.g. Penck 1924, 1925). In more than 400 pages it concentrates mainly on the broad folds of the Sierra de Fiambalá and the Famatina Range, together with the intervening downwarped Bolson of Fiambalá (plate 10.4). Summits of the ranges of the southern Puna de Atacama rise generally to 4,000–5,000 m or more, with associated bolson floors at between 1,000 and 2,000 m. Although faulting, much of it strike in character, is important structurally, Penck believed that the topography of this, and other, basin and range regions is dominated by anticlinal flexures of the broad fold type, which are not necessarily bounded by faults. Broad folds (*Grossfalt*) are produced by compression and become progressively higher and narrower with time (Penck 1920a: 341–58). Faults and thrusts were believed by Penck to occur in the later stages of folding but not to be dominant in the process of formation of the ranges (see figure 10.5). This contrasts strongly with Davis' (1901B, 1903G, 1905B) elaboration of Gilbert's ideas (see volume 1, pp. 548–9; and also Spurr 1901) on the fundamental block-faulting origin of such regions. Penck further believed that broad folds are usually superimposed on the larger structural features of regional domes (*Gewölbes*) (Simons 1962: 5–6). The folds in the Puna de Atacama were viewed as asymmetrical with the western flanks exhibiting greater flexure with dips of 5–11°, whereas the eastern slopes are more gentle. This tectonic difference was seen by Penck as helping to explain the topographic differences observed on opposite sides of the ranges. In his area of geological mapping (see figure 10.1) rocks as old as the Palaeozoic and pre-Cambrian are present and the region clearly has had a long and complex history culminating in Tertiary folding of various ages. The Sierra de Fiambalá, cored with pre-Mesozoic granites and gneisses, appears to be an older range, whereas the Famatina Range, with a post-Mesozoic core, is younger.

The Bolson of Fiambalá (Penck 1920a: 280–6) has a sedimentary fill of Upper Cretaceous (Calchaqui) sandstone and Tertiary (Puna) sandstone and

Plate 10.4 Example of a broad fold: the Sierra de Fiambalá, western Argentina, from San Salvador. The Famatina Range is in the background

Source: Photograph by W. Penck (from Penck (1953), Plate X(2), facing p. 261). By permission of Macmillan, London and Basingstoke

Figure 10.2 The relations between upland denudation surfaces and nearby basin sedimentation in the Puna de Atacama

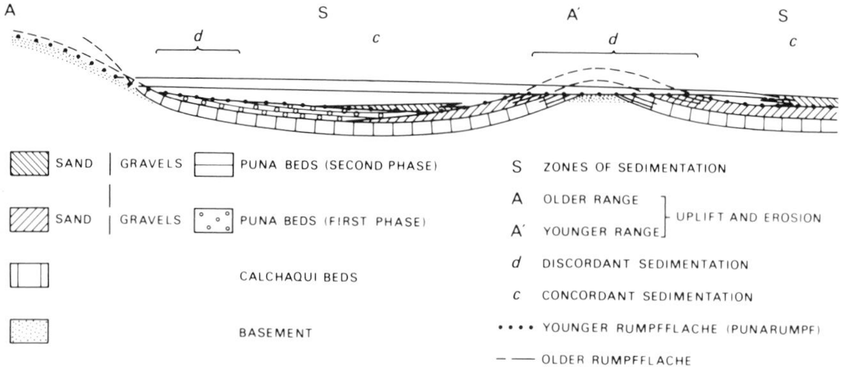

SAND	GRAVELS	PUNA BEDS (SECOND PHASE)
SAND	GRAVELS	PUNA BEDS (FIRST PHASE)
		CALCHAQUI BEDS
		BASEMENT

S   ZONES OF SEDIMENTATION

A   OLDER RANGE ⎤
                              ⎬ UPLIFT AND EROSION
A´ YOUNGER RANGE ⎦

d   DISCORDANT SEDIMENTATION

c   CONCORDANT SEDIMENTATION

•••• YOUNGER RUMPFFLACHE (PUNARUMPF)

— — OLDER RUMPFFLACHE

Source: After Penck (1920a), Figure 14, p. 367

gravel facies divided into earlier and later phases (Penck 1920a: 175–93). To the west, near the base of the east flank of the Famatina Range, are Late Tertiary and Quaternary gravels, and Recent alluvium floors the centre of the bolson (Penck 1920a: 216–56). Penck (figure 10.2) makes it clear that there is a link between surfaces of erosion on, or projected above, the broad folds and the bolson sediments, the former being preserved as unconformities on the basin margins and passing into stratification planes farther out towards the centre of the bolson. The summit, or primary, erosion surface is correlated with the Lower Tertiary (Calchaqui) Beds (figure 10.2). There has been continuous bolson sedimentation since the Upper Cretaceous, when the first gentle broad folds began to rise producing only finer sediments in the basins of sedimentation. Subsequently the resulting bolson sediments have become progressively coarser and more extensive, indicating increasingly rapid uplift of the ranges which continues to the present time (Simons 1962: 7). Penck believed that this facies distribution suggested continuous uplift of the ranges and not intermittent rises. It is interesting that in 1917 Joseph Barrell of Yale had demonstrated that the character of such a sedimentary record is determined as much by the nature of basin subsidence as by the behaviour of the adjacent source area, although the two are clearly linked. However, this view did not receive wide approval until the 1930s.

Penck postulated that an increased rate of stream incision (i.e. waxing development) causes valley-side slopes to steepen with depth in successive units of time. Thus waxing development, convex slopes, high erosion rates and high relief go together (plate 10.5). Conversely, a decreased rate of stream incision (i.e. waning development) produces concave slopes as the

Plate 10.5 Walther Penck in convoy in one of the gorges of the Puna de Atacama

Source: Courtesy Helmut Penck

Figure 10.3 The longitudinal and transverse profiles of a valley cut into the flank of a rising range

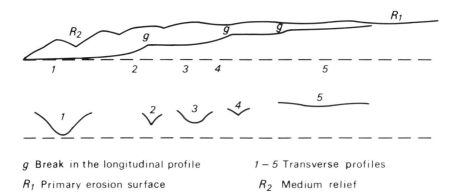

g Break in the longitudinal profile   *1 – 5* Transverse profiles

R₁ Primary erosion surface   *R₂* Medium relief

Source: After Penck (1924); see Penck (1953), Figure 13, p. 200. By permission of Macmillan, London and Basingstoke

individual slope elements retreat parallel to themselves, leaving behind a concavity which is growing progressively towards the divide at the expense of the retreating steeper convex slopes. The lower parts of the developing concavities become flats or benches. By this means a given location may pass successively through conditions of low, medium, high and low relief. On the flanks of well-developed accelerating broad folds a migrating zone of waxing development was thought by Penck to have produced a flight of extending benches by a process which he did not make clear. These benches are not, as Davis would suggest, of radically differing age and caused by intermittent uplift, but due to continuous, accelerating uplift and the ability to continue to develop more-or-less simultaneously (Simons 1962). The extension of any given bench is the result of the erosional backwearing of a steep scarp of waxing erosional development and the bench persists until a lower waxing slope element retreats sufficiently to destroy it. The inception of the waxing slope elements is successive, with the highest being the oldest and those associated with the lowest steps being the youngest. However, all may continue to extend back into the range simultaneously. The highest part of the range is formed of the oldest summit or primary erosion surface (*Rumpffläche*), whereas the lowest and steepest part of the flank of the range is the most recently dissected. Thus a valley extending from the main divide down to the adjacent basin exhibits the oldest features near the valley head and the youngest near the mouth (figure 10.3) – *the reverse of the Davis model*.

Different sides of the ranges of the Puna Andes, as well as different ranges, have different benching. The western flanks of the ranges have been more steeply warped than the eastern and are generally deeply dissected. These flanks either show little evidence of benching, as on the west side of the Sierra de Fiambalá, or limited benches on the restricted divides between the valleys, in places connected with valley-like benches extending eastward into the ranges. The more gently tilted eastern flanks of the ranges exhibit the most complete staircases of benches backed by retreating scarps and dissected by steep valleys connected to lower benches. The higher, older ranges, such as the Sierra de Laguna Blanca, have many benches separated by parallel-retreating convex waxing slopes. Slopes are especially steep near the base of the range where a pronounced concavity connects them to the basin floor. The younger ranges have fewer flats, and these are located near the mountain foot. The youngest ranges have none at all. Figure 10.4 shows a typical profile of the middle and lower part of the eastern flank of an Andean range, possessing different types of characteristic relief:

Relief 1: (not shown) a flat summit erosion surface formed by waxing slope retreat, at the crest of which may still exist inselbergs whose tops preserve evidence of an even earlier surface. Relief 1 has been almost entirely removed over much of the Famatina Range and the Sierra de Fiambalá.

Relief 2: below Relief 1, or forming the tops of some ranges, is a concave (waning) profile leading down to a flat a few kilometres broad, into which valley troughs as much as 500 m deep have been incised. Relief is moderate and there are signs of tectonic warping. This surface may have replaced Relief 1. At its outer edge there is a dissected (waxing) zone of steep, convex slopes up to 500 m high penetrated by deep valleys.

Relief 3: below this dissected zone, and penetrating it along the valleys, is a zone of concave slopes commonly below elevations of about 3,500 m extending down to lower flats and terraces. Like Relief 2, this zone has undergone recent uplift and tilting.

Figure 10.4 Typical profile of the middle and lower eastern slope of a broad fold in the southern Puna de Atacama

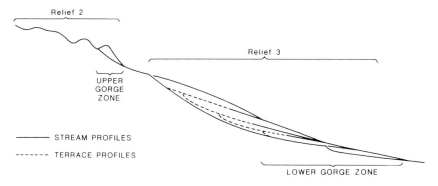

Source: After Penck (1920a), Figure 16, p. 399 By permission of Macmillan, London and Basingstoke

As the subtitle of Penck's (1920a) long article made clear, the primary intention of his work on the southern Puna de Atacama was to enquire into the different mountain types and into the nature of mountain building. His interest in geomorphic and sedimentary features was mainly directed to this end and thus his approach differed radically from the American, French and British traditions, if not from the Russian.

Der Südrand der Puna de Atacama appeared in 1920 and was generally inadequately reviewed for geomorphologists (e.g. Ogilvie 1923). It was left to Isaiah Bowman to attempt to resolve what was clearly developing into a radical departure from the Davis model. In 1915 Bowman had been appointed Director of the American Geographical Society and soon acquired a wide reputation for his scholarly writings, especially on South America and political problems of the world. Late in 1918 he became the chief territorial specialist with the American commission concerned with the peace negotiations in Paris (Martin 1980: 81–97). However, he always retained his interest in physiography and his warm relations with his many geomorphological friends, including his former tutor and mentor, W. M. Davis, and the Pencks. Most of his correspondence with the latter has already been published (Martin 1974). We read that Bowman wrote to Albrecht Penck on 26 July 1923 protesting against Walther's stance on the cycle of erosion and Davisian methodology, as expressed in *Der Südrand der Puna de Atacama*:

> Permit me to say that I have just gone through your son's book on the Puna of Atacama and find it a most admirable piece of work. It is one of the best regional studies I know of anywhere. I only regret that he should have misinterpreted (as it seems to me) the idea of the cycle of erosion. I do not agree with him in his statements about the cycle theory ..., for I think the cycle idea as developed by Davis has a place for every kind of topographic fact which your son gathered together with such skill and success in Argentina. The idea no doubt should be amplified and extended and modified in detail, but this all of its supporters themselves would wish to see done. I hope that in our criticism of this matter we shall not fall into a narrowly nationalistic attitude. It would be a great misfortune to civilization if we should do so ...
>
> (Quoted in Martin 1974: 9–10)

On the same day Bowman wrote to W. M. Davis referring to the book and especially to its adverse Davisian criticisms:

> I have been through it very carefully myself from cover to cover, and I have great respect for the work which he has done and much admiration for his mode of presentation, but when it comes to the erosion cycle, he appears to me to be very 'young and foolish' .... It seems to me that he is going out of his way to take a crack at the idea of the cycle, while yet being compelled to employ it in explanation of the landscapes he found.
>
> (Quoted in Martin 1974: 29–30)

Davis replied to Bowman on the following day (27 July 1923) expressing gratification that someone besides himself thought Walther Penck 'rather askew':

> He seems first to misunderstand a general scheme for a very limited, rigid, special scheme; and second equally determined to show how the work of an earlier time than his ought to have been done.
> You do not mention A. Penck, ... the father is as exasperating as the son.
>
> (Quoted in Martin 1974: 31–2)

Davis goes on to complain that Albrecht Penck had adopted the cycle idea in 1906 or 1908 without any personal credits to its originator.

It seems that now Bowman had decided to act as a negotiator in trying to bring peace in the Davis–Penck disagreement. On 22 August 1923 Albrecht Penck replied to him on behalf of his son who was severely ill in Stuttgart and had not written any letters since Easter:

> He [Walther] was grateful that he was able to procure funds for its publication [*Der Südrand der Puna de Atacama*, 1920], through the medium of Professor Boas, whom I expect to see here shortly, ... I am so familiar with his [Walther's] views that I know exactly what his attitude is toward the geographical cycle. In principle it coincides with mine. If by cycle of erosion is meant ... quite generally the theory that mountains can be eroded and uplifted, we are of course both agreed .... But here in Germany the geographical cycle is quite generally understood to mean ... that erosion takes place in a definite sequence .... For a long time I have believed in this logical sequence. But it is not always the case.
>
> (Quoted in Martin 1974: 10)

Bowman replied to Albrecht on 25 September 1923, the German's sixty-fifth birthday, and explained at great length that Davis' scientific theory was necessarily at first expressed in simple terms and one would expect it to be enlarged, amplified and elaborated later. It appeared to him that Penck and Davis were, in fact, practically in agreement in the controversy. Before Albrecht had replied Bowman had heard of the death of Walther Penck and wrote to his father expressing the deep grief of the American Geographical Society at the sad news:

> I hope also that you will let me have any posthumous works of his that may be published, as, for instance, the first part of his 'Morphology of the Earth's Crust' which you mention.
>
> (Quoted in Martin 1974: 20)

Eventually, on 30 October 1923, Albrecht Penck replied at equally great length to Bowman's earlier letter (of 25 September) requesting an exchange of views about Davis' cycle theory. The reply is friendly and expresses

> agreement with you when you say that Davis's way of consideration (conception) never fails when one uses it in the analysis of forms. In these it has always been of

invaluable help, and, I believe, that if one regards it as a method of procedure one will always derive the greatest benefit from it.

(Quoted in Martin 1974: 21)

But, Penck continues, many of his more academic colleagues in Germany see in Davis' theme a definite theory rather than a method, proposing a universal change in form from 'young' to 'old'. Although this could be seen frequently in the erosional valleys of the Alps, Albrecht's recent studies in the German Highlands (*Mittelgebirge*) have revealed valleys that actually reverse the postulated Davisian sequence. This theory, developed by Walther (1920a), is not to be designated merely as an expansion of Davis' theory because it includes a step far in advance of Davis who did not consider crustal movement from its inception:

> My new investigations in Mittelwald had just been ended when the news of Walther's death reached me. I thereupon hurried to Stuttgart and then went to Leipzig to help my daughter-in-law. From there the boxes of Walther's manuscripts have just arrived, and new lines of work are opening up for me here. I am also anxious as to what my daughter-in-law and her children can live on. Times are hard for us here. An American dollar is [30 October 1923] worth 60 billion marks.
>
> (Quoted in Martin 1974: 25–6)

On 28 November 1923 Bowman thanked Penck for his long and careful statement which he was very pleased to have. He expressed dismay at the financial plight of Walther's family and offered to purchase his collection of South American literature on behalf of the American Geographical Society. Bowman thought it best to postpone any further discussion of the Davis–Penck controversy until 'your son's book has been printed, so that I may study it very closely and carefully'.

The Southern Black Forest

After his return to Germany in 1918 Walther Penck was employed by the University of Leipzig and, with his wife, occupied much time walking in, and contemplating, the German Highlands (Penck 1953: viii). On 13 and 20 December 1921 he delivered two lectures at Leipzig which were published posthumously by his father under the title *The Piedmont Flats of the Southern Black Forest* (Penck 1925). This work was directed towards more long-lived, larger-scale and tectonically varied features than the Andean broad folds, and it was not surprising that Davis (1932G) chose to concentrate a detailed attack upon it under the guise of providing a 'translation' (see volume 2, pp. 699–714). A more proper translation had to wait until that by Martin Simons in 1961; from which the following extracts have been taken.

Penck believed that the area between the Alps and the North German Plain contained evidence of upheaval at different dates and varying rates. In those of stronger uplift and deeper dissection Variscan (Hercynian) structures are exposed (e.g. Harz, Thuringer Wald, Erzgebirge, Black Forest). However, the Mesozoic Central German Scarplands are another such uplifted dome, and the relatively depressed other Mesozoic areas have simply not been elevated as much as the intervening mountains. In contrast with the previously accepted view, that of fault-bounded horst and graben topography, Penck held that the mountain units are not generally bounded by faults, except on one or two sides, and that on the southern side of the Harz and in parts of the Thuringer Wald and Black Forest the Mesozoic sedimentary strata lap up on to the substructures in the manner in which they were formerly deposited (except for some later distortion). The formation of the present Middle Mountains was largely coincident with that of the Alps but was not effected by the same lateral forces. The former were produced by the uparching of individual domes (*Gewölbes*) due to magmatic intrusion from beneath. Under this mechanism the domes expanded upward and outward without the development of marginal synclines, but with some mutual interference by adjacent forms – e.g. between the Black Forest and the Central German Scarplands. The domes thus differ from the broad folds (*Grossfalt*) which are compressed laterally and only expand longitudinally (figure 10.5) (Simons 1962). However, broad folds are commonly superimposed on domes.

The eastern and southern flanks of the Black Forest afford a particularly promising field of research as they are free of faults and, especially on the

Figure 10.5 The progressive development (1–4) of a broad fold (Grossfalt) (A) and of a doming or uparching (*Gewölbes*) (B)

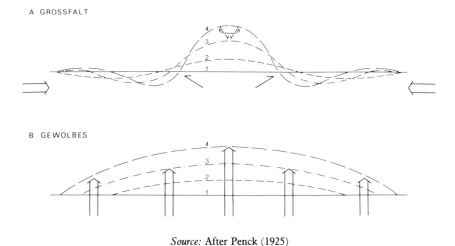

A GROSSFALT

B GEWÖLBES

Source: After Penck (1925)

east, without any marked irregularities. This gradient declines to the foot of the Malm scarp in the Swabian Alb, which, in distinction from other scarps, does not owe its existence to the greater resistance of the Malm Rock. When it comes to evidence from the drainage pattern, the middle and lower course of the Neckar are parts of a very old stream, antecedent even to the arching of the southern Odenwald, and indicative of an early tectonically created gradient which shows that the northern parts of the Swabian Alb never formed part of the Black Forest–Odenwald slope (Penck 1925: 85). To the evidence of structural relations and drainage must be added that of morphological analysis, which 'offers the only method of solving ... multifarious geological problems', and more particularly of deciding how far tectonically the Black Forest has been independent of its surroundings, and if it is a dome, or tilted block, or fold, and what has been the history of its tectonic development. Any investigation faces the great difficulty that strata correlated with Black Forest uplift are entirely absent, so that progress depends on the study of individual relief forms.

Conventional Davisian treatments of the Black Forest topography had associated it with a south- and east-tilting peneplain (*Rumpffläche*) whose uplift had initiated a new cycle of valley cutting. The surface had been variously dated from exhumed Permian, supported by Hettner, to Tertiary. Penck believed that the Upper Permian was a time of considerable relief in the region, but that a Black Forest erosion surface could be traced as far east as the Keuper outcrop and was therefore of post-Triassic inception. At this point, however, Penck departed radically from earlier views. He saw an important difference between the northern and southern parts of the Black Forest. In the north the greatest heights lie on the outcrop of thick Bunter Sandstone and the summit is a broad erosion surface which sinks eastward, below which is a sharp step. Probably no landform elements here are older than the summit surface. The Southern Black Forest is quite different, as its crystalline substructure was lifted up powerfully to form high mountains relatively early and simultaneously the younger strata became impoverished. Lower Bunter Sandstone is absent and the Middle Sandstones are thinner and do not form scarps. The flat erosional surfaces of the crystalline areas pass over on to the Middle Bunter Sandstones and Muschelkalk without their form being affected. This southern part of the Black Forest was highland as early as the Lower Trias and the elevation of its crystalline region was not simply a result of post-Triassic uplift. Moreover, field survey shows that the surface which truncates the Trias over such an extensive area is not the only one.

Penck believed that domes, such as those of the Middle Mountains, developed by means of accelerating uplift followed by deceleration, although to this day it is not clear whether he recognized the distinction between truly continuously accelerated uplift or, alternatively, continuous uplift containing discrete, intermittently accelerated phases of uplift (Simons 1962: 9, 11). This ambiguity was later seized upon by Davis who naturally would not allow breaks of slope to be produced as a result of continuously accelerated uplift. These breaks of slope were central to Penck's thesis. Rivers flowing off the dome have broadly convex profiles but these are made up of a sequence of concave reaches between sharp convex breaks of slope. The upper reaches occupy broad, shallow upland valleys with concave slopes, whereas the lower reaches are dominated by steep-sided deep valleys. The breaks of stream slope (nicks) are independent of structural control and occur at similar elevations on different streams. On the dome margins and interfluves and extending up the valleys are erosional (piedmont) flats (*Piedmontfläche*). The river valleys of the Black Forest possess terraced, valley-in-valley forms with a series of alternating convex and concave side slope features one above the other. These, Penck stressed, were the result of waxing and waning phases of *intensity of erosion* (not of phases of *uplift*, as Davis later postulated) formed under continuously accelerating uplift (Simons 1962: 2). Penck further believed that the gradient of any river on the dome (neglecting lithological differences) was a function of the rate of tectonic uplift and everywhere equal to that local rate.

The initiation of more-or-less horizontal valley-side slope elements (or 'flats' – *geneigten Flächen*) was held to be due to the rates of stream incision:

> The development of valley slopes is the fundamental natural measure of erosion during the incision of a stream . . . . The amount of denudation which takes place while a stream cuts down to a given depth increases as the rate of erosion decreases; . . . conversely, if a river's rate of erosion increases, so the valley slopes being formed above it become steeper in successive units of time. *Convex slope profiles result.* As erosion intensity weakens, the reverse is found, *concave* profiles result.
>
> (Penck 1925: 88)

Once formed, the steeper slope elements retreat faster than the less steep ones such that over a protracted period of time the *slope as a whole* tends to become concave as the lower foot flat grows upward, providing basal river incision ceases (figure 10.6) (i.e. there is waning development). This form of slope retreat implies that in time the rounded convex (waxing) slope crests are transformed into sharp convex breaks of slope (Simons 1962: 3):

> The denudation on all inclined flats [slope units – *geneigten Flächen*] proceeds in such a way that the flats retreat parallel to themselves at a constant gradient, and at their foot a flat of smaller gradient grows upward increasingly, so long as no river erodes deeper there, hindering the formation of the gentler foot-flats. Every rock wall shows the operation of this law of the denudation process. The retreating flats [slope units] finally become crowded out and are replaced by the upward-growing lower foot-flats.
>
> (Penck 1925: 88–9)

Figure 10.6 The unequal parallel retreat of slope elements (*geneigten Flächen*) of different angle during the time units 1 to 20, leading eventually to a completely concave slope profile (34)

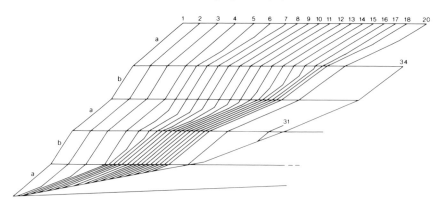

Source: After Penck (1924); see Penck, (1953), Figure 11, p. 172. By permission Macmillan, London and Basingstoke.

After that, the flattening and lowering of the land continues as long as there is no further erosion in depth, . . . and *the slope profile is preserved* as long as there is no newly-growing slope unit eating through, from below upwards. A valley slope which first experiences increase, then decrease and standstill of erosion has a succession of convex and concave forms . . . the first indicating increase, the second decrease of the rate of erosion.

(Penck 1925: 89)

Under otherwise equal circumstances . . . convex slope profiles signify acceleration of upheaval.

(Penck 1925: 89)

Extensive and prolonged slope retreat of this kind produces a surface of low relief dominated by low-angle, concave (waning) slopes, which Penck termed the *Endrumpf*.

Similarly, accelerated upheaval was seen by Penck to produce breaks of slope in longitudinal stream profiles which develop in much the same manner as breaks in valley-side slopes. The more voluminous lower course of a river will be able to erode more powerfully than the upper course and so can counteract more forcibly the increase of gradient produced by accelerated uplift. So in the lower course a *convex* nick is formed in the long profile, which moves headward and withdraws the upper reach of the river from the renewal of gradients caused by further upheaval of the mass. The erosional intensity of the upper course is weakened because its local base of erosion moved relatively upstream and concave slope profiles come to be formed there:

Except for the gradient, rate of erosion depends on the fixed maximum water volume [of a river]. And the gradient is, under otherwise equal circumstances, a function of the crustal movements. Uniform upward movement of a land mass does not alter the gradients of the rivers flowing on it. On the contrary, these establish, each and all, a profile of equilibrium and incise themselves exactly as much as the land rises. The curve of equilibrium is steeper where the amount of water flowing is less.

(Penck 1925: 89)

If the upheaval, which in time units 1–3 was uniform (thus, itself uniform, over-all the same amount of incision of the long profile), now becomes more rapid, so inevitably the water-rich lower course immediately erodes deeper than before, in each unit of time, but not the feebler upper course. A convex nick is formed in the long profile of the stream. This nick eats back in accordance with the laws of headward erosion, and forms the erosion base for the upper reach of the course. This, therefore, is no longer tributary to the erosion base at the edge of the rising mass, but to a point in the valley course which is not sinking relatively, but has become upheaved and, moreover, moved up valley. The upper reach of the river is from now on withdrawn from the renewal of gradients caused by the upheaval of the land mass, the nick, local base of erosion for the upper course, has become lifted up, and in addition moved up valley and thus rises relatively to the upper course section. In this, consequently, erosion intensity weakens and concave slope profiles come to be formed there. Continuous acceleration of the upheaval leaves in the long profile of the stream one convex nick after the other, all move up-valley, below each one there begins a narrow, steep course reach with convex valley slopes, above each there is a broader reach with concave slope profiles. That is typical of the Black Forest valleys.

(Penck 1925: 89–90)

The growth of the Black Forest dome would imply to Penck that low-angle concave slopes near the centre of the dome would be separated from rapid stream erosion by the flanking breaks of slope which were retreating and being blurred into concavities in the process. Thus an *Endrumpf* of low relief and concave (waning) slopes would expand outward from the centre by the merging and addition by headward retreat of more peripheral lower benches. At the outer margins of the dome the initial uplift would be slow and the first erosional effect would be to produce a surface of low relief with gentle convex (waxing) slopes (a *Primärrumpf*). With the passage of time the uplift of a given location would accelerate and the expanding dome would be etched by a flight of roughly concentric erosional benches (*Piedmonttreppen*). Each bench would be backed by a dissected scarp of convex–concave slopes retreating in the manner shown in figure 10.6 under the control of the erosional baselevel produced by the bench itself:

When the rate of uplift increases and the dome broadens, the existing piedmont-flat comes into the sphere of livelier upward movement and becomes dissected, while

Figure 10.7 The piedmont flats and major river terraces of the southern Black Forest. The highest (sixth) flat is too limited to be shown, being present on the tops of a few inselbergs situated on Flat V

Source: After Penck (1925), Figure 7, p. 92

the out-pushed edge of the dome now finds itself in the zone of slowest upheaval, in which, outside the edge of the old, dissected piedmont-flat, a newer, lower one is formed. The two flats are separated from one another by steeper gradients, always by a concave, lower, and convex, upper, break of gradients. Such gradients, and the general arrangement of piedmont-flats one above the other, signify not in the slightest that the upheaval was intermittent, but only becoming continually more rapid.

(Penck 1925: 91)

There are five major erosional surfaces on the south and east sides of the Black Forest (figure 10.7):

5 The highest and oldest surface, Mesozoic in age, which is enclosed on all sides by level 4.
4 A piedmont bench, also Mesozoic in age, situated at about 1,300 m on which are located inselbergs crowned with eroded remnants of surface 5.
3 The main piedmont bench with valleys leading down to it which are cutting back and dissecting level 4.
2 The second youngest level, probably Eocene in age.
1 The lowest piedmont flat some 20–30 km wide separating the Black Forest from the upper margins of the Rhine and Wutach valley floors. It is possibly of Upper Miocene age.

Penck stresses that all the piedmont benches show no signs of structural control and that they resulted from doming of increasing acceleration which began in the Mesozoic. The flanking river Neckar flows in a tectonic trough, already perceptible in the Eocene, and the Alb in a separate, analogous dome.

We have already treated in detail Davis' (1932G) description and commentary relating to Penck's Black Forest paper (see volume 2, pp. 699–714), and it is only necessary to remind ourselves of its major critical elements. Davis' treatment shows skilful advocacy, great artistic ability – a quality obviously lacking in Walther Penck – and, in spite of the author's talk of brevity, an utter indifference to the economy of printed space. He is delighted with its strong deductive bias, a joy to one who has always criticized Albrecht Penck and other outstanding German geographers for being far too inductive. Walther Penck's principles, he says, range from truisms to elaborate ideas, some of which are very controversial – if not altogether wrong. Davis misinterprets Penck's concept of the parallel retreat of slope *elements* and claims that this condition can only occur in some desert mountains. He misinterprets slope convexity and concavity to imply acceleration and deceleration of *uplift* and then attacks the implications of this. Moving on to firmer ground, Davis (1932G: 411–15) will not accept that continuous uplift can produce discontinuous longitudinal stream profiles, but even here he confuses his argument by introducing his own concept of grade.

Davis similarly introduces confusion by trying to relate the *Primärrumpf* and the *Endrumpf* to the initial and penultimate (not final) stages of his own cycle of erosion. Going back to a more secure position, he attacks the notion that continuous, expanding upheaval can produce an irregular landscape of piedmont benches, and interprets each bench as a partial peneplain. Supplying a superb diagram which convincingly displays his own concept (see volume 2, Figure 148, p. 701), he adds in a truly Davisian strain:

> How could serious students of physiography ever have persuaded themselves that continuous upheaval would or could cause intermittent erosion?
>
> (Davis 1932G: 440)

Indeed, the Achilles' heel of Penck's work was undoubtedly his inability to illustrate his ideas with convincing diagrams, partly because of their relatively complex nature. Davis, with his superb graphical skills, was able to reinterpret and corrupt many of Penck's ideas with the aid of new diagrams. However, Davis was undoubtedly influenced by Penck's notions more than he himself realized, particularly in his (1930E) 'Rock floors in arid and humid climates' (see volume 2, pp. 669–675). Davis' major mistake was not to study more closely *Die morphologische Analyse*.

Die morphologische Analyse

The term 'morphological analysis' was coined by Walther Penck in a paper which he published in 1921. He outlined a projected larger work on the topic in a letter written in the following year, clearly stating his prime objective by predicting that in its culminating chapter:

> wide and surprising vistas will appear leading towards the *causes* of crustal movement. It is to this fundamental problem of geology that I intend to devote the whole of my labours, possibly to the end of my life.
>
> (Penck 1953: vi)

These words were only too prophetic and a year or so later he was dead. After a further interval of only a year *Die morphologische Analyse* was published – incomplete, hastily assembled and poorly edited by his grieving father, to remain as an enigmatic monument to his brilliant son. Almost 30 years later the first English translation of the whole work (Penck 1953) clarified for a wider audience some of the misunderstandings of Penck's ideas, but it by no means dispelled completely the intellectual fog which emanated from the monument itself.

The first five chapters, comprising some 40 per cent of the book, introduce the work and treat the tectonic, weathering, mass movement and denudational processes in a general way. Chapter 1 reiterates that the significance of landform studies lies in their assistance in solving the geological problems of crustal movement; that landforms result from the opposing tendencies of exogenetic processes (i.e. weathering and transport) and endogenetic (crustal) processes; that deposits correlated with uplift throw light on the latter; and that Davis' tectonic assumption is a special case and is generally inadmissible. Penck also believed that rate of denudation increases with gradient (and thus usually with height); that the only exogenetic processes which are of geomorphic importance are those dependent on gradient; and that the simultaneous action of non-equal opposing forces (i.e. exogenetic and endogenetic) must be studied through the development of a series of forms by an incremental method similar to the calculus. Chapter 2 identified two types of relief-producing tectonic processes:

1 Simple warping or arching in continental areas such as the Appalachians and Urals.
2 Folding in less stable areas in the form of:
   (a) Compressional folding which leads to uplift, as in the Alpine–Himalayan belt;
   (b) Broad folding (*Grossfaltung*).

There is often a coincidence between arching and folding, as in the Mesozoic–Tertiary mountain belts. However, in Anatolia and the Pamirs the present mountain ranges have little to do with lines of recent folding.

The treatment of weathering in chapter 3 is reasonably standard but stress is laid on the relative rates of weathered debris production *in situ* and debris removal. The latter is held to depend on the ratio of gradient and debris mobility produced by weathering which determines the intensity of denudation. For a given weathering rate, gradient is the dominant controlling factor over denudation. This theme is taken up under the topic of mass movement in chapter 4 where rock resistance is ascribed to cohesion, internal friction, external (i.e. basal debris) friction and root binding. Mass movement must overcome this resistance and is then proportional to the sine of the slope angle, down to a limiting angle of less than 5°. If the slope angle increases towards the crest and/or the weathered mass is thinner there (giving less external frictional resistance), the rate of mass movement is greater and the crest is rounded. Weathered particles move more freely at gradients steeper than 25–30° and collect at the base of the steeper slope elements producing a low-angle subtalus and lower slope element termed the *Haldenhang*. In this accumulation zone movement continues until a lower limiting gradient of 2 or 3° is reached.

Chapter 5 concentrated on the distinction which Penck drew between denudation and erosion, and on the developmental effect of baselevels. Denudation occurs over the general land surface under the processes of mass wasting where gradients exceed a lower minimum value (i.e. 0–5°), and its rate is predominantly proportional to gradient. Erosion is a linear activity of rivers (and also ice, etc.) which can affect the developing lower parts of slopes, for example by progressively increasing the angles of new basal slope elements. Slopes are thus dominated by denudation in their upper parts, by erosion near their bases and by a combination of the two in the middle. Departing somewhat from his father's ideas (see chapter 12), Walther held that similar landforms are produced under all fluvial conditions and that only rates of development differ under different climates. Penck's use of the concept of *absolute baselevel* (i.e. sea level) was in accord with Davis', but he recognized *immediate baselevels* separating areas developed under different endogenetic influences (e.g. piedmont steps), as well as *local baselevels* such as are associated with tributary junctions, stream nicks, breaks of valley-side slope and resistant rock beds. Penck believed that above a given baselevel slopes comprise a single system of units which are capable of developing independently of those below the baselevel.

A long, central Chapter 6 detailed Penck's vision of the development of slopes (*Hangentwicklung*) which was at the core of his view of geomorphology. Here absence of clear definitions and difficulties of translation have generated continuing problems for the scholar but, in this connection, the unpublished analysis by Kesseli (1940) proves invaluable. For Penck, slope development was dominated by mass movement, wherein the largest size of debris which is mobile on a slope element varies with its gradient and the development of the slope is controlled by this debris size and by its rate of comminution (being exponentially faster for coarse debris than for fine). He used a graphic/differential method to predict slope transformations, not dissimilar from that employed by Osmond Fisher (1866; see volume 1, pp. 440–1), but considered consecutive time units which were so small that it was possible to assume the *successive* operation of weathering, removal, comminution, basal erosion and deposition which, in reality of course, occur *concurrently*.

Penck's simplest erosional case assumes a stable baselevel, uniform lithology, a uniform weathering rate throughout which is exceeded by the transport rate, and complete removal of basal debris in each time unit. An initial, straight, steep cliff slope at some 45–90°, dependent on debris characteristics, will retreat by the removal of parallel slices leaving beneath another straight slope element (*Haldenhang*) at a lower gradient controlled by the calibre of the weathered debris shed from the retreating initial slope (figure 10.8(i)). This retreat makes the *Haldenhang* progressively longer and the initial slope progressively shorter as the latter develops above its rising baselevel (C), is removed from the controls at the base of the slope as a whole and proceeds towards its final consuming by the mutual backcutting of opposing slopes. Below C, as soon as the *Haldenhang* element begins to form, it, too, is subject to a similar parallel retreat conditioned by the size and comminution of its debris cover, leaving yet another successively lower

Figure 10.8 Kesseli's representation of five of Walther Penck's major slope development models

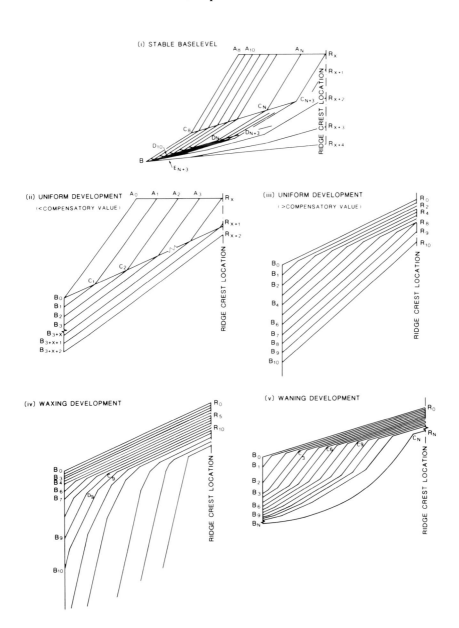

Source: After Kesseli (1940), Figures 3, 5, 6, 7 and 8

gradient slope element below it. Thus a complex queue of retreating and successively consuming slope elements develops leading in time to complete valley-side slopes of the marked concavity which was previously noted (see figure 10.6). This process involves a decrease both in relief and in the average gradient per unit length of the total slope which Penck termed *waning development*.

A second case is one in which basal stream incision occurs such that the slope can evolve without the development of a *Haldenhang*, the gradient of which would be entirely conditioned by the character of the debris shed from a higher slope element. This implies that stream erosion does not permit a lower debris-dominated slope element to form, such that the newly formed slope retreats parallel to itself ultimately producing constant relief and constant average gradient per unit length of slope (i.e. *uniform development*). Of course, this could occur if the rate of basal stream downcutting were just sufficient to produce valley-side slope elements just equal in gradient to that of the initial slope (i.e. achieving what Penck termed a 'compensatory value') – but this would be a remote possibility. Penck therefore considered the two possibilities of stream erosion producing a basal slope angle either less than or greater than that required by the compensatory value. The former case is shown in figure 10.8(ii) where a straight basal slope composed of new elements retreats parallel to itself consuming the initial slope by about time unit $3+X+1$ (see subscripts). It should be stressed that, unlike the *Haldenhang*, the gradient of which is controlled by the character of the debris shed from *above*, the lower slope element here is dominantly influenced by the rate of stream cutting from *below*. Where this downcutting produces a slope gradient greater than that required by the compensatory value a different geometrical sequence evolves (figure 10.8(iii)) with uniform development occurring after time unit 8.

If stream downcutting accelerates beginning at the compensatory level (time unit 3 in figure 10.8(iv)) progressively steeper slope elements are created which would retreat parallel and mutually consume each other to give progressively steeper average gradients and increasing relief (i.e. *waxing development*). Conversely, if the rate decreases (after time unit 3 in figure 10.8(v)) the production of lower gradient slope elements gives a decreasing average gradient through time and consequently a concave slope profile (i.e. *waning development*). In a similarly incremental manner, Penck theoretically and graphically worked out what he considered to be the possible effects of basal slope alluviation and of differences of lithology on slope development. The latter he regarded as of only transient importance during the earlier phases of slope development when weathering differences were most evident.

The final, long chapter deals with 'linking of slopes, form associations and sets of landforms' and is mainly devoted to a recapitulation and elaboration of Penck's ideas of piedmont flats and benchlands (1925) and on broad folds

(1920a). In it there is concentration on the Variscan blocks of central Europe and on the Andes, but there is also treatment of scarplands, inselbergs, the Alps and Anatolia. Widespread world concave slopes are held to be dominantly denudational but also to be derivative from the convex and straight erosional slopes associated with stream incision accompanying accelerated uplift. Concave slope elements must therefore connect upward with convex ones, unless the latter have been obliterated by the retreat of the former. Main river valleys, the floors of which form stable local baselevels, may be flanked by concave valley-side slopes, while the sides of their still actively eroding tributary valleys may be convex.

The German scarplands between the rivers Danube and Wutach (see figure 10.7) provide good examples of these slope patterns. The main valleys have sections of active stream erosion with steep valley sides, alternating with reaches flanked by slopes with concave bases where erosion has almost ceased until the encroachment by the retreating next-lower active reach revitalizes it. In the retreating active zone valleys take the form of canyons: higher up they are V-shaped and higher still, where the zone of active erosion has not yet reached, they are wide and shallow as they merge with the next-higher erosion surface or outward-sloping piedmont bench. Between the main rivers scarp lobes jut out, marking the active zone of slope retreat. Inselberg landscapes are dominated by the concave slopes which are associated with a drainage network whose erosive activity is weakening. This waning development, like the waxing, spreads upvalley, but at any given point it is unaffected by developments downstream below the controlling local baselevel until a new waxing impulse works headward and engulfs it by the supervention of the influence of a new baselevel. It is only when active uplift has ceased that waning development can, working headward, dominate the landforms of the whole region. Concave-sided inselbergs are distinctive features of zones of waning development and are not restricted to any special climatic regime. Piedmont benchlands are formed of stairways of piedmont flats leading up to a summit surface composed of concave, waning slopes with wide shallow valleys and commonly crowned with unconsumed concave residuals. Each Piedmont flat continues back into the valleys which are cutting into the next, older flat above (figure 10.9). Thus each piedmont flat forms the baselevel for the destruction of the flat above it as the retreat of the originally convexo-concave slopes of the riser consumes the latter. These slopes originated as the valley sides of the major streams flowing on to the lower flat, which have retreated. The *Endrumpf* is an extensive, high, old surface of low relief with concave (waning) slopes formed by such retreat. The Atlantic slope of the Appalachians is a possible example of piedmont benchlands.

The final part of chapter 7 deals with an extension and refinement of the concept of broad folds also set out by Penck in 1925. The form of the *Gipfelflur* surface shows that the Alps are a broad, undulating arch with the

Figure 10.9 Schematic profile from the older centre to the younger periphery of an expanding dome showing piedmont flats (P) and the development of a Primärrumpf and an Endrumpf

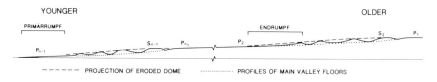

Source: After Penck (1924), Figure 16, p. 176; see also Penck (1953), Figure 16, p. 214

major longitudinal rivers following the undulations. Much of the original fluvial terrain has been destroyed by glaciation but piedmont flats are evident in the eastern Alps, as they are in the mountains of Anatolia. In the Andes, the Sierra de Achala possesses the highest (2,100–2,200 m) and oldest denudation surface with concave slopes (probably an *Endrumpf*) leading up to inselbergs, with a more extensive piedmont flat 500 m below this, and so on. The Puna de Atacama is made up of broad folds, the edges and ends of which exhibit a *Gipfelflur* of individual peaks (figure 10.10). One of the older of these folds is the Sierra de Fiambalá (about 5,000 m) which possesses, in descending order of progressively younger age:

1 A summit surface of low relief extending up to an *Endrumpf* crowned by tors and down to a zone of convex (waxing) slopes indicative of increased uplift.
2 A surrounding, lower belt of medium relief and concave slopes.
3 A zone of steep, convex forms extending down to narrow valleys.
4 Steep relief extending from about 3,500 m to the mountain foot.

Figure 10.10 North–south profile along the longitudinal axis of an older broad fold in the Puna de Atacama, with the vertical scale twice that of the horizontal. The marginal *Gipfelflur* is shown, together with the summit level II erosion surface surmounted by an inselberg and the associated valley level II correlating southward with a lower set of summit levels

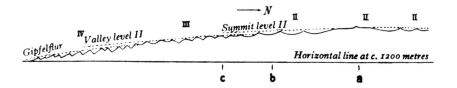

Source: After Penck (1924); from Penck (1953), Figure 20c, p. 263. By permission of Macmillan, London and Basingstoke

Figure 10.11 Topographic and geological cross-section showing the broad folds of the Sierra Narvaez and the Sierra de Fiambalá in the southern part of the Puna de Atacama (see figure 10.1). Here Penck correlated the range summits (c, p₁ and p₂) with the deposition of the Calchaqui, Lower Puna and Upper Puna Beds, respectively

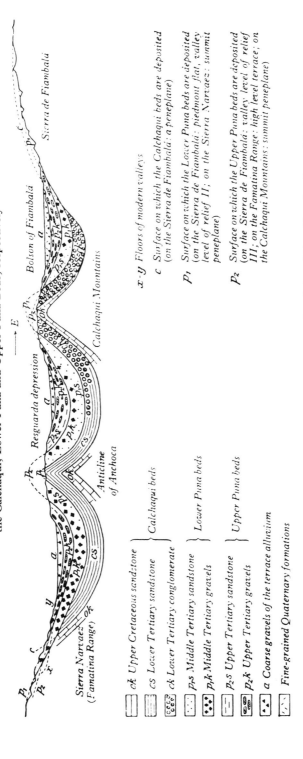

ck Upper Cretaceous sandstone ⎫
cs Lower Tertiary sandstone   ⎬ Calchaqui beds
ck Lower Tertiary conglomerate ⎭

p₁s Middle Tertiary sandstone ⎫ Lower Puna beds
p₁k Middle Tertiary gravels   ⎭

p₂s Upper Tertiary sandstone ⎫ Upper Puna beds
p₂k Upper Tertiary gravels   ⎭

a Coarse gravels of the terrace alluvium

Fine-grained Quaternary formations

x–y Floors of modern valleys

c Surface on which the Calchaqui beds are deposited (on the Sierra de Fiambalá: a peneplane)

p₁ Surface on which the Lower Puna beds are deposited (on the Sierra de Fiambalá: piedmont flat, valley level of relief II; on the Sierra Narvaez: summit peneplane)

p₂ Surface on which the Upper Puna beds are deposited (on the Sierra de Fiambalá: valley level of relief III; on the Famatina Range: high level terrace; on the Calchaqui Mountains: summit peneplane)

Source: After Penck (1924); from Penck (1953), Figure 21A, p. 274. By Permission of Macmillan, London and Basingstoke

The broad folds of the Puna de Atacama began by growing in height, length and, subsequently, width, during which time piedmont benches formed. The youngest ranges have not yet greatly increased in width and, consequently, do not possess well-marked piedmont flats. Later the broad folds, while continuing to grow in height and length, became compressed and were narrowed in width, such that their erosion surfaces (including benches) were warped. Thus the older ranges are higher, possess more piedmont flats and exhibit greater warping. It is possible to date these surfaces by correlating them with the deposits of the adjacent bolson (figure 10.11).

The Aftermath

On 11 February 1925 Albrecht Penck informed Isaiah Bowman that a copy of *Die morphologische Analyse* had been sent to him, and Bowman later agreed to review it himself, hinting that this would take some time! (Martin 1974: 27). Bowman's (1926) review was published early the following year, after W. M. Davis had been consulted, and we have previously referred to its main points (see volume 2, pp. 694–7). The review began with a general complaint against German criticisms of the Davisian cycle:

> During the past fifteen years increasing criticism of the idea of the topographic cycle has appeared in foreign literature. New methods of attack have been proposed in the avowed belief that the alleged simplicity of the topographic cycle was illusory. Too many facts were thought to be in disharmonious relation to the simple scheme of uplift and subsequent erosion through youth, maturity, and old age. The objections seem to spring ... in part from an inexplicably persistent misunderstanding of *stage* to mean *age* ... , [and] the refusal to describe two parts of a single valley as mature where excavated in weak rock and young when excavated in more resistant rock, again overlooking the distinction between *time* and *stage*.
>
> (Bowman 1926: 123)

However, Penck's book was 'commended for thorough reading to every American student of physiography' as it provided 'the most profound and revolutionary criticism of the American school of physiography yet to appear'. Stressing Penck's concentration on the importance of crustal mobility, Bowman considered the work to be only a more explicit and detailed rendering of the complications of the cycle that Davis had already discussed. This point was later disputed by Albrecht Penck (1926) who insisted that his son's work was far from being an expanded Davisian theme and that it posed an entirely different problem, namely to deduce from different types of landforms the intensity of crustal movements.

In describing Penck's view of slope development it is clear that Bowman is already making the error of directly relating slope geometry to manner of uplift:

Penck claims to find through his detailed analysis of the mode of retrogression of the slopes of the land an explanation for the forms of old age independent of the idea of the cycle. Certainly he has brought into the literature a number of useful terms and has challenged prevailing explanations at a number of quite critical points. His recognition of the piedmont step; the brow of the uplifted block upon which an old erosion surface may be identified; the sharp topographic unconformity that exists between the old erosion surface and the residuals that rise above it and the remarkable localization of this last feature on a line almost as definite as a strand line; the importance of studying the effects of piedmont stripping upon an upland border where accumulation of sediments was made before an old erosion surface was developed upon the adjacent uplifted mass; the agencies which effect a change in the form of a landscape from concave to convex or from convex to concave; the constant challenge with which he meets a new grouping of forms – these are among the most important contributions of his book. A far-reaching influence is attributed to *intensity and degree of uplift* (as opposed to climate) in the modelling of slopes. If convex, the slopes are a response to rapid uplift; if concave, they bespeak a slower rate. Obviously the full test of the idea must come with the inspection of a far larger number of examples under the most divergent conditions.

(Bowman 1926: 128)

Bowman, in criticizing Penck's disregard for American researches on 'basin and range' landforms, claimed that 'front-range physiography' is too specialized a field to be used as a basis for a revision of the whole science of landforms:

In his attempt to harmonize the topographic cycle of mountain areas such as the Sierra de Fiambalá and the corresponding cycle of sedimentation in the adjacent basin floor, Penck has found the key not to systematic physiography as a whole but to that special chapter of it which might be called front-range physiography.

(Bowman 1926: 130)

Making use of the work of Louderback (1904), Bowman supports the idea of sudden uplift and differential movement of discrete fault blocks:

one is led to conclude that there is ample justification for treating the mountain range in just the simple manner that Penck supposes Davis uses on all occasions, that is, rapid uplift into its present position .... .

(Bowman 1926: 131)

The review ends prophetically:

The next great steps in physiography may well be the application of Penck's argument to landscapes in critical or classical topographic regions and its use in the fuller discussion of cases of variable rates of upheaval which have been only briefly treated in earlier studies. When this is done it seems fairly clear that the idea of the topographic cycle as developed by Davis will still be the most important part of interpretive generalization.

(Bowman 1926: 132)

In 1925 Bowman and Douglas Johnson failed to persuade John Wiley to publish a translation of Penck's *Die morphologische Analyse* (Martin 1974: 6–7), and consequently for many years afterwards the English-speaking world had to rely on reviews of a distinctly Davisian character. However, in 1929 a version of Penck's analytical morphology was applied to the Peninsular Range of California by Carl Sauer. Sauer moved to Berkeley from the University of Michigan in 1923 (see chapter 4), taking with him John Leighly, who like himself was fluent in German. Later they were joined by John Kesseli (1895–1980), of German–Swiss background, who completed a notable trio of Penckian scholars. Their students used to study the original German texts and went on frequent field investigations in geomorphology which were undertaken partly to test the relative merits of Davis' and Penck's methodology. Typically in 1925 Sauer in his *The Morphology of Landscape* advocated replacing genetic geomorphology by a descriptive and classificatory study of surface forms – a suggestion he later reversed. Sauer's opposition to Bowman and Davis was long continued and ironically in 1952, nearly 2 years after Bowman's death, he delivered the Bowman Memorial Lectures in which, according to S. W. Wooldridge, a devoted Davisian of London University:

> he went out of his way to claim that W. M. Davis delayed somewhat our learning about the physical earth by his system of attractive but unreal cycles of erosion. In these references there is a distinctly bitter or rancorous note which many of us, still heavily in debt to Davis, must resent and, indeed, flatly repudiate.
>
> (Wooldridge 1955: 90)

Sauer remained in charge of geography at Berkeley until 1957 and lived until 1975 by which time Davis' popularity had slumped and Bowman's (1926: 132) prophecy had long been falsified.

Early in his stay at Berkeley Carl Sauer (1929) (see chapter 7) set out to confound the suggestions by critics that Walther Penck's method of landform analysis was as yet far too restricted in application – to a few Variscan blocks in the Old World and a frontier range in South America – for it to be accepted as more than a special case. With this criticism in mind, Sauer investigated a relatively limited area in San Diego county, California, and applied to it 'a trial of the Penck method' that suggested that the surface complex is 'almost in its entirety, an expression of intensity of diastrophism to weathering of rock, denudation of slopes, and accumulation of detritus'. This study was used to generalize the geomorphology of the Peninsular Range, extending 300 miles south into Baja California, a conspicuous block inclined towards the Pacific Ocean on the west and dropping sharply to the Colorado desert and Gulf of California on the east. The western flank bears a so-called Eocene marine terrace – its gravels, once considered fluviatile, were now considered marine – at about 2,500 feet altitude with a series of sedimentary terraces at and below about 1,500 feet. The basal rock is granitic and was intruded into gneisses and schists that have been largely eroded away. The intrusion is generally thought to be a large batholith of the time and kind that formed the Sierra Nevada. Sauer believed that the range mass, rather than being a great fault block, is a broad fold (*Grossfalt*) and that the faulting was secondary to the Jurassic batholithic intrusion. The summit areas, the Mesa Grande (4,000–3,000 feet) and Julian Mesa (5,000–3,500 feet), were thought by most Americans to have been part of a continuous peneplain surface that has now been dislocated and eroded into separate blocks. However, these were regarded by Sauer as parts of a pre-uplift *Primärrumpf* which is in the process of being continually reduced in relief at a presently high elevation. Sauer thus broke with what he termed the 'conventional base-levelling thesis' by allowing uplift and erosion to take place with reference to the baselevel provided by the retreating edges of mesa-like masses. Stressing the importance of intensity and duration of tectonic movement on surface relief, he held that concave slope profiles indicate the effect of fixed baselevels and convex ones of increasing rates of uplift and denudation.

Sauer's (1929) work on the Peninsular Range brought a traditionally critical review from Kirk Bryan and Gladys Wickson (1931), eliciting a spirited response from Sauer (1932). The reviewers attacked Sauer's lack of regard for the implications of the marine terraces, questioned his views on baselevel and the possibility of the continued development of surfaces of low relief at high elevations, and ascribed differences in slope geometry to regional climatic differences. However, Sauer raised important geomorphic questions regarding this tectonically active region, such that more than 30 years later Thornbury (1965: 550) wrote:

> It is not surprising that many California geologists regard peneplains and erosion cycles with considerable skepticism. In an area that has been as active diastrophically as California has been (since and) during Tertiary . . . time there has been little opportunity for baselevelling to take place except very locally . . . . Under such conditions the kind of geomorphic interpretation that may have validity in the eastern United States does not work too well.

Before the Second World War, despite Sauer's 'trial' study, the American view of Walther Penck's work was dominated by Davis' (1932G) adverse and misleading review. However, a nagging feeling remained that tectonic movements might have left a more subtle landform record than Davis allowed. Naturally in Germany there were a number of reviews and assessments (Henkel 1926, Priem 1927, Morawetz 1932, 1937, Spreitzer 1932 (see also 1951), Louis 1935, Würm 1935–6, Braun 1938). Surprisingly, these were not entirely laudatory and special criticism was levelled at Penck's simplistic disregard of such processes as creep and the rate of basal debris removal in slope development (Morawetz 1932). In France, Henri Baulig pursued a moderate Davisian line critical of most Penckian concepts. His articles on the profile of equilibrium (1925, 1950), on piedmont benches (1939), and on the equilibrium profile of slopes (1940, 1950) were widely read. Typically he warned (1939) that the convexity and concavity of slopes most frequently depend on lithology; that a graded slope cannot exist except through the action of a loose mass in motion and so it is wrong to believe that a graded valley side consists of several distinct units developed successively; that it is impossible to deduce with reasonable definition the past rate of river-bed erosion and so of landmass uplift; and that a break of slope (nickpoint) in a river's long profile does not have any special properties involving the 'indispensable fixity' of a baselevel.

Late in 1939 several American geomorphologists met to discuss Walther Penck's contribution to geomorphology (Symposium 1940) as a result of an, in places, misleading questionnaire drawn up by O. D. Von Engeln (see volume 2, pp. 715–717). None of the contributors supported Penck's theory of the origin of piedmont benchlands, especially with regard to the formation and function of nickpoints. However, Kirk Bryan thought the new radical uplift doctrine had greatly stimulated the American school of geomorphology especially in the study of slopes (p. 254). He came close to Penck's slope concept in stating that, once developed, slope units persist in their inclination as they retreat, and disappear only when all rocks above the encroaching foot-slope (pediment) have been removed (p. 266). He discussed textbook diagrams of the retreat of slopes but reproduced only one (p. 258) and that the erroneous diagram devised by Davis. John Leighly averred that Davisian ideas on processes generating slopes lacked 'an adequate foundation' (p. 225). He, and some other participants, considered that the Penckian and Davisian main systems of landform analysis were not mutually exclusive; that Davis was the more comprehensive and Penck a fragment but enough 'to constitute an entire system by itself'; and that the future ideal system might well incorporate only part of each (p. 224). With regard to tectonics, Meyerhoff thought the Davisian assumption of long still-stands a handicap for American geomorphologists and he wanted to replace its tectonic premises and those of Penck – which seemed to him diametrically opposed – by the idea of 'a restive earth in differential but intermittent motion'. This debate clearly suggested that the main body of geomorphologists in the United States had not yet grasped the fact that Walther Penck was a revolutionary who aimed at a new diastrophic system but also that, except among a few staunch followers, the unquestioned acceptance of the Davis method had already been broken there. In the meanwhile students at Berkeley were using Kesseli's (1940) analysis of chapter 6 on slopes in *Die morphologische Analyse*.

In the light of Penck's obscurity and Davis' misinterpretations, it is interesting to observe some of the treatments of Walther Penck's ideas in popular textbooks. Cotton (1941, 1948: 205, 232–3, 360) repeated Davis'

interpretation of Penck, questioned the latter's concept of slope development, referred to the 'discredited theory of intermittent erosion during
continuous upheaval' and criticized the piedmont benchland concept. He
also mentioned Sauer's (1929) work on the Peninsular Range but preferred
the traditional interpretation involving faulting. Martonne (1948) noted
criticisms of Penck at the International Congress of Geography in Amsterdam
(Symposium 1938), considered the *Primärrumpf* concept purely hypothetical,
condemned Penck's ignoring of climatic influences and expressed the view
that the piedmont benchland theory, although successful in Germany, 'is not
founded on a secure basis' (Martonne 1948: 614). However, by far the most
influential textbook of Penck's work was by Von Engeln (1942, 1948) who
devoted a whole chapter to the system of Walther Penck ('the Challenger'), as
opposed to that of Davis ('the Master'). In reality, many of the views ascribed
to Penck originated in Davis' (1932G) misleading analysis and were supported by skilful diagrams, for example figure 10.B which clearly links slope
form directly to uplift. Having set up this straw man, Von Engeln (1948: 257)
demolished Penck's 'generally discredited' correlations between slope form
and diastrophism. He also questioned the value of the concepts of the
Endrumpf and *Primärrumpf*, and believed that the development of piedmont
benchlands through parallel slope retreat requires very special conditions.
Nevertheless, Von Engeln concluded that Penck's ideas on the possible
relations between landforms and tectonic uplift should be kept in mind.
Although Thornbury's (1954) text appeared after the English translation of
Penck's book, its publication was too soon for it to profit from the latter.
However, Thornbury was critical of Davis' tectonic assumptions and considered that, in stressing crustal mobility, Penck came nearer than anyone
else to offering a philosophy of geomorphology different from that of Davis.
Although Penck was probably wrong in his conclusions on the diastrophic
significance of varying cross-valley profiles, his most significant contribution
to geomorphology might well be:

> that he has caused geomorphologists to realize how inadequate have been the usual
> explanations of how slopes develop and recede.
>
> (Thornbury 1954: 200)

We are told with regard to the parallel retreat of slopes that:

> the consumption of a land mass is conceived by Penck as largely the result of
> *backwasting* and by Davis as the result of *downwasting*. It is probably too much to
> say that one is right and the other is wrong.
>
> (Thornbury 1954: 201)

Moreover, there may well be an element of truth in the idea that slopes, once
'graded', maintain a constant angle as they retreat. This suggestion had
clearly not been refuted by recent studies in arid lands and in South Africa.

Figure 10.12 Representation of the system of interaction between processes and form properties discussed by Walther Penck (1924). The star denotes a site of error in that the relationship here is not direct but differential

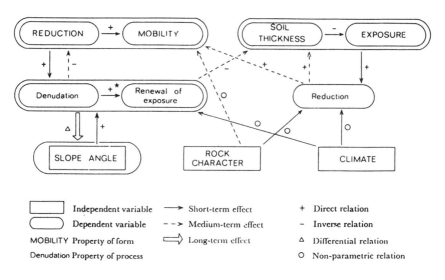

Source: From Young (1972), Figure 9, p. 31

The *Primärrumpf* and *treppen* concepts of Penck were dealt with at some length by Thornbury (1954: 201–5) who considered it difficult to distinguish between a Penckian *Primärrumpf* and a Davisian old-from-birth peneplain. The piedmont benchland concept was viewed as relatively unpopular and, unfortunately, the explanatory diagram given was that concocted by Davis to explain the effect of intermittent uplift.

Around 1960, partly under the influences of the English translation of *Die morphologische Analyse* (1953), two important explanatory works appeared which made Penck's ideas more clear to an English-speaking audience. As we have already seen (volume 2, pp. 704 and 717–8), Tuan (1958) clarified Penck's ideas on slopes and Simons (1962) performed a similar service in respect of his broader notions. Some time later Young (1972) devoted considerable attention to Penck's 'slope replacement theory' which, although as highly deductive and as divorced from measurement as that of Davis, was recognized to have certain intrinsic value. Penck's examination of the processes of mass movement was considered to be in advance of its time and his deductive model of slope development had continuing value. The latter, based on seven dependent variables and three independent variables, was expressed as a process–response model by Young (1972: 31) (figure 10.12). One Penckian assumption had been widely criticized (Morawetz 1932, 1937,

Figure 10.13 Three interpretations of landform development through time, showing schematic slope profiles and relief. C gives a more faithful representation of the Walther Penck model than does B

A   DAVIS (1899 *ET SEQ.*)

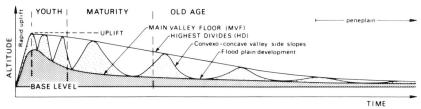

B   PENCK AFTER DAVIS (1932G) AND VON ENGELN (1948)

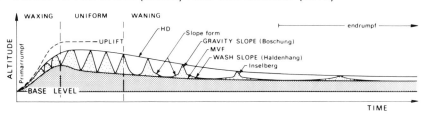

C   PENCK AFTER THORNES AND BRUNSDEN (1977)

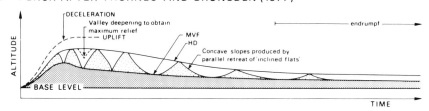

Source: After Thornes and Brunsden (1977), Figure 6.2, p. 122

Louis 1935, Birot 1949, Beckett and Webster 1968). Penck assumed the *instantaneous* removal of rock waste (on a slope at a given angle) immediately it attained a mobility susceptible to denudation, whereas in fact it then had to be transported *across all parts of the slope below it*. This would create a progressive downslope decrease in the degree of 'exposure'. Thus in figure 10.12 the star denotes a differential and not a direct relation. In his deductive model of slope evolution Penck began with a steep cliff of homogeneous rock topped by a level surface and footed by a river which removes all debris supplied to it but does not actively erode (see figure 10.8(i)). The whole of the rock face retreats parallel to itself, except the basal element which lacks a slope angle adequate for its removal at that degree of mobility. Thus a basal slope of lesser and uniform gradient develops at the expense of the rock face or cliff, which result, as noted above, invalidates one of the assumptions of the model. Young avers that:

> Penck's system ... stands in its own right as a powerful deductive model, applicable to cases where the manner of action of denudation processes simulates instantaneous removal of material, and where there is unimpeded basal removal.
>
> (Young 1972: 34)

Thus difficulties in explaining Penck's ideas by the use of diagrams remained, despite the welcome clarification provided by Thornes and Brunsden (1977) (figure 10.13C).

After mid-century much of world geomorphology became increasingly taken up with studies of process, as will be described in our volume 4. In this context Penck's work appeared just as stylized, dated and irrelevant as that of Davis. However, recent research into current tectonic processes promises a revival in Penck's fortunes.

Had Penck enjoyed a normal lifespan he might well have changed Hutton's famous dictum to read – 'While there is a sea and a mobile crust, there is that which is required for the system of the world'. As it was, he became the only contemporary scholar to loosen the Davisian stranglehold on landform concepts and to ensure that:

> The old order changeth, yielding place to new, ...
> Lest one good custom should corrupt the world.
>
> (Tennyson: *Morte d'Arthur*)

References

Aigner, A. (1925–6) 'Die geomorphologischen probleme an den Ostrand Alpen', *Zeitschrift für Geomorphologie* 1: 29–44.

Annaheim, H. (1936) 'Die Landschaftsformen des Luganer Seegebietes', *Geographische Abhandlungen* 8, 1–148.

Baulig, H. (1925) 'La notion de profile d'équilibre', *Comptes Rendus, Congrès International de Géographie (Cairo)* 3: 51–63 (also in (1950) *Essais de Géomorphologie*, Paris, pp. 43–77).

—— (1939) 'Sur les "gradins de piedmont"', *Journal of Geomorphology* 2: 281–304.

—— (1940) 'Le profile d'équilibre des versants', *Annales de Géographie* 49: 81–97 (also in (1950) *Essais de Géomorphologie*, Paris, pp. 125–47).

—— (1950) *Essais de Géomorphologie*, Paris: Société d' Edition Les Belles Lettres.

—— (1952) 'Surfaces d'aplanissement', *Annales de Géographie* 61: 161–83, 245–62.

Beckett, P. H. T. and Webster, R. (1968) 'Soil formation and slope development: 1. A new look at Walther Penck's Aufbereitung concept', *Zeitschrift für Geomorphologie* 12: 1–24.

Beckinsale, R. P. (1974) 'Albrecht Penck 1858–1945', *Dictionary of Scientific Biography*, vol. 10, New York: Scribners, pp. 501–6.

Birot, P. (1949) *Essai sur Quelques Problèmes de la Morphologie Generale*, Lisbon: Centro d'Estudos Geograficos.

Bowman, I. (1909) 'The physiography of the Central Andes: I The Maritime Andes, II The Eastern Andes', *American Journal of Science* series 4, 28: 197–217, 373–402.

—— (1916) *The Andes of Southern Peru*, New York: American Geographical Society.

—— (1924) *Desert Trails of Atacama*, New York: American Geographical Society.

—— (1926) 'The analysis of landforms: Walther Penck on the topographic cycle', *Geographical Review* 16: 122–32.

Braun, G. (1938) 'Zum problem der Piedmontreppen', *Comptes Rendus, Congrès Internationale de Géographie (Amsterdam)* 2: 125–32.

Bryan, K. and Wickson, G. G. (1931) 'The W. Penck method of analysis in Southern California', *Zeitschrift für Geomorphologie* 6: 287–91.

Chorley, R. J. (1974) 'Walther Penck, 1888–1923', *Dictionary of Scientific Biography*, vol. 10, New York: Scribners, pp. 506–9.

Cotton, C. A. (1941) *Landscape: As developed by the processes of normal erosion*, London: Cambridge University Press (2nd edn 1948).

Daly, R. A. (1905) 'The accordance of summit levels among alpine mountains', *Journal of Geology* 13: 195–205.

Engeln, O. D. von (1942) *Geomorphology*, New York: Macmillan (second printing 1948).

Festband (1918) *Albrecht Penck: Zur Vollendung des sechzigsten Lebensjahrs gewidmet von seinen Schulern und der Verlagsbuchhandlung*, Stuttgart: Engelhorn.

Fisher, O. (1866) 'On the disintegration of a chalk cliff', *Geological Magazine* 3: 354–6.

Heim, A. (1927) 'Die Gipfelflur der Alpen', *Neujahrsblatt der Naturforschenden Gesellschaft in Zürich*, No. 129, 1–25.

Henkel, L. (1926) 'Einwände gegen wichtage punkte in W. Penck's erosionstheorie', *Petermanns Geographische Mitteilungen* 72: 263–4.

Hettner, A. (1913) 'Rumpfflächen und Pseudorumpfflächen', *Geographische Zeitschrift* 19: 185–98.

Kesseli, J. E. (1940) *The Development of Slopes: A summary of studies by Walther Penck and Sieghart Morawetz*, unpublished manuscript.

Krebs, N. (1928) *Die Ostalpen und das heutige Osterreich*, 2nd edn., 2 vols, Stuttgart: Engelhorn.

Leutelt, R. (1929) 'Die Gipfelflur der Alpen', *Geologische Rundschau* 20: 330–7.

Louderback, G. D. (1904) 'Basin range structure of the Humboldt region', *Bulletin of the Geological Society of America* 15: 289–346.

Louis, H. (1935) 'Problems der Rumpfflächen und Rumpftreppen', *Verhandlungen über Wissenschaft Abhandlungen Jahr 25, Deutschen Geographentages zu Bad Neuheim (1934)*, pp. 118–37.

—— (1958) 'Albrecht Penck und sein Einfluss auf Geographie und Eiszeitforschung', *Die Erde* 89: 161–82.

Machatschek, F. (1919) *Geomorphologie*, Leipzig (2nd edn 1934, 6th edn 1954, 9th edn 1968), (English translation by D. J. Davis, 1969, Edinburgh: Oliver & Boyd).

Martin, G. J. (1974) 'A fragment on the Penck(s)–Davis conflict', *Geography and Map Division: Special Libraries Association Bulletin* 98: 11–27.

—— (1980) *The Life and Thought of Isaiah Bowman*, Hamden, CT: Archon, Shoe String Press.

Martonne, E. de (1948) *Traité de Géographie Physique*, tome 2, *Le Relief du Sol*, 8th edn, Paris: Colin.

Mojsisovics, E. von (1879) *Die Dolomitriffe von Südtirol*, Vienna.

Morawetz, S. (1932) 'Eine Art von Abtragungsvorgang', *Petermanns Geographische Mitteilungen* 78: 231–3.

—— (1937) 'Das problem der Taldichte und Hangserschneidung', *Petermanns Geographische Mitteilungen* 83: 346–50.

Ogilvie, A. G. (1923) 'Argentine physiographical studies: A review', *Geographical Review* 13: 112–21.

Penck, A. (1887) 'Ueber Denudation der Erdoberfläche', *Schriften des Vereines zur Verbreitung naturwissenschaftlicher Kenntnisse in Wien* 27: 431–57.

—— (1894) *Morphologie der Erdoberfläche*, 2 vols, Stuttgart: Engelhorn.

—— (1919) 'Die Gipfelflur der Alpen', *Sitzungberichte der Preussischen Akademie der Wissenschaften zu Berlin* 17: 256–68 (English Trans. by Mrs G. Wills).

—— (1926) 'Reply to the review by I. Bowman', *Geographical Review* 16: 350–2.

Penck, A. and Brückner, E. (1901–9) *Die Alpen im Eiszeitalter*, 3 vols, Leipzig: Tauchnitz.

Penck, W. (1920a) 'Der Südrand der Puna de Atacama (Nordwestargentinien): Ein Beitrag zur Kenntnis des andien Gebirgstypus und zu der Frage der Gebirgsbildung', *Abhandlungen der sächsischen Akademie der Wissenschaften, (Leipzig), Math.-Phys. Klasse* 37: (1) 420.

—— (1920b) 'Wesen und Grundlagen der morphologischen Analyse', *Bericht der sächsischen Akademie der Wissenschaften, (Leipzig), Math.-Phys. Klasse* 72: 65–102.

—— (1921) 'Morphologische Analyse', *Verhandlungen deutsch. Geographentages* pp. 122–8.

—— (1924) 'Die morphologische Analyse: Ein Kapitel der physikalischen Geologie', *Geographische Abhandlungen*, 2, Reihe, Heft 2, Stuttgart.

—— (1925) 'Die Piedmontflächen des südlichen Schwarzwaldes', *Zeitschrift der Gesellschaft für Erdkunde zu Berlin* pp. 83–108 (English trans. by M. Simons, 1961).

—— (1933) Durch Sandwüsten auf Sechstausender, Stuttgart: Engelhorn (second edn. 1938).

—— (1953) *Morphological Analysis of Landforms*, London: Macmillan (English trans. of Penck (1924) by H. Czech and K. C. Boswell).

Piotrovsky, M. V. (1984) *Towards an Understanding of the Cognitive Laws of the Earth*, Moscow, (in Russian).

Priem, A. (1927) 'Uber die Merkmale und den Entwicklungsgang der Piedmonttreppen', *Geographischer Anzeiger* 28: 373–81.

Richter, H. (1928) 'Zum Problem der alpinen Gipfelflur', *Zeitschrift für Geomorphologie* 4: 149–60.

Sauer, C. O. (1925) 'The morphology of landscape', *University of California, Publications in Geography* 2: 19–35.

—— (1929) 'Land forms in the Peninsular Range of California as developed about Warner's Hot Springs and Mesa Grande', *University of California, Publications in Geography* 3: 199–290.

—— (1932) 'Land forms in the Peninsular Range', *Zeitschrift für Geomorphologie* 7: 246–8.

Simons, M. (1962) 'The morphological analysis of landforms: A new review of the work of Walther Penck', *Transactions of the Institute of British Geographers* 31: 1–14.

Spreitzer, H. (1932) 'Zum Problem der Piedmonttreppe', *Mitteilungen der Geographischen Gesellschaft in Wien* 75: 327–64.

—— (1951) 'Die Piedmontreppen in der regionalen Geomorphologie', *Erdkunde* 5: 294–304.

Spurr, J. E. (1901) 'Origin and structure of the Basin Ranges', *Bulletin of the Geological Society of America* 12: 217–70.

Staff, H. von and Rasmuss, H. (1912) 'Zur Morphologie der Sächsischen Schweiz', *Geologische Rundschau* 2: 373–444.

Symposium (1938) 'La question de l'escalier de Piedmont', *Comptes Rendus du Congrès International de Géographie (Amsterdam)*, vol. 2, section 2, pp. 98–204.

Symposium (1940) 'Walther Penck's contribution to geomorphology', *Annals of the Association of American Geographers* 30: 219–80.

Thornbury, W. D. (1954) *Principles of Geomorphology*, New York: Wiley.

—— (1965) *Regional Geomorphology of the United States*, New York: Wiley.

Thornes, J. B. and Brunsden, D. (1977) *Geomorphology and Time*, London: Methuen.

Tuan, Y.-F. (1958) 'The misleading antithesis of Penckian and Davisian concepts of slope retreat in waning development', *Transactions of the Indiana Academy of Science* 67: 212–14.

Wooldridge, S. W. (1955) 'Review of "Geographical Essays" by W. M. Davis', *Geographical Journal* 121: 89–90.

Würm, A. (1935–6) 'Morphologische Analyse und Experiment, Stangentwicklung, Einebung, Piedmontreppen', *Zeitschrift für Geomorphologie* 9: 76–87.

Young, A. (1972) *Slopes*, Edinburgh: Oliver & Boyd.

PART IV

Regional Geomorphology

CHAPTER ELEVEN

Regional Classifications

Problems of Landform Classification

Classification is a necessary, but early, stage in the development of any science (Grigg 1967), but regional classification presents such special taxonomic problems to the spatial sciences that concern for it has had a particularly long history. The most satisfactory classifications are those conducted with a specific purpose in mind, rarely serving two purposes equally well. In this respect, the difficulty of defining the role of the study of landforms within physiography, geomorphology, geology and geography (see chapter 4) has created problems of definition and purpose which have led to the proliferation of landform classification and regionalization. In addition, not least among the challenges facing regional taxonomy are those associated with the definition of spatial individual units and how to group these units into regional classes, the hierarchical scales of which are distinct but do not impose any progressive change in the fundamental basis of the classification.

Were regional classifications of landforms concerned with comparable morphometric features alone, they would still present a variety of problems. In particular, the conflict between the genetic basis favoured by Davis and the generic followed by German geomorphologists provided grounds for heated debate during the first three decades of the present century. However, the problems of the role and relations of the study of landforms particularly in association with climate (see chapter 12), pedology (see particularly the Russian work described in chapter 6 and the 'catena' concept of Milne (1935)), vegetation (Bourne 1931, Veatch 1933), geography (Herbertson 1905) and human utility have led to increasing difficulties as the twentieth century has progressed. Indeed, the classificatory association of landforms with other spatial features in a complex functional association has introduced into geomorphology elements of the debate regarding the 'nodal', 'organic', 'real' and 'natural' character of regions which so occupied geographers before the Second World War. Even apparently simple attempts to depict landforms artistically (Lobeck 1923, Raisz 1938, 1946) have raised methodological issues mostly to do with spatial scales of generalization. Before the last decade of the nineteenth century detailed topographic surveying was providing a flood of information at ever more refined scales for the student of landforms and it should not be forgotten that Davis' studies of the evolution of

Appalachian landforms (e.g. 1889D, 1890K) were only made possible by the advent of suitable topographic maps of Pennsylvania and New Jersey (see volume 2, p. 232). After the First World War improvements in aerial photography swelled this flood.

It is appropriate to treat the regional taxonomy of landforms in terms of the various strategies which have been adopted to attack the problems referred to above. These strategies have involved the encyclopedic approach, subdivision, accretion, the use of drainage basin hierarchies, practical considerations and complex regionalizations.

Encyclopaedic Classifications

In his historical account of the growth of physiographic geology during the second half of the nineteenth century, Von Zittel (1901) selected for special praise Oscar Peschel (1826–75), Baron Ferdinand von Richthofen (1833–1905) and Albrecht Penck (1858–1945). Peschel's *Neue Probleme der vergleichenden Erdkunde: als Versuch einer Morphologie der Erdoberfläche* (1870) was based on oversimplified and crude classificatory assumptions. A fuller and more adequate effort to group together types of landforms which have a similar genetic history was made by Richthofen in 1886 in his *Führer für Forschungsreisende* (Beckinsale 1975; see Volume 1, p. 601). The second part of this handbook, or guide, for scientific travellers discussed the interrelations of underlying geology and surface forms and provided appreciable detail of the processes involved. The third and final part (pp. 652–85) dealt with soils, rocks and mountain structures and attempted to classify the chief kinds of landforms (*Bodenplastik*) as far as possible according to the process that was dominant in their formation. The ideas, illustrated by more than 100 small line blocks (see figure 12.3), are at times remarkably percipient and demonstrate climatic influences:

> In its classification of material and approach it was the first truly successful compilation of genetic geomorphology. It immediately became the standard work in Germany for the systematic treatment of landforms. . . .
>
> (Beckinsale 1975: 440)

Richthofen's volume strongly influenced the content of a larger and superior work produced in 1894 by Albrecht Penck who

> accomplished the difficult task of arranging our present knowledge of surface-configuration upon the basis of leading genetic principles. In his *Morphologie der Erdoberfläche* Penck has presented the chief results of the special literature of physiography in a clear, concise form.
>
> (Von Zittel 1901: 185)

This compilation is outstandingly important as the first unified text of geomorphology. It is often said to be of two volumes but is in fact two main parts bound within one cover. The first part deals in 486 pages with agents of erosion and deposition. Its introduction refers to Peschel's work of 1870, which carried in its subtitle the title of Penck's volume. It expresses also Penck's indebtedness to Suess (1883–1908), Richthofen (1886) and Dana (1863). The second part, of 696 pages, discusses landforms under descriptive headings, such as plains; hills; valleys and their relation to various forms of composite relief; basins and lakes; and depressions. Throughout the whole volume the approach is encyclopedic, concise and precise, being a close-knit amassment of facts and formulae, suggestions and measurements, more reminiscent of Dana's *Manual of Geology* (1863) than of Richthofen's *Führer für Forschungsreisende*. Most of the principles given were already known but some of the information was new and the presentation was unified and scientific.

In 1895, the year following the publication of the monumental *Morphologie der Erdoberfläche*, Albrecht Penck delivered a notable address to the Sixth International Geographical Congress in London at a meeting attended among others by Lapparent, Passarge and the brothers Archibald and James Geikie. Entitled *Die Geomorphologie als genetische Wissenschaft: eine Einleitung zur Diskussion über geomorphologische Nomenclatur*, it was accompanied with a summary in English called *The morphology and terminology of land forms*. This brilliant outline provided the quintessence of German morphological classification at a time when geomorphology was about to be invaded by rival Davisian cyclic concepts. At the meeting, Harry Seeley, professor of geography at King's College, London, expressed the British attitude:

> The elaborate classification of the origin of similar physical features of the earth by Professor Penck is indispensable for the scientific teaching of geography, and in giving names to these phenomena he has gone beyond what has hitherto been considered necessary in this country. Professor Penck has laid necessary stress upon the nature and intensity of the forces concerned in denudation. In this country we are accustomed to regard the nature of the material upon which the force works as not less important. Both the form of force and the time for which it works have to be considered in relation to each variety of rock.
>
> (Seeley in Penck 1896: 736)

The following ideas which Albrecht Penck developed regarding the geomorphic nomenclature and classification around the turn of the century were later somewhat modified under the full impact of Davis' influence but by the First World War, under the influence of Passarge and of his son's ideas (see chapter 10), he rid himself of cyclic tendencies.

During the early 1890s Penck considered that the fundamental morphological problems involved the forms or shapes of terrain features (stereometry), the processes involved in their formation, and the grouping of associated forms into distinctive assemblages. Changes in the form of the earth's surface were held to result from *true erosion* (the linear action of rivers) and from *denudation* (by distributed weathering and mass movements), the former being generally more rapid than the latter; from accumulation; and from dislocation (figure 11.1). All forms of the surface could be assigned to a comparatively few fundamental types which were themselves arrived at by a systematic grouping of the elements of form. Fundamental surface forms may be classified into:

1 The plain; virtually horizontal.
2 The escarpment; steeply sloping surface.
3 The valley; two more-or-less steeply sloping surfaces facing each other across a narrow strip of even surface, which usually declines in the direction of its downhill slope.
4 The mount or mountain; which slopes away in every direction.
5 The cup-shaped hollow; surface falls in a circle towards a fixed point.
6 The cavern; almost enclosed by land surface.

As a rule these six fundamental forms occur in association, with either one form predominating or different forms occurring regularly side by side, for example escarpment and plain. Larger groupings of associations occur, as in mountain systems. So, on the earth's surface six categories of units may be found in ascending order of magnitude:

1 Form element.
2 Fundamental form.
3 Group of forms or landscape.
4 Extended area of equal elevation.
5 A system, or grouping of such areas.
6 Continental block and abyssal deep.

1–3 are minor categories which have arisen in a variety of ways, while 4–6 are major categories which are always exclusively due to dislocations of portions of the earth's crust. Any fundamental form can be created in three ways (by erosion, accumulation, or dislocation) but the nature of the topographic surfaces developed depends heavily on the underlying geological structure; on whether the structure consists of horizontal strata, slightly folded or bent strata (*Verbiegungsland*), intensely folded strata (*Faltungsland*), faulted blocks (*Schollenland*), lava flows (*Ergussland*), or intrusive igneous masses (*Intrusivland*).

If fundamental forms have the same outward appearance they may be called *homoplastic*, and if of the same origin, *homogenetic*. A combination of homoplastic and homogenetic creates a *homonymous* form, whereas homoplastic and heterogenetic creates an *analogous* form. Penck pursued other relationships with remarkable persistency and then turned to outstanding tasks. The first task was to describe and name individual forms and to determine their origin. At that time even the nomenclature of the six

Figure 11.1 Albrecht Penck's (1896: 750) presentation of the genetic relations of six fundamental geomorphic forms

Fundamental Form.	Erosion.		Accumulation.	Dislocation.	
	True Erosion.	Denudation.		Faulting.	Folding.
Plain	Littoral } abrasion Fluviatile	Subaerial	Æolian (loess) Fluviatile (river plain) Lacustrine Marine (coast plain) (sea bottom) Organic (peat moss)		
Escarpment	Littoral (cliff) Fluviatile	Subaerial (strata escarpment)	Subaerial (detritus esc.) Fluviatile (delta slope) Lacustrine (lake cliff) Marine (sea cliff)	Fault-scarp	Flexure-scarp
Valley	Fluviatile (normal valley)	Subaerial (strata valley)	Æolian (dune valley) Glacial (moraine valley) Volcanic Organic (reef valley)	Trough valley Monoclinal valley	Synclinal valley
Mount	Fluviatile (ridge)	Subaerial (strata ridge)	Æolian (dunes) Fluviatile (gravel heaps) Glacial (moraines) Marine (banks) Volcanic (volcanoes)	Horst-mount	Anticlinal mountain
Hollow	Subaerial (doline) Æolian (wind hollow) Fluviatile (river-hollow) Glacial (glacier-hollow) Marine (tide-hollow) Volcanic (crater of explosion)	Æolian } removal Glacial	Subaerial Æolian Fluviatile } Ring-deposit Littoral } and formation Volcanic } of a dam Organic	Monoclinal hollow Trough hollow	Synclinal hollow
Cavern	Fluviatile	Subaerial (earthfall)	Thermal (sinter-cave) Volcanic (lava-cave)	Fault-cave	Fold cave

Source: From Penck (1896), p. 750

fundamental forms had still to be found; for example, the fifth form (cup-shaped hollow) had neither clear definition nor standardization, being *basin* in English, *Becken* in German and *bassin* in French, all of which are ambiguous. However, because each fundamental form, except the plain, embraced three groups of homogenetic features, which themselves may be subdivided on the basis of the kind of erosion, or accumulation, or dislocation that operated, its plastic and genetic relations could be expressed by two nouns, for example, plain of accumulation. The addition of an adjective would give precision to the mode of origin, for example plain of fluviatile accumulation. A further requirement of this geomorphological nomenclature was to limit the use of special terms to features of similar origin. Thus mounts of aeolian accumulation are dunes and the term must not be applied to any other kind of hill. Modern geomorphologists will find many weaknesses in Penck's scheme but he was well aware that the geomorphological knowledge of his time was too limited to produce a natural classification into 'families' based on the grouping together of all homogenetic forms without reference to their outward appearance. We discuss elsewhere (chapter 12) his great contribution to climatic geomorphology.

It is thus apparent that in the late nineteenth century German scholars were pre-eminent in devising detailed encyclopedic classifications of land-forms, introducing a spatial element by grouping individual forms into assemblages. One noteworthy contemporary British classification of processes appeared in Archibald Geikie's *Textbook of Geology* (1882, 1903, vol. 1, book 3) where the description was arranged under the following heads, 'the reader being warned that the subdivision is employed merely for convenience':

A. HYPOGENE ACTION
    1. Volcanic
    2. Earthquake
    3. Secular Upheaval and Subsidence
    4. Other Crustal Changes

B. EPIGENE ACTION
    I. Atmospheric (Destructive; Reproductive)

    II. Water
        1. Rain (Chemical; Mechanical)
        2. Underground Water (Chemical; Mechanical)
        3. Brooks and Rivers (Chemical; Mechanical)
           (a) Transport
           (b) Excavating
           (c) Reproductive

    III. Terrestrial Ice
        1. Frost (Destructive)

    2. Snow (Conservative; Destructive)
    3. Ice-Sheets and Glaciers (Transport; Erosion; Deposition)

IV. Oceanic Waters
    1. Erosion (Chemical; Mechanical)
    2. Transport
    3. Reproduction

V. Life (ORGANISMS)
    1. Destructive
    2. Conservative
    3. Reproductive

However, it is generally recognized that the so-called epigene agents are in reality complex and capable of further subdivision; that, whereas a simple classification of forces may be sufficient for geological needs, for geography, where emphasis is placed on the evolution of topographic features, the systems of subdivision should be those into which individual landforms can readily be fitted. Many landforms are the result of joint action of various activities or forces; they are polydynamic and in that eventuality, if it is possible, the form in question should be classified under the particular force that has predominated in its production. In Geikie's classification, some of the activities enumerated do not produce definite surface forms, while others seem capable of further subdivision.

The belief that Albrecht Penck had explored the limits of Germanic encyclopedic taxonomy were soon dispelled by the work of Passarge. Siegfried Passarge (1866–1958) (plate 11.1) started his academic life as a pupil of Richthofen at Berlin University and produced his first publication in 1891 (on the New Red Sandstone of Thuringia) and his last, a review of morphological studies near Aswan, in 1955. In the meanwhile he spent some years at Jena and Freiburg and became fully qualified in both geology and medicine. He attended the International Geographical Congress at which Albrecht Penck spoke in 1895 and then left to study the deserts of South-West Africa. Later he moved successively to the chair of geography at Breslau and, during 1908–36, to the newly established Kolonial Institut at Hamburg. He wrote prolifically under the stimulus of the geomorphological ideas of Richthofen and Penck, and after about 1909 under the influence of his abhorrence of the cyclic concepts of Davis. In geography he was often influenced by his professional knowledge of medicine and psychology so that his 'regional geography' frequently seems to be too all-embracing, but in his geomorphology he found a unifying theme in his dislike of Davisian stages.

Passarge's *Physiologische Morphologie* was published in 1912 as a section of the Hamburg Geographical Society's journal and as a separate book. In it Passarge set out to establish German geomorphology in the tradition of Richthofen and Penck and to dissuade readers from embarking on Davis'

Plate 11.1 Siegfried Passarge

Source: Courtesy of Professor G. Sendler and University of Hamburg Archives

educational and descriptive methods. Spatial variations in the earth's surface could not be imputed to any single cause because local forces (geology) and genetic forces (climate) were both involved. Davis' simplified 'structure, process, stage' concept should be replaced by 'region, place [local forces], and form'. The morphological study of landforms should be derived from the superimposition of maps – on the same scale – showing relief, lithology, and vegetation. Such a method would largely avoid, according to Passarge, the unreality and superficiality of Davis' scheme which is riddled with abstractions (or deductions) and rather neglects the climatic side of the genetic processes involved. Passarge analyses and classifies the form elements in groups according to their genetic bases, which may be monodynamic (one dominant force) or polydynamic. He sets up two great types of surface forms, landforms and coastal forms or landscapes and coastscapes, and then, using a classification of 'forces', establishes, in agreement with it, groups of landforms representing classes, orders, families, genera and species or special forms. The following outline omits the many species, as well as orders and smaller units for coastal forms and polydynamic forms:

TYPE A. LANDFORMS

  CLASS A. ENDOGENOUS FORMS
    ORDER I. TECTONIC FORMS
    Family 1. Flexure Forms
    Genus   a. Convex Forms
             b. Concave Forms
        2. Fracture Forms
             a. Block and ridge lands
             b. Sunk lands
        3. Fold Forms
             a. Symmetrically folded mountains
             b. Asymmetrically folded mountains
             c. Overthrust and folded mountains

    ORDER II. VOLCANIC FORMS
        1. Intrusive Forms
             a. Without change of Form
             b. With change of Form
        2. Eruptive Forms
             a. Explosive accumulations
             b. Effusive accumulations

  CLASS B. EXOGENOUS FORMS
    ORDER I. ACCUMULATION FORMS
        1. Water Accumulation Forms
             a. Mechanical Accumulation Forms

        b. Chemical Accumulation Forms
        c. Superficial Mass Movements
    2. Ice Accumulation Forms
        a. Bottom or Ground Ice
        b. Glacier Ice
    3. Glacier Accumulation Forms
        a. Moraines
        b. Fluvio-glacial accumulations
    4. Aeolian Accumulation Forms
        a. Dune lands
        b. Loess Deposits
    5. Sea Accumulation Forms
        a. Wave
        b. Tidal
        c. Current
    6. Vegetable Accumulation Forms
        a. Autochthonous accumulations
        b. Allochthonous accumulations
    7. Animal Accumulation Forms
        a. Accumulations due to land animals
        b. Accumulations due to marine reefs
        c. Pelagic accumulations
    8. Accumulations due to Temperature Changes
        a. Insolation accumulations
        b. Frost accumulations

ORDER II. DENUDATION FORMS

    1. Water Excavation Forms
        a. Erosion (sub-Genera: Fluvial; Pluvial; Mass
           Movements)
        b. Solution
    2. Ice Erosion Forms
        a. Glacier
        b. Snow-water
    3. Aeolian Erosion Forms
        a. Mechanical Erosion
    4. Disruption Forms due to Temperature Changes
        a. Insolation
        b. Frost
    5. Disruption Forms due to the Action of Lightning
        a. Mechanical
        b. Chemical
    6. Disruption Forms due to the Action of the Sea
        a. Wave

      b. Tidal
   7. Disruption Forms due to Vegetable Action
      a. Mechanical
      b. Chemical
   8. Removal Forms due to the Action of Animals
      a. Mechanical
      b. Chemical

TYPE B. COASTAL FORMS

CLASS A. DETERMINED BY THE FORM OF THE ADJACENT
LAND
CLASS B. DETERMINED BY THE FORM OF THE ADJACENT
SEA FLOOR
CLASS C. DETERMINED BY NOW-ACTIVE PROCESSES OF
ACCUMULATION
CLASS D. DETERMINED BY NOW-ACTIVE PROCESSES OF
EROSION

Passarge did not wait for any testing of the practical value of his classificatory scheme and produced from 1919 to 1930 a large literature on it. His first major exposition was *Die Grundlagen der Landschaftskunde*, a work of four volumes, the first of which (*Beschreibende Landschaftskunde*, 1919) discussed with considerable thoroughness the elements or individual forms that may contribute to a landscape; the second dealt with the causes of landscape phenomena, including oceans, atmosphere, plants and animals; the third *Die Oberflächengestaltung der Erde* (1920) surveyed at great length the processes creating landforms, and especially climate; the fourth provided an account of man's influence on the surface of the earth. Passarge developed and modified the same general theme in his *Vergleichende Landschaftskunde* which appeared in five parts or volumes from 1921 to 1930. Here he expanded the concept of landscape elements combining to form a hierarchy of sizes successively from a simple grouping of contiguous elements, to a 'section', a 'region', major regional groupings, and finally great zonal or latitudinal belts (*Landschaftsgürtel*), which we will mention again under climatic geomorphology. But much of these later volumes tends to be geographical rather than geomorphological and the earlier volumes (*Die Grundlagen der Landschaftskunde*) are more interesting for our present concern.

    Davis was well aware of Passarge's *Physiologische Morphologie* (1912) which he obviously distrusted but he decided not to advertise it by a specific review, referring to it in passing (Davis 1915K: 72). His reaction to *Die Grundlagen der Landschaftskunde* was much more violent and he immediately reviewed at length the first volume, *Beschreibende Landschaftskunde*, in 1919 (Davis

1919D). This reaction has already been summarized (see volume 2, pp. 502–4) and here we will only outline that summary. The aim, Davis states is:

> To serve as a guide to the investigation and description of the visible features of geographical areas in their natural combinations.
>
> (Davis 1919D: 266)

He continues:

> It is furthermore held to be a necessity that the scientific study of landscapes . . . should begin with an empirical description of facts of observation, uninfluenced by theoretical or explanatory preconceptions.
>
> (Davis 1919D: 266)

The volume consists largely of a systematic analysis of landforms in which the fundamental features are first classed as individual and as composite forms. These composite forms are grouped into larger regions and zones:

> The classification is thorough, sometimes to the point of being labored . . . . The variety of forms treated is great, and a knowledge of them is profitable.
>
> (Davis 1919D: 267)

Inevitably Davis finds certain minor landforms, such as the cusps and concaves of terrace fronts, had not been included in the text. But his chief complaint is the dominance of empirical definitions and descriptions from which all rational, explanatory treatment ('the refreshing juice of explanation') has been excluded. He deplores the 'heavy laboriousness' of such an approach and compares its educational principles with those of Gilbert, 'our American master of geomorphology', who believed in improving the deductive faculty and enlarging the logical side of his subject rather than deterring readers by a surfeit of tedious descriptive content.

Davis, however, had been premature in attacking Passarge's relative lack of explanation and genesis because the German author's third volume (*Die Oberflächengestaltung der Erde*) went into considerable detail on sculpturing agencies and the resultant surface forms. In his inevitable review (see volume 2, pp. 504–8), Davis admits its strength on soils, but regrets that:

> the soil forms are insufficiently related to their physiographic environment and are almost wholly divorced from the various stages of the cycle of erosion in which they belong.
>
> (Davis 1923G: 601)

In spite of reservations, Davis also admits its strength on wind action and climatic effects, including climatic changes. As would be expected, he finds some errors and ambiguities but his main objection seems to be to Passarge's hostility to deduction. Part of Passarge's third volume attempted to classify landforms according to their manner of origin in a natural system. As in the 1912 classification, the system proposed is extremely systematic. It includes types, classes, orders, families, genera and subdivisions of genera and thus becomes so elaborate that its general adoption in landscape description is altogether improbable.

(Davis 1923G: 605)

Passarge's volume concluded with attacks on the cycle of erosion and especially on the widespread idea of peneplains developed under so-called 'normal conditions'. Passarge's explanation of 'plains of degradation', which is placed near the end of the book and printed in spaced letters 'as a sort of dénouement to the scheme', leans heavily on one of Richthofen's ideas that extensive and deeply penetrating planation is the consequence of repeated climatic changes. Davis suggests that this seemed to be deduction and lacked sufficient evidence to replace his own scheme of peneplanation accomplished by the action of erosional processes within a 'single climate'. Davis (1923G: 607) concluded his review with a warning:

it is feared that, in spite of the insistence [Passarge] places on direct observation and empirical description in his first volume, a distraction from explanatory description toward analytical explanation will be the result of his third volume. In geomorphology, ... such analytical explanation is appropriate enough, ... but in the explanatory description of the visible landscape it is not the development of land forms but the developed forms that should stand in the foreground; ... *Landschaftskunde*, learnedly analytical as it is, leaves much to be desired.

Passarge's 1912 classificatory system was introduced to the British audience by J. D. Falconer (1914a), who followed this with a very detailed encyclopedic classification of his own (1914b, 1915). John Downie Falconer (1876–1947) seems to have slipped undeservedly into the shadows. He was educated at the University of Glasgow where he graduated in 1897 and, later, at Edinburgh University where he became a lecturer in petrology and assistant to James Geikie. From 1911 to 1917 he acted as lecturer in geography at Glasgow University and, from 1914 to 1921, as lecturer in geology at the British Museum. At the same time, from 1904 onward, he was closely associated with the geological survey of Nigeria and from 1918 to his retirement in 1927 directed it. In his account of the classification of landforms, Falconer affirmed that:

Emphasis may be laid either upon the external shape or upon the internal structure or upon the processes through which the various forms have originated and such classifications may be termed respectively morphological, structural, and physiological.

(Falconer 1915: 244)

He suggested that as a morphological classification the reader should use that by A. J. Herbertson (1911: 25–56), based upon certain 'elementary' forms:

The ridge; the mount; the table; the furrow; the hollow; the depression; the col; the plain; and the cavern.

These forms might then be subdivided according to the variation of shape and size. For example:

> A mount may be conical and either peaked, truncated, or rounded; dome-shaped and either symmetrical, elliptical, or crested; columnar and either angular or rounded, peaked or truncated; pyramidal and either triangular or polygonal, truncated or stepped.

(Falconer 1915: 244)

Herbertson's classification, drawing heavily on Supan's *Grundzuge der Physischen Erdkunde* (1884, 1899), was recognized as having, in common with other purely morphological schemes, the obvious weakness that its various groups included forms which differed widely in origin and that it emphasized superficial form to the exclusion of genetic aspects. Structural classifications, based on the petrological character and structural arrangement of the underlying rocks, such as igneous or sedimentary or metamorphic, do, according to Falconer, usefully emphasize the dependence of surface form upon the natural structure of the local geology, but they appeal more to geologists than to geographers who regard structure as one of several fundamental factors in landform development. Physiographic classifications based on the results of the various processes being tabulated together in separate groups need careful handling. Too often, in textbooks of physical geography and geology current about 1915, the detailed description of processes was not accompanied by a classification of landforms, or the authors proceeded to discuss landscapes under headings such as mountains, valleys, plateaus, and plains, as if they did not clearly distinguish between landforms and landscapes.

Falconer's (1915) own classification, similar to that of Passarge, was based on class, order, family, genus, subgenus and specific landform (number of divisions in each group in parenthesis):

CLASS (2) — endogenetic
             exogenetic

ORDER (4) — endogenetic negative
             endogenetic positive
             exogenetic degradation (denudation)
             exogenetic aggradation (accumulation)

FAMILY (26) — e.g. endogenetic positive landforms due to superficial volcanic activity

GENUS (58) — e.g. endogenetic positive landforms due to superficial volcanic activity: effusion forms; explosion forms

SUBGENUS (?) — e.g. endogenetic positive landforms due to superficial volcanic activity: solidification forms; precipitation forms

SPECIFIC FORM (?) — e.g. endogenetic positive landforms due to superficial volcanic activity: solidification forms: lava plains, slopes, scarps, etc. By extension, lava cones, domes and flows

Falconer (1915: 394) believed that the ideal landform was a surface or slope which is monodynamic, produced by the activity of a single force. In reality, however, he recognized that most surfaces are polydynamic and the surface or slope may then be assigned either to the predominant force, when several forces have acted simultaneously, or to the proximate force when several forces have acted more or less in succession. Falconer distinguished between *landforms* and *landscapes*, the latter made up of more-or-less orderly repetitions of forms. Davis (1915K: 71–2; see volume 2, pp. 410–11) clearly regarded the practicality of such classifications as highly questionable and referred specifically to Falconer's scheme:

> in which he classifies over 60 specific forms under various orders, families, genera and subgenera. So far as I have seen, the schemes have not yet been put to use in the description of actual landscapes or regions, although such a test of their practical value is evidently desirable.

Characteristically, Davis goes on to advocate his own classificatory scheme based on structure, process and stage, which could be made fivefold by the addition of 'relief of surface or local measure of vertical inequality, and texture of dissection or spacing of stream lines'.

Classification by Subdivision

Linton (1951: 217) outlined alternative strategies for the regional classification of landforms:

> Morphological delimitation must proceed from the consideration of one or other of the two natural units of the lands, the *site* and the *continent*, by the recognition of the characteristic groupings of the one and the characteristic subdivisions of the other, carrying both processes to the point where their results converge

The practical limits of regional subdivision have been explored by Linton (1951) and by Mabbutt (1968), the former using a helpful chemical analogy:

> The smallest subdivision of a substance that can exist and still retain the physical and chemical properties of that substance is the molecule. The smallest physiographic subdivision of a land mass for which description is simplified by reference to its morphological evolution is the section. Subdivision of the chemical molecule yields atoms of the constituent elements whose properties are unrelated each to the other or either to the compound whole. Subdivision of the physiographic section yields unit areas of the constituent morphological elements, closely related to the subjacent geological formations, but unrelated to each other or to the compound whole.
>
> (Linton 1951: 205)

Mabbutt (1968: 13–4) notes an inherent problem in subdivision in that it

> cannot proceed beyond the level of unit appropriate to the genetic bond itself . . . .
> As the range of criteria widens in order to encompass the increasing heterogeneity
> of the lower units, so the applicability of each factor becomes more localized and its
> relationships more complex. For each lower level of subdivision there is an
> increasing number of alternative higher groupings; the arrangement becomes
> increasingly *ad hoc* and the simple hierarchic structure is lost.

Linton identifies the strategy of subdivision with the United States and
Britain:

> In fact, it can even be urged that a predilection for morphological subdivision may
> be considered as a characteristic of British geographical method . . . the predilection
> is one which we seem to share with our American colleagues – if indeed we have not
> derived it from them – and in which we differ from the French.
>
> (Linton 1951: 199)

Apart from the early mapping by Lewis Evans in the eighteenth century
and the more detailed work by Joseph Peter Lesley in the Appalachians a
century later (see volume 1, pp. 236–8, 346–54), the first important regional
classification of North American landforms appeared in John Wesley
Powell's *Physiographic Regions of the United States* (1895) wherein:

> The regions here delineated are held to be natural divisions, because in every case
> the several parts are involved in a common history by which the present
> physiographic features have been developed.
>
> (Powell 1895: 66)

Later W. M. Davis (1899 A, B) improved and popularized Powell's regions
and in 1911 Isaiah Bowman produced the first great book in English on the
subject of the regional classification of landforms. Although entitled *Forest
Physiography*, its subtitle, *Physiography of the United States and Principles of
Soils in Relation to Forestry*, reveals its true nature. It is a fine book, amply
illustrated, including two large, pull-out, coloured maps, physiographic and
geological, and just over four-fifths of it concerns physiographic regions
which Bowman (1911: 108–10) defined (after Davis) as having uniform
topographic expression resulting from uniform geological structure, physiographic process and stage of development. A more restricted regional
physiographic treatment of Kansas had previously been conducted by Adams
(1902) who used the contact between the Upper and Lower Carboniferous
rocks to delimit accurately the western boundary of the Ozark Mountains.

Interest in problems of regional geomorphology now rose to a crescendo in
academic circles in the United States. In 1913 W. L. G. Joerg (1914),
pointing to the lack of harmony among geographers in delimiting regions,
took twenty-one existing national schemes and combined them to produce a
generalized national map of 'natural regions', or of portions of the earth's

Plate 11.2 Nevin M. Fenneman, aged 73

Source: Photograph by Nancy Ford Cones. Courtesy University of Cincinnati Archives

surface 'whose physical conditions are homogeneous'. At the same meeting, at Princeton, N. M. Fenneman suggested a subdivision or classification that was more essentially geomorphic in quality, although it had the dual purpose of discussing and explaining the physical features of the country and also of assisting planning and discussion of matters of human concern. Fenneman postulated that 'unity or similarity of physiographic history' is a formula which *almost* designates the basis here in mind for the delimitation of provinces' (1914: 86) but, although he stated that the influence of broad geological structures on boundary delimitation was subsidiary, it was clearly more important than he was prepared to admit. His conclusion was:

> The ever-present considerations . . . are topography and soils. They are, on the one hand, the end product and record of physiographic history, and, on the other, the beginning of geographic development. These therefore are the most fundamental criteria in determining physiographic provinces.
>
> (Fenneman 1914: 85)

Nevin Melancthon Fenneman (1865–1945) (plate 11.2), the son of an Ohio Calvinist minister, came under the influence of W. M. Davis at a Harvard summer course in 1895 and was subsequently encouraged by Isaiah Bowman. He was professor of geology and geography at the University of Cincinnati from 1907 to 1937. A cultured man of wide literary interests, he projected an ambiguous image – to his peers he was a modest and diffident scholar but to certain of his students he appeared as 'a gruff, tyrannical, relentless, rigorous, sarcastic and frightening teacher', doubtless a legacy of his own strict upbringing (Ryan 1986).

At the December 1914 meeting of the Association of American Geographers in Chicago a whole session was devoted to physiographic regionalism and a committee (M. R. Campbell, F. E. Matthes, D. W. Johnson and E. Blackwelder) was set up under Fenneman to prepare a revised map of the major physiographic divisions of the United States and to give a brief account of the salient physical characteristics of each (Matthes 1915). The first result of the committee's work appeared 2 years later (Fenneman 1916) when *form* was identified as the primary basis for subdivision (p. 24) into major divisions, provinces, sections and districts. Indeed, Fenneman (1916: 27) went so far as to write:

> The basis on which the major divisions are distinguished is not essentially different from that which is assumed in distinguishing the smaller units.

A later version appeared in 1928 based on more detailed work, such as that by Fenneman (1922) himself on the delimitation of the High Plains border, and on consultations with the US Geological Survey. This version became a standard document which was applied to a wide range of regional classifications of human as well as physical significance. Fenneman's influence was considerable, and Thayer (1916, 1918) extended the former's regional classification south into Mexico and north into Canada. Revisions and extensions were also conducted for the Interior Low Plateau (Flint 1928), Alabama (Johnston 1930), Hawaii (Wentworth 1936), Idaho (Anderson 1941) and California (Jenkins 1943).

In his critical analysis Linton (1951) pointed out that, because the United States covers more than 3 million square miles and the Fenneman method employed aims essentially at finding the 'Highest Common Factor of as wide an area as possible', the subdivisions tend to be enormous. The major divisions rest on a few basic characteristics, usually those reflecting general rock structure or structural history, while the provinces generally embrace wider considerations, and the sections still more. But so far the problem of classification has dealt only with macrodivisions, and it is reasonable to assume that physiographic sections can be further subdivided into their component parts, at which stage physiographic history usually ceases to be 'a unifying factor' and the 'unit areas of the constituent morphological elements' yielded by the subdivision become 'closely related to the subjacent geological formations, but unrelated to each other or to the compound whole' (Linton 1951: 205).

From at least 1928 onward the use of broad physiographic divisions dominated American geography. In 1931 Fenneman produced *The Physiography of Western United States* in which he aimed to provide a consistent frame for the addition of extra knowledge of smaller units, but which was heavily cyclic in its approach:

> Such local treatises will be vastly more useful if all conform to some previously accepted scheme of physical units for the entire country. They will thus conform with one another.
>
> (Fenneman 1931: vii)

A student concerned with a small area bounded, or traversed, by any of Fenneman's generalized boundaries was advised to delineate them 'with a degree of detail suited to the scale on which he works'. The accompanying map of the physical divisions of the United States (see figure 7.1) was considered 'to be subject to alteration and refinement as knowledge increases' (Fenneman 1931: viii). This fine book was followed in 1938 by a larger volume on *The Physiography of Eastern United States* which supplied details of relevant provinces and sections. Inevitably one-volume summaries of the physiography of the whole United States or North America also appeared, notably those by F. B. Loomis (1937) and W. W. Atwood Sr (1940). The latter

> was written more from the viewpoint of the geographer than from that of the geologist; its treatment of most geomorphic units is rather brief, and it departs somewhat from the classification set up by the Fenneman committee.
>
> (Thornbury 1965: 8)

The divisions used by Atwood in this and other textbooks were 'natural regions' rather than strictly physiographic provinces but they had great public appeal. According to Preston James (1972: 396), Atwood's various textbooks were used by something like 50 million children and it may be that 'no American has ever brought geography to so many people'.

Classification by Accretion

In the early 1930s there was considerable interest among geographers and geomorphologists in regional classification based on the amalgamation of supposedly basic spatial units. Unstead (1933) used a so-called 'synthetic method' to combine *stows* – small regions exhibiting a unity of relief – into the successively higher orders of *tracts* and *regions*. Bourne (1931) identified the 'soil site' as a basic morphological unit, groups of which combined into distinctive spatial assemblages or patterns, and Milne (1935) used soil slope catenas as landscape components. More important, Wooldridge (1932, 1935) proposed that landform regions should be assembled from the accretion of 'the basic physiographic atoms out of which the matter of regions is built' (Wooldridge 1932: 33). These atoms were defined as the landscape *facets* of 'flats' and 'slopes' forming the intersecting erosional or depositional elements believed to be characteristic of every landscape (Wooldridge 1932: 31–3). In reality, the existence of these facets was based on the assumption that landscapes were composed of Davisian peneplains, partial peneplains, straths, terraces and floodplains (i.e. the 'flats') connected by intervening 'slopes'. Thus this classification from its inception confused generic and genetic criteria. Linton (1951) elaborated the idea that flats and slopes form the ultimate units of relief by suggesting their hierarchical combination into *site, stow, tract, section, province,* and *continental subdivision*. In this way region building by accretion merged with Fenneman's subdivisions. Savigear (1956) termed Wooldridge's facets (or areas between significant breaks of slope (Waters 1958)) 'morphological units', contributing the concept of curved slope segments (facets) in 1966 which could also be added to the regional classes (Gregory and Brown 1966). The notion of landscape facets was developed into a military concept by the British Military Engineering Experimental Establishment (MEXE), initially in the context of a test area around Oxford. Here facets such as gravel terrace units were defined in terms of 'physiographic unity' and combined into *landscape patterns* (e.g. clay valley or limestone scarp) based on 'common or related morphogenesis' (figure 11.2A) (Beckett and Webster 1962, 1965; see Mitchell 1973).

A more objective American terrain classification by accretion for military purposes was devised by the US Army Engineer Waterways Experiment Station (USAEWES), Vicksburg, Mississippi, using four terrain factors (characteristic slope, characteristic relief, occurrence of slopes greater than

Figure 11.2 Landscape classifications. A: The pattern of a mature river valley, the Upper Thames, developed on the Corallian Limestone, the Oxford Clay and the Oolite, showing the following facets: 1, high gravel terrace; 2, spring line; 3, clay crest; 4, clay slope; 5, clay footslope; 6, unbedded glacial drift; 7, river and banks; 8, local bottomland; 9, floodplain alluvium; 10, old alluvium, not flooded; 11–12, scarp slope; 13, Dipslope. B: A component landscape defined in terms of four terrain factors, showing the relation between a component (top) and a gross landscape (below)

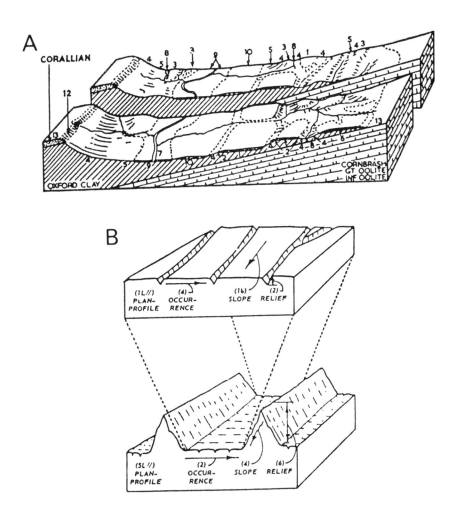

Source: A, from Beckett and Webster (1962); B, from Van Lopik and Kolb (1959)

26.5°, and characteristic plan profile involving topographic 'highs') to identify *component landscapes* which could be combined into *gross landscapes* (figure 11.2B) (Van Lopik and Kolb 1959). However, although Aitchison and Grant (1967) produced landform components which could be successively combined into units and patterns, terrain classifications of this type languished both in Britain and the United States after the 1960s.

Drainage Basin Hierarchies

It has long been recognized that the erosional drainage basin presents a convenient and unambiguous topographic unit available in a nested size hierarchy (see above all Woldenberg, 1968). During the eighteenth century many European scholars became interested in the general pattern of river systems and in the idea of river basins demarcated by drainage divides. The writings of a British physician Christopher Packe (1686–1749) are few and rare but provide significant straws in the wind. In 1743 he wrote a brief exposition of the 'origin, cause and insertion, extent, elevation and congruity of all the valleys and hills, brooks and rivers of East-Kent'.

One of Packe's French contemporaries, Philippe Buache (1700–1773) had a much wider impact, particularly in cartography. He was elected assistant geographer to the French Royal Academy of Sciences in 1730 and presented to it in 1752 both his *Essai de géographie physique* (1756) and a *Carte physique et profil du canal de la Manche* or 'physical map and profile of the (English) Channel'. In the latter he made use of underwater contours (isobaths) in the manner of Nicolas Cruquius, a Dutch engineer, who in 1729 had devised a contour map of the submarine relief of the mouth of the Meuse. Buache's map became widely known and he was commonly (and wrongly!) considered as the first to make use of isobaths for the sea. However, his map with 10 fathom intervals did lead to better underwater contouring and eventually stimulated the application of contour techniques to land surfaces. On land the heights for contours were at first measured *downward* from known summits but this method was soon replaced by measuring *upward* from sea level (Dainville 1958). Exact contouring has proved an essential boon to geomorphologists; it was ideal for assessing the shape of landforms and after about 1858 was occasionally made more pictorial by the addition of hachures or lines of hill shading to express the steepness and length of surface slopes. In his *Essai de géographie physique* Buache (1756) discussed his ideas on global structure, in which mountain chains were the 'bones', and on submarine basins and river basins. These suggestions were at times crudely overgeneralized as, for example, in his account of river basins which were shown diagrammatically as divided from each other by very strong watersheds (figure 11.3). This watershed exaggeration strengthened concepts on the general topographic unity of the drainage basin, and some academic geographers seemed to find herein:

Figure 11.3 Parts of northern France and southern Britain draining directly into the English Channel (Canal de la Manche) and the North Sea

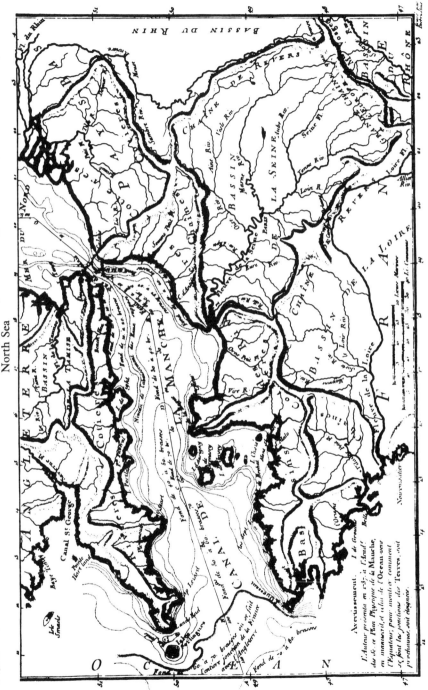

Source: From Buache (1756), Plate XIV

a concrete and 'natural' unit which could profitably replace political units as the areal context for geographical study.

(Smith 1969: 101)

Since Buache, the drainage basin concept has never ceased to be important in landform studies. In 1802 Playfair stressed the obvious unity of river drainage systems (see volume 1, pp. 61–3) and more recently Davis wrote:

Although the river and the hill-side waste sheet do not resemble each other at first sight, they are only the extreme members of a continuous series, and when this generalization is appreciated, one may fairly extend the 'river' all over its basin and up to its very divides. Ordinarily treated, the river is like the veins of a leaf; broadly viewed, it is like the entire leaf.

(Davis 1899H: 495)

However, it should be noted that, in setting up his physiographic subdivisions of the United States, Powell (1895: 65) discarded the

old custom of describing great physiographic regions in units of basins ... [partly because it] fails to exhibit the association of great features that are intimately connected in physiographic history.

This view was held strongly by other American regional subdividers, particularly Fenneman.

The unity of drainage basins also acquired support from a hydrophysical viewpoint when the basin was used to study the nature of the branching of the river systems draining it. This method, the hierarchical ordering of streams in a drainage basin, can be traced back at least to 1834 when Colonel Julian Jackson published an article entitled 'Hints on the subject of geographical arrangement and nomenclature'. In this he produced a system for rivers which involved identifying orders of stream branching or ramification based on working headward and following stream alignment (figure 11.4):

By a system, I mean only such a methodized arrangement of facts and objects already known as shall serve to render our acquaintance with them more complete, our notions more concise.

(Jackson 1834: 72)

By order of ramification, I refer to the order of the recipient as being primary, secondary, etc., reckoned from the sea. Thus I say, a river falling directly into the sea is the primary recipient of the system to which it belongs and all rivers falling immediately into this recipient form ramifications of the first order whatever may be their number.

(Jackson 1834: 72)

Jackson (1790–1835) made other interesting observations on rivers, having spent some time in Russia (Goudie 1978), but his remarks on stream ordering seem to have failed to create any real impression for some 80 years.

Figure 11.4 Systems of stream ordering by Jackson (1834), Gravelius (1914), Horton (1945) and Strahler (1952; see 1957)

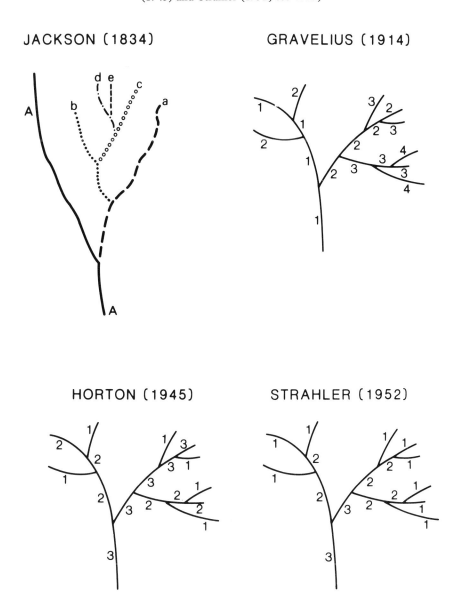

JACKSON (1834)

GRAVELIUS (1914)

HORTON (1945)

STRAHLER (1952)

Source: Mostly after Woldenberg, 1968, figure I-1

Gravelius (1914: 1–3) employed the same reasoning as Jackson in developing a system of stream ordering in which the whole main stream was assigned to order 1 with the smaller tributaries of progressively higher order. Once generated at the lower end, the segment of given order extended to the watershed, implying that there could be no nested hierarchy of basins (figure 11.4). The same was true of the ordering system of Robert E. Horton (1945) who designated each 'finger-tip' tributary as a first-order stream and the result of the combination of two first-order streams as a second-order stream, and so on. Once the initial ordering in a basin was completed, reordering took place in which the highest-order stream was projected back to the head stream which showed the least deviation from the higher-order's direction until the divide was reached. This was then repeated, with each of the next-highest-order streams being reordered back to the watershed, and so on (figure 11.4). After the Second World War, Arthur N. Strahler (1957) devised a system of stream ordering based on the first part of Horton's procedure which, for the first time, allowed a nested basin hierarchy to emerge and for drainage basins of given order to become fundamental and comparable geomorphic units within a region. This development will be treated in detail in our volume 4, as it formed much of the stimulus for the 'quantitative revolution' in geomorphology.

The association of geomorphology with geography and economic planning has meant that, for much of the twentieth century, river basins have been studied in the context of human and economic geography. Notable uses were made by C. B. Fawcett (1917) in his 'natural divisions' of England and by Jean Brunhes (1920) in his major territorial divisions in his *Géographie Humaine de la France*. After the close of the First World War, the concept of river basin management was pursued intensely. In 1921 the French government laid down the principles for the economic development of the Rhône and in the 1930s the American Tennessee Valley Authority completed a highly developed drainage basin management scheme. After the Second World War numerous large schemes, national or international, were undertaken for multipurpose river basin development and the desirability of studying complete river basins, irrespective of their political subdivisions, was universally fostered.

Practical Classifications

Although the practical applications of regional landform classifications were not common until after the middle of the twentieth century, they originated in earlier times, particularly in the Russian classifications described elsewhere in this chapter and in chapter 6. It has long been a basic tenet of Russian scholarship that regions possess tangible reality, as distinct from the theoretical or artistic view of regions adopted by many western scholars.

Morphological, or geomorphic, maps are those which contain geomorphic information representing an addition to, and an improvement on, contours (e.g. breaks of slope (Savigear 1965)) which, together with other information, allow their conversion into purpose-oriented maps containing regional information relevant to the needs of management, planning, land classification, military operations, etc. (Mabbutt 1968, Cooke and Doornkamp 1974: 352–3). St-Onge (1968) has noted that the only effective geomorphological map in existence before the Second World War was that produced by Passarge in 1914, who developed his ideas in 1919. In the early 1950s geomorphological mapping began in earnest in Poland (Klimaszewski 1956, 1963) (figure 11.5), followed by France (Tricart 1965), Russia (Bashenina *et al.* 1960) and Britain (Waters 1958, Mitchell 1973). In 1960 the International Geographical Union in Stockholm established a subcommittee on geomorphological mapping, but these studies were soon to languish.

The mapping of land system regions for management purposes began in the interwar period with an attempt at land evaluation conducted in 1922 by the Michigan Economic Land Survey, followed by a land classification in that state by Veatch (1933). In East Africa Milne (1935) used his catena concept to identify soil sampling units which would enable a 'complex entanglement' to be mapped at small scale (Mitchell 1973: 47). However, it was not until 1946 that serious work began in Australia under the Comonwealth Scientific and Industrial Research Organization (CSIRO) to provide the basis for rapid reconnaissance surveys of land capability. Subsequent reports by Christian and Stewart (1953) and Christian (1958) identified *land units* (or facets) located on a single rock type which 'should be sufficiently uniform for a prudent arable farmer to manage the whole extent of one unit in the same way' (Cooke and Doornkamp 1974: 333). Land units might be composed of two or more elements and could themselves be combined into a *land system* within which there is a recurring pattern of topography, soils and vegetation, correlated with geology, geomorphology and climate, respectively (Mabbutt 1968, Mitchell 1973).

The identification of terrain regions for strategic military purposes was carried out retrospectively by Douglas Johnson (1921) with regard to First World War operations in France; whereas tactical uses were only introduced during the Second World War but were not developed until later (Hunt 1950). In the 1950s terrain features affecting visibility, cross-country mobility, construction, availability of water, etc. were quantified to produce regional classifications of military utility. Chief among these have been the parametric terrain maps and regional terrain analogues produced by the US Army Engineer Waterways Experiment Station, Vicksburg, Mississippi (Van Lopik and Kolb 1959) (Figure 11.2); the slope maps based on statistical point sampling developed by the Quartermaster Research and Engineering Command, Natick, Massachusetts (Wood and Snell 1957, 1959, 1960); and the regional classifications developed in Britain by the Military Engineering Experimental Establishment (Beckett and Webster 1962, Mitchell 1973). However, as the 1960s advanced concerns regarding possible nuclear war caused a shift of military interest to a much larger regional scale. The quantitative work in the 1950s and 1960s by Arthur N. Strahler and his team at Columbia University, New York, sponsored by the US Office of Naval Research (ONR), had such far-reaching implications for the whole field of landform study that it will form a major focus of interest in our volume 4.

Figure 11.5 Geomorphological map of a part of the Upper Silesian industrial district. The key to the symbols follows

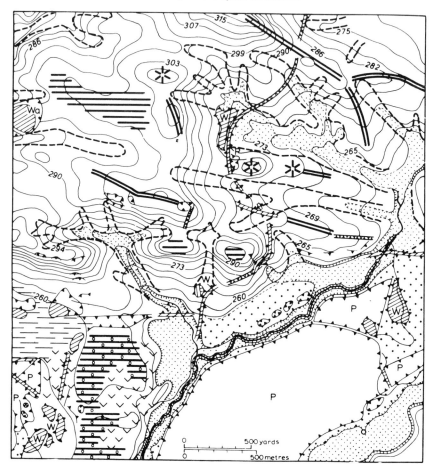

Source: After Klimaszewski; from Cooke and Doornkamp (1974), Fig. 14.1, p. 357. By permission of the Polish Academy of Sciences

Key:

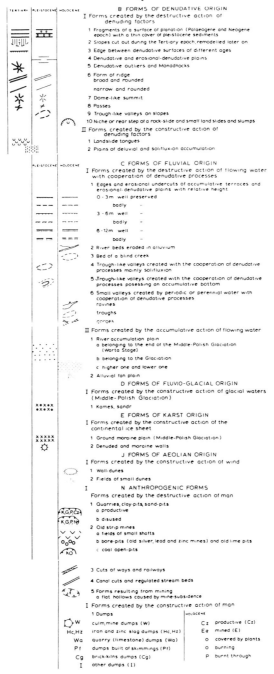

B FORMS OF DENUDATIVE ORIGIN
I Forms created by the destructive action of denuding factors

1 Fragments of a surface of planation (Palaeogene and Neogene epoch) with a thin cover of pleistocene sediments
2 Slopes cut out during the Tertiary epoch, remodelled later on
3 Edge between denudative surfaces of different ages
4 Denudative and erosional-denudative plains
5 Denudative outliers and Monadnocks
6 Form of ridge
broad and rounded

narrow and rounded
7 Dome-like summit
8 Passes
9 Trough-like valleys on slopes
10 Niche or rear step of a rock-slide and small land slides and slumps

II Forms created by the constructive action of denuding factors
1 Landslide tongues
2 Plains of deluvial and solifluxion accumulation

C FORMS OF FLUVIAL ORIGIN
I Forms created by the destructive action of flowing water with cooperation of denudative processes

1 Edges and erosional undercuts of accumulative terraces and erosional-denudative plains with relative height
0 - 3 m well preserved
badly "
3 - 6 m well "
badly "
6 -12 m well "
badly "
2 River beds eroded in alluvium
3 Bed of a blind creek
4 Trough-like valleys created with the cooperation of denudative processes mainly solifluxion
5 Trough-like valleys created with the cooperation of denudative processes possessing an accumulative bottom
6 Small valleys created by periodic or perennial water with cooperation of denudative processes
ravines
troughs
gorges

II Forms created by the accumulative action of flowing water
1 River accumulation plain
a belonging to the end of the Middle-Polish Glaciation (Warta Stage)
b belonging to the Glaciation
c higher one and lower one
2 Alluvial fan plain

D FORMS OF FLUVIO-GLACIAL ORIGIN
I Forms created by the constructive action of glacial waters (Middle-Polish Glaciation)
1 Kames, sandr

E FORMS OF KARST ORIGIN
I Forms created by the constructive action of the continental ice sheet
1 Ground moraine plain (Middle-Polish Glaciation)
2 Denuded and moraine walls

J FORMS OF AEOLIAN ORIGIN
I Forms created by the constructive action of wind
1 Wall-dunes
2 Fields of small dunes

I   **N ANTHROPOGENIC FORMS**
Forms created by the destructive action of man
1 Quarries, clay-pits, sand-pits
a productive
b disused
2 Old strip mines
a fields of small shafts
b bore-pits (old silver, lead and zinc mines) and old lime pits
c coal open-pits

3 Cuts of ways and railways
4 Canal cuts and regulated stream beds
5 Forms resulting from mining
a flat hollows caused by mine-subsidence

I Forms created by the constructive action of man
1 Dumps

W	culm, mine dumps (W)
Hc,Hz	iron and zinc slag dumps (Hc,Hz)
Wa	quarry (limestone) dumps (Wa)
Pf	dumps built of skimmings (Pf)
Cg	brick-kilns dumps (Cg)
I	other dumps (I)

HOLOCENE
Cz productive (Cz)
Eq mined (E)
o covered by plants
o burning
p burnt through

Complex Regionalization

The concept of complex associations of factors based partly, but by no means wholly, on physiographic considerations has been a persistent regional theme during the twentieth century. It has is roots in the Continental European scholarship of Russia (Dokuchayev), Germany (Supan and A. Penck) and France (Vidal de la Blache, Brunhes and Demangeon). Work in the English-speaking world has been much less important and almost entirely derivative. The Germanic tradition was introduced widely into Britain by A. J. Herbertson who in 1905 produced an account of 'The major natural regions: an essay in systematic geography'. A natural region, we are told, should have 'a certain unity of configuration, climate and vegetation', but the resultant world map – based largely on existing atlas maps – was so crude that it was not well received in London. However, the point must be made that Herbertson had spent some time (1889–90) in the University of Freiburg, where he took a doctorate and was familiar with the work of Penck and Supan and no doubt also with the ideas of Passarge who was a contemporary of his at Freiburg. In the United States Veatch (1933) used vegetation to define terrain units, as had Bourne (1931) in the British colonies, but, as Linton (1951) has pointed out, most so-called 'natural regions' have been *morphologically* delimited. In this regard Russian regional studies have been highly distinctive in that they have been firmly based on the concept of landscape.

Whereas in the United States many of the major physiographic divisions show a strong north–south or meridional trend, in Eurasia the same features often take on an east–west or zonal alignment. Without being overdeterministic it does seem almost inevitable that Russian earth scientists would tend to favour a zonal basis for dividing the earth's surface. At the same time they brought the importance of soils and of sedimentation to the forefront in classifications of the physical landscape. Recently A. G. Isachenko (1965) has given a fine review of Russian concepts of physical–geographical differentiation (see also Solntsev 1962) and set forth the principles underlying 'the classification and regionalization of terrain'. His book has been translated into English by R. J. Zatorsky as *Principles of Landscape Science and Physical–Geographic Regionalization* (1973) and produced under the skilled editorship of John S. Massey. Here we can do no more than provide a brief summary of its sections which seem most relevant to geomorphological aspects of regionalization and append a few asides.

The earth's envelope, we are told, develops regular and territorially limited entities called geographic or natural territorial complexes, hereafter referred to as *geocomplexes*. Landscape science aims at studying this set of geocomplexes, particularly their morphological structure – that is, the relationship between the various units which constitute a landscape – and their development and distribution. These geocomplexes, which range in size from small and simple to large and complex, are examined both as individual entities and as members of a typological scheme.

The study of geocomplexes and landscape zones, which is practically coincident with regional physical geography, has roots dating back many centuries but, if any individual can be said to be its originator, it would be the great Russian scientist V. V. Dokuchayev (1846–1903) (see chapter 6). He regarded soils as a function of all geographical factors – climate, bedrock, the organic world, age of the land, and topography – and as the clearest expression of the geocomplex. In his new classification of genetic soil types (1888–1900) he related soil belts to vegetation and climatic belts or zones:

> thanks to the known position of our planet relative to the sun, thanks to the rotation of the earth, its spherical shape – climate, vegetation, and animal life are distributed on the earth's surface from north to south, in a strictly determined order ... which allows the division of the earth's sphere into belts: *polar, temperate, subtropical, equatorial*, and so forth. And since ... soil formers, which are subject to known laws in their distribution, are distributed by belts, their results also – the soil – must be distributed on the earth's sphere in the form of definite *zones* going more or less (only with certain deviations) parallel to the circles of latitude.
>
> (Dokuchayev 1876–1900: VI: 407)

Isachenko (1973: 26–7) expresses it thus:

> Dokuchayev first treated zonality as a natural law: every natural or natural history zone constitutes a regular natural complex in ·which the living and non-living aspects of nature are closely associated one with another.

During the early twentieth century the chief Russian propagator and popularizer of landscape science was Lev Simonovich Berg (1876–1950) (plate 11.3) who in 1913 defined natural zones as 'regions in which a specific landscape predominates' and named them *landscape zones*. Berg, who from 1917 onward was professor of geography at Petrograd University, continued to be interested in geographical zones or in the classification of the earth's surface according to climatic, soil, biological and other natural phenomena. He distinguished ten different zones, several of which were subdivided and most of which had a latitudinal (east–west) sequence in lowlands and a longitudinal (south–north) sequence in mountainous terrain. The geographical zones were distinguished on the principle of uniformity of landscape; that is, a combination of relief, climate, plants and soil into a uniform harmonious whole which typically duplicates itself over an entire geographical zone. In addition, Berg also examined the evolution of landscapes by means of reconstructing their palaeogeographic conditions and tracing their transformation into the modern landscape. For example, he was convinced that a change in the geographical landscape led to a change in the character of its sedimentary processes. In 1931 Berg published the first volume of *The Landscape–Geographic Zones of the USSR* which went into a third edition of two volumes (1947–52) (Tikhomirov 1970: 621–4).

Plate 11.3 Lev Simonovich Berg (1876–1950)

Source: By permission of *American Geographical Society, Occasional Publication* No. 1, 1962, p. 29

However, as Isachenko points out, there was a duality in Berg's concept of landscape. One aspect of it led to the decisive role in landscape research of a typological approach, whereby landscape was considered a type of territory which grouped together all similar, homogeneous sections of the earth no matter where they be. The other aspect was regional, under which, according to L. G. Ramyenski, a landscape is assumed to be 'a relatively complex territorial system consisting of diverse yet mutually consistent, interrelated elementary natural complexes which develop as a single entity' (Isachenko 1973: 34). This approach demanded an understanding of the dynamics or fundamental processes affecting the landscape.

Anyone reading Isachenko's volume will soon be aware that his ideas represent a culmination of decades of Russian landscape science. In elaborating on the general principles underlying the physical–geographical differentiation of a territory he progresses from zonal to local influences. The effect of geographical zonality is shown to depend mainly on the distribution of solar energy at the earth's surface and also to include climate, soil and the deposition of sediments. At the same time there are azonal influences, especially of relief, geological structure, and climate, which mainly arise from 'the history of the tectonic development of our planet, resulting from the action of its internal energy' (p. 73). These azonal phenomena manifest themselves against or within the zonal background. Present-day landscape differentiation expresses the interaction of zonal and azonal factors during the historical evolution of the existing geographical envelope. This differentiation reveals itself in a system of units of physical–geographical regionalization which rank hierarchically downward from *zone* to *landscape* (figure 11.6).

The landscape is the principal physical–geographical unit and the principal object of the study of terrain, within which zonal and azonal conditions are uniform or are considered non-differentiable. It has a uniform geological base and developmental history and characteristic relief and climate. For classification purposes it is viewed both 'from below', that is, zonally and azonally wherein it is uniform, and morphologically or internally. So for purposes of classification, landscape has a multilevel character based on a taxonomic system of subordinate units. The higher level of these classificatory units is the landscape *type*, such as the taiga. Most of these types are further divided into subtypes such as northern taiga, middle taiga and southern taiga. On the next level are *classes* such as highland and lowland landscape which also may be subdivided, for example low-lying plains and elevated plains. The most basic category of landscape is the *species* which groups together landscapes with a similar genesis, structure and morphology.

In the classification hierarchy below landscape the differentiation depends no longer on zonal and azonal factors but on more specific internal or local factors which are effective over a relatively small area and are designated as internal or morphological (figure 11.6). It is true that these morphological

Figure 11.6 A system of units in physical geography

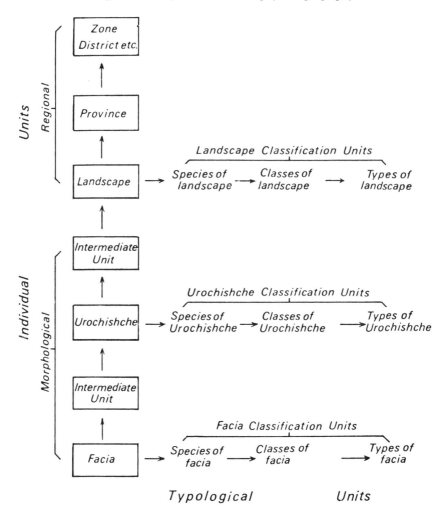

Source: From A. G. Isachenko (1973) *Principles of Landscape Science and Physical–Geographic
Regionalization*, Melbourne University Press, Fig. 1, p. 7. Translated from the original Russian
edition of 1965 by R. J. Zatorski and edited by J. S. Massey

differences must act upon a background of specific zonal–azonal conditions
but at this level these latter are insignificant, the differences being caused by
actual variations within a landscape and in the evolution of a landscape.
These differences may result from erosion, or the redistribution of moisture,
or surface deposition, or competition between plant communities, and so on.
The process of exogenous dissection of the earth's surface and the creation of

an enormous number of small and medium-sized topographical forms or landforms are especially important in the development of internal physical–geographical differences. The presence of such forms allows the discernment of, and differentiation into, localities, which are areas distinguished by their position in the orographic profile, for instance flats, peaks, parts of a slope, and so on. Within a single locality the physical–geographical conditions have a uniform character, which is tantamount to stating that in it every geographical component is represented by its smallest territorial unit: that is, relief is represented by a single element, climate by a single microclimate, soil by a single phase, and vegetation by a single phytocenosis. The complete group of components in a single locality forms the simplest geocomplex, incapable of further subdivision and known in Russia as the *facia*.

Between the facia and the landscape lies the *urochishche* (figure 11.7), an association or assemblage of facia exhibiting uniformity in geology, drainage, transfer of solid materials and chemicals, etc. (Ramyenski 1938, Solntsev 1962, Prokayev 1962, Vinogradov *et al.* 1962). In English nomenclature, facia is roughly equivalent to site and urochishche to facet or land unit. Some earth scientists consider that, strictly speaking, neither of these smallest morphological units are true regionalization units because they sometimes fail to preserve the typical characteristics of all the higher regional complexes and do not provide information on the complete natural conditions in a territory. On the other hand, some geographers maintain that all geocomplexes, including facia, constitute categories in regionalization. Not surprisingly there is an incredible variety and inconsistency in the writings of authors interested in the problem.

Isachenko leaves no doubt about the gradation and essential interrelations of this system of units of physical geography. He also has interesting comments on regionalists in non-Russian countries. For example he praises Herbertson's studies of geocomplexes (natural regions) which, as we have seen, were not rapturously received in London, and discusses Passarge's theoretical studies of surface differentiation, including major land zones and landscape regions, and landscapes. The latter category was subdivided into 'constituent landscapes' and 'landscape parts', 'which correspond approximately to the urochishcha and the facia of Soviet geographers' (Isachenko 1965: 38). Passarge felt that the method of investigating landscapes was a question of individual preference, or purely subjective, an approach which was criticized by Berg (1931: 1: 130) who also considered that the German scientist's concept of landscape excluded the need to express the mutual interdependence of its components. With regard to A. Penck, Berg asserted that in addition to ignoring the significance of soils, he viewed landscape as a mechanical collection of various unconnected or independent components. Isachenko expresses his own opinion when he says that in most capitalist countries, landscape science had been little developed because the *region* had been regarded as an intellectual concept rather than a reality.

Figure 11.7 Schematic distribution of urochishcha types in a region of mid-taiga lacustrine–glacial plain. 1. Gently hilly sandy plain; spruce-pine; well-drained mod.

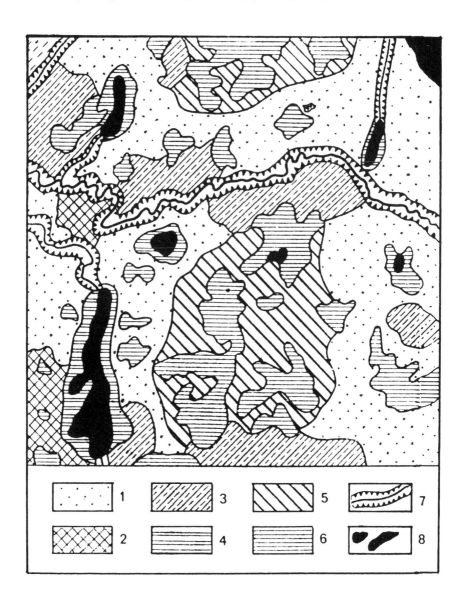

Source: After Dashkevich. From A. G. Isachenko (1973) *Principles of Landscape Science and Physical-Geographic Regionalization*, Melbourne University Press, Fig. 31, p. 156. Translated from the original Russian edition of 1965 by R. J. Zatorski and edited by J. S. Massey

podsolized soils. 2. Kames; bilberry–cowberry pine stands; well-drained weakly podsolized soils. 3. Small-hill plain with gentle slopes; moss-cowberry and bilberry; med. to strongly podsolized soils. 4. Lake basins; spruce forest, podsolized soils. 5. Flat sandy plain; sphagnum spruce-pine forest; swampy soils. 6. Highland and transitional moors. 7. River valleys with poorly defined floodplains and elevated terraces; spruce forests. 8. Lakes.

Here, in leaving Isachenko's excellent survey, we must reiterate that we are concerned with geomorphology rather than with geography, with natural elements and physical landscape patterns rather than geographical regionalization in general. In geomorphology the higher divisions will probably remain within relatively generalized boundaries and will be used for two main purposes: to study world patterns and to investigate their internal conditions. The smaller divisions will be dealt with from a physical rather than geographical viewpoint although the environmental influence of humanistic agents will be included. It seems odd that Isachenko did not stress the strongly environmental approach adopted by the 'climatic geomorphologists' of the German and French schools and latterly of American academics. This theme has provided 'a uniquely geographical approach to genetic geomorphology' (Butzer 1973: 40) and has lessened the powerful roles of structure and history in the geomorphology of the west.

References

Adams, G. I. (1902) 'Physiographic divisions of Kansas', *Bulletin of the American Geographical Society* 34: 89–104.

Aitchison, G. D. and Grant, K. (1967) 'The P.U.C.E. programme of terrain description, evaluation, and interpretation for engineering purposes', *Proceedings of the 4th Regional Conference for Africa on Soil Mechanics and Foundation Engineering*, Cape Town.

Anderson, A. L. (1941) 'Physiographic subdivisions of the Columbia Plateau in Idaho', *Journal of Geomorphology* 4: 206–22.

Atwood, W. W., Sr (1940) *The Physiographic Provinces of North America*, Boston: Ginn.

Bashenina, N. V. *et al.* (1960) *Legend for the Geomorphological Map of the Soviet Union*, Department of Geography, Moscow University (in Russian).

Beckett, P. H. T. and Webster, R. (1962) 'The storage and collation of information on terrain (An interim report)', *Military Engineering Experimental Establishment, Christchurch, Hampshire* (mimeo.).

—— (1965) 'A classification system for terrain: Units and principles', *Military Engineering Experimental Establishment, Christchurch, Hampshire, Report* 872.

Beckinsale, R. P. (1975) 'Ferdinand von Richthofen', *Dictionary of Scientific Biography*, vol. XI, New York: Scribners, pp. 438–41.

Berg, L. S. (1913) 'A proposed classification of Siberia and Turkestan into landscape and morphological regions', in *Collected Papers in Honour of Professor D. N. Anuchin's 70th Birthday*, Moscow, pp. 117–51 (in Russian).

—— (1931) *The Landscape–Geographic Zones of the USSR*, Moscow (3rd edn, 2 vols, 1947–52) (in Russian).

Bourne, R. (1931) 'Regional survey and its relation to stocktaking of the agricultural resources of the British Empire', *Oxford Forestry Memoirs* 13: 16–8.

Bowman, I. (1911) *Forest Physiography*, New York: Wiley.

Brunhes, J. (1920) *Géographie Humaine de la France*, vol. 1, Paris: Société de Libraire Plon.

Buache, M. (1756) 'Essai de géographie physique', *Mémoires de Mathématique et de Physique, Académie Royale des Sciences, 1752* pp. 399–416.

Butzer, K. W. (1973) 'Pluralism in geomorphology', *Proceedings of the Association of American Geographers* 5: 39–43.

Christian, C. S. (1958) 'The concept of land units and land systems', *Proceedings of the 9th Pacific Science Congress, 1957*, vol. 20, pp. 74–81.

Christian, C. S. and Stewart, G. A. (1953) 'General report on the survey of the Katherine–Darwin region, 1946', *CSIRO Australian Land Resources Series* no. 1.

Cooke, R. U. and Doornkamp, J. C. (1974) *Geomorphology in Environmental Management*, Oxford.

Dainville, F. de (1958) 'De la profondeur à l'altitude', in M. Mollat (ed.) *Le navire et l'economie maritime du moyen-âge au XVIIIe siècle*, Paris, pp. 195–213.

Dana, J. D. (1863) *Manual of Geology: Treating the principles of the science*, Philadelphia: Bliss.

Dokuchayev, V. V. (1876–1900) *Collected Works*, 9 vols, Moscow (latest edn 1944–61) (English transl. by V. A. Yesakov).

Falconer, J. D. (1914a) 'The progress of physical geography', *Scottish Geographical Magazine* 30: 537–42.

—— (1914b) 'The classification of land forms and landscapes', *Geographical Journal* 43: 424–7.

—— (1915) 'Land forms and landscapes', *Scottish Geographical Magazine* 31: 57–71, 143–51, 169–80, 244–53, 393–406.

Fawcett, C. B. (1917) 'Natural divisions of England', *Geographical Journal* 49: 124–34.

Fenneman, N. M. (1914) 'Physiographic boundaries within the United States', *Annals of the Association of American Geographers* 4: 84–134.

—— (1916) 'Physiographic divisions of the United States', *Annals of the Association of American Geographers* 6: 19–98.

—— (1922) 'Physiographic provinces and sections in western Oklahoma and adjacent parts of Texas', *US Geological Survey Bulletin* 730-D: 115–34.

—— (1928) 'Physiographic divisions of the United States', *Annals of the Association of American Geographers* 18: 261–353.

—— (1931) *The Physiography of Western United States*, New York: McGraw-Hill.

—— (1938) *The Physiography of Eastern United States*, New York: McGraw-Hill.

Flint, R. F. (1928) 'Natural boundaries of the Interior Low Plateau physiographic province', *Journal of Geology* 36: 451–7.

Geikie, A. (1882) *Textbook of Geology*, London: Macmillan (4th edn 1903, 2 vols).

Goudie, A. S. (1978) 'Colonel Julian Jackson and his contribution to geography', *Geographical Journal* 144: 264–70.

Gravelius, H. (1914) *Flusskunde: Grundriss der gesamten Gewässerkunde*, Berlin: Goschenesche Verlagshandlung.

Gregory, K. J. and Brown, E. H. (1966) 'Data processing and the study of land form', *Zeitschrift für Geomorphologie* 10: 237–63.

Grigg, D. (1967) 'Regions, models and classes', in R. J. Chorley and P. Haggett (eds) *Models in Geography*, London: Methuen, pp. 461–509.

Herbertson, A. J. (1905) 'The major natural regions, an essay in systematic geography', *Geographical Journal* 25: 300–12.

—— (1911) *Handbook of Geography*, Vol. 1, London: Nelson.

Horton, R. E. (1945) 'Erosional development of streams and their drainage basins: Hydro-physical approach to quantitative morphology', *Bulletin of the Geological Society of America* 56: 275–370.

Hunt, C. B. (1950) 'Military geology', in S. Paige (ed.) *Applications of Geology to Engineering Practice*, Geological Society of America, Berkey Volume, pp. 295–327.

Isachenko, A. G. (1965) *Principles of Landscape Science and Physical–Geographic Regionalization*, Moscow: Vysshaya (English transl. by R. J. Zatorski, Melbourne University Press, 1973).

Jackson, J. (1834) 'Hints on the subject of geographical arrangement and nomenclature', *Journal of the Royal Geographical Society* IV: 72–88.

James, P. E. (1972) *All Possible Worlds*, New York: Bobbs-Merrill (2nd edn with G. J. Martin, 1981, New York: Wiley).

Jenkins, O. P. (1943) 'Geomorphic provinces of California', *California Division of Mines, Bulletin* 118, pp. 83–88.

Joerg, W. L. G. (1914) 'The subdivision of North America into natural regions', *Annals of the Association of American Geographers* 4: 55–83.

Johnson, D. W. (1921) *Battlefields of the World War: A Study in military geography*, New York: Oxford University Press.

Johnston, W. D., Jr (1930) 'Physical divisions of northern Alabama, *Bulletin of the Geological Survey of Alabama* 38.

Klimaszewski, M. (1956) 'The principles of the geomorphological survey of Poland', *Przeglad Geograficzny* 28: 32–40 (suppl.).

—— (1963) 'Problems of geomorphological mapping', *Polish Academy of Science, Institute of Geography, Geographical Studies* no. 46 (in Polish).

Linton, D. L. (1951) 'The delimitation of morphological regions', in L. D. Stamp and S. W. Wooldridge (eds) *London Essays in Geography*, London: Longman, pp. 199–218.

Lobeck, A. K. (1923) *Physiographic Diagram of Europe*, New York: Geography Press, Columbia University.

Loomis, F. B. (1937) *Physiography of the United States*, New York: Doubleday.

Mabbutt, J. A. (1968) 'Review of concepts of land classification', in G. A. Stewart (ed.) *Land Evaluation*, Melbourne: Macmillan, pp. 11–28.

Matthes, F. E. (1915) 'Conference on the delimitation of physiographic provinces in the United States', *Annals of the Association of American Geographers* 5: 127–9.

Milne, G. (1935) 'Some suggested units of classification and mapping, particularly for East African soils', *Soil Research* 4: (3): 183–98.

Mitchell, C. (1973) *Terrain Evaluation*, London: Longman.

Packe, C. (1743) *Convallium Descriptio*, Canterbury.

Passarge, S. (1912) 'Physiologische Morphologie', *Mitteilungen der Geographischen Gesellschaft in Hamburg* 26: 133–337.

—— (ed.) (1914) *Morphologischer Atlas*, vol. 1, *Morphologie des Messtischblattes Stadtremda*, Hamburg: Friedrichsen.

—— (1919–20) *Die Grundlagen der Landschaftskunde*, Hamburg: Friedrichsen, (vol. 1) (1919) *Beschreibende Landschaftskunde*, vol. 2 (1920) *Die Oberflächengestaltung der Erde*.

—— (1921–30) *Vergleichende Landschaftskunde*, 5 vols, Berlin: Reiner.

Penck, A. (1894) *Morphologie der Erdoberfläche*, 2 vols, Stuttgart: Engelhorn.

—— (1896) 'Die Geomorphologie als genetische Wissenschaft: eine Einleitung zur

Diskussion über geomorphologische Nomenclatur', *Comptes Rendus, Sixth International Geographical Congress (London), section C* pp. 735–52.

Peschel, O. (1870) *Neue Probleme der vergleichenden Erdkunde: als Versuch einer Morphologie der Erdoberfläche*, Leipzig: Duncker and Humblot.

Powell, J. W. (1895) 'Physiographic regions of the United States', *National Geographic Society, Monograph* 3, pp. 65–100.

Prokayev, V. I. (1962) 'The facies as the basic and smallest unit in landscape science', *Soviet Geography, Review and Translation* 3: (6): 21–9.

Raisz, E. J. (1938) 'Developments in the physiographic method of representing the landscape on maps', *Proceedings of the 15th International Geographical Congress, Amsterdam*, vol. 2, section 1, pp. 140–9.

—— (1946) 'Landform, landscape, land use, and land type maps', *Annals of the Association of American Geographers* 36: 102–3.

Ramyenski, L. G. (1938) *An Introduction to Complex Soil and Geobotanical Investigations*, Moscow: Selhozgiz (in Russian).

Richthofen, F. F. von (1886) *Führer für Forschungreisende*, Hanover: Oppenheim.

Ryan, B. (1986) 'Nevin Melancthon Fenneman (1865–1945)', *Geographers' Biobibliographical Studies*, vol. 10, London: Mansell, pp. 57–68.

Savigear, R. A. G. (1956) 'Technique and terminology in the investigation of slope forms', *First Report of the Commission for the Study of Slopes, International Geographical Union, (Amsterdam)* pp. 66–75.

—— (1965) 'A technique of morphological mapping', *Annals of the Association of American Geographers* 55: 514–38.

Smith, C. T. (1969) 'The drainage basin as an historic basis for human activity', in R. J. Chorley (ed.) *Water, Earth and Man*, London: Methuen, pp. 101–10.

Solntsev, N. A. (1962) 'Basic problems of Soviet landscape science', *Soviet Geography, Review and Translation* 3: (6): 3–15.

St-Onge, D. A. (1968) 'Geomorphic maps', in R. W. Fairbridge (ed.) *The Encyclopedia of Geomorphology, vol. III of The Encyclopedia of Earth Sciences Series*, New York: Reinhold, pp. 388–403.

Strahler, A. N. (1957) 'Quantitative analysis of watershed geomorphology', *Transactions of the American Geophysical Union* 38: 913–20.

Suess, E. (1883–1908) *Das Antlitz der Erde*, Wien: Tempsky, 3 vols.

Supan, A. G. (1884) *Grundzuge der Physischen Erdkunde*, Leipzig: Verlag Von Veit (2nd edn 1899, 5th edn 1911).

Thayer, W. N. (1916) 'The physiography of Mexico', *Journal of Geology* 24: 61–94.

—— (1918) 'The northward extension of the physiographic divisions of the United States', *Journal of Geology* 26: 161–85, 237–54.

Thornbury, W. D. (1965) *Regional Geomorphology of the United States*, New York: Wiley.

Tikhomirov, V. V. (1970) 'Lev Simonovich Berg', *Dictionary of Scientific Biography*, vol. I, New York: Scribners, pp. 621–4.

Tricart, J. (1965) *Principes et Méthodes de la Géomorphologie*, Paris: Masson.

Unstead, J. F. (1933) 'A system of regional geography', *Geography* 18: 185–7.

Van Lopik, J. R. and Kolb, C. R. (1959) 'A technique for preparing desert terrain analogs', *US Army Engineer Waterways Experiment Station, Vicksburg, Mississippi, Technical Report*, 3–506.

Veatch, J. O. (1933) 'Agricultural classification and land types of Michigan', *Michigan Agricultural Experiment Station, Special Bulletin* no. 231.

Vinogradov, B. V., Gerenchuk, K. I., Isachenko, A. G., Ramans, K. G. and Tsesel'chuk, Yu. N. (1962) 'Basic principles of landscape mapping', *Soviet Geography, Review and Translation* 3: (6): 15–20.

Von Zittel, K. (1901) *History of Geology and Palaeontology*, London: Scott.

Waters, R. S. (1958) 'Morphological mapping', *Geography* 43: 10–17.

Wentworth, C. K. (1936) 'Geomorphic divisions of the island of Hawaii', *University of Hawaii Bulletin* 16, no. 8.

Woldenberg, M. J. (1968) 'Hierarchical systems: Cities, rivers, alpine glaciers, bovine livers, and trees', *Harvard Papers in Theoretical Geography*, No. 19.

Wood, W. F. and Snell, J. B. (1957) 'The dispersion of geomorphic data around measures of a central tendency and its application', *H.Q. Quartermaster Research and Engineering Command, Natick, Mass., Report* EA-8.

—— (1959) 'Predictive methods in topographic analysis: 1. Relief, slope and dissection on one inch-to-the-mile maps in the USA', *H.Q. Quartermaster Research and Engineering Command, Natick, Mass., Technical Report* EP-112.

—— (1960) 'A quantitative system for classifying landforms', *H.Q. Quartermaster Research and Engineering Command, Natick, Mass., Technical Report* EP-124.

Wooldridge, S. W. (1932) 'The cycle of erosion and the representation of relief', *Scottish Geographical Magazine* 48: 30–6.

—— (1935) 'The "facet" as the ultimate unit of geographical analysis', *Nature* 135: (3403): 119.

CHAPTER TWELVE

Climatic Geomorphology

Introduction

The term climatic geomorphology was introduced by Martonne (1913) and, as elaborated by Holzner and Weaver (1965), has come to imply that the geomorphologically significant processes associated with a number of distinctive world climatic belts are capable of producing in each belt over a prolonged period a characteristic and distinctive assemblage of landforms. These belts have been termed *morphoclimatic* by Büdel (1948a, b) (see also *Formkreisen* (Büdel 1944)), in contrast to the smaller-scale and geologically controlled *morphostructural regions* (Cholley 1950). A second aspect of climatic geomorphology, and one which has grown in relative importance during the twentieth century, is based on the assumption that landform assemblages exhibit features resulting from a succession of past (or 'fossil') climates such that, in an extreme interpretation, 'we should find in the morphology of a region traces of as many systems of erosion as it has experienced climates, and even traces of the passage of one system or climate to another' (Cholley 1950). The broad landform regions so defined have been termed *morphogenetic* (Cotton 1958) or *climatogenetic* (Büdel 1963); see also *Formkreisen* (Büdel 1944, 1948b).

It is clear from the previous chapter that climatic geomorphic regions occupy larger areas than those of other regional geomorphic classifications. Passarge (1920, 1921) showed their position in a hierarchy:

Landschaftsteile: landform elements (e.g. slopes, valley bottoms);
Teillandschaft: characteristic groups of elements;
Landschaft: basic units (e.g. the Harz Mountains);
Landschaftsgebeit: larger structural units having climatic unity (e.g. the North German Plain);
Landschaftsgürtel: the highest order of units related to major climatic provinces.

The idea of the geomorphologically significant *landschaftsgürtel* was generated by the great soil regions of Dokuchayev (1883) which were advertised in the western world by the work of Glinka (1924, 1927), by the climatic classifications of Supan between 1884 and 1908 and of Voyeykov (1885–7), and by Herbertson's (1905) natural regions. However, the greatest impetus to climatic geomorphology came from the climatic classification of Köppen (1901, 1923, 1936) (figure 12.1). Although based primarily on regional

Figure 12.1 World climatic regions after W. Köppen

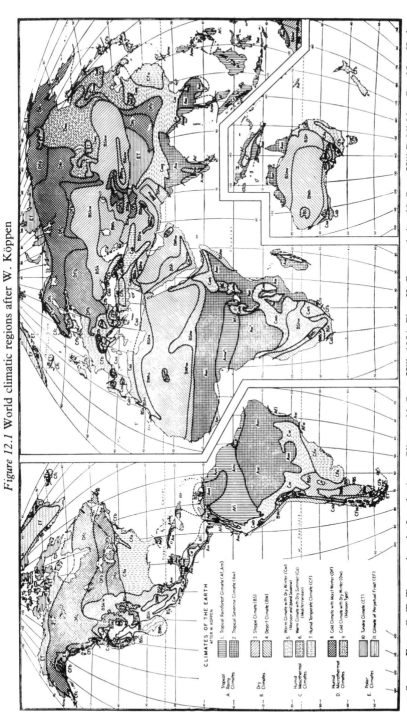

Source: From G. T. Trewartha, *An Introduction to Climate,* McGraw-Hill, 1954, Plate II. By permission McGraw-Hill Publishing Co., New York.

delimitations of vegetational significance, the combination of temperature considerations and seasonality of precipitation seemed also to give climatic expression to geomorphologically significant processes. Thus five major climatic regions (single letters in figure 12.2), subdivided first into twelve regions (double letters) and second into seventeen regions (double and triple letters), came to be viewed as forming the basis of global classifications of morphoclimatic regions numbering between three and nine.

Figure 12.2 Köppen's climatic nomenclature

TROPICAL RAINY	A	TROPICAL RAINFOREST	RAIN ALL YEAR	Af
			MONSOONAL	Am
		TROPICAL SAVANNA: DRY WINTER (WET/DRY TROPICS)		Aw
DRY	B	STEPPE BS	HOT	BSh
			COLD	BSk
		TROPICAL DESERT BW	HOT	BWh
			COLD	BWk
HUMID MESOTHERMAL	C	WARM: DRY WINTER (MONSOONAL AND UPLAND SAVANNA)		Cw
		WARM: DRY SUMMER (MEDITERRANEAN)		Cs
		HUMID TEMPERATE Cf	WARM SUMMER	Cfa
			COOL SUMMER	Cfb
HUMID MICROTHERMAL	D	COLD: WET WINTER Df	LONG, WARM SUMMER	Dfa
			LONG, COOL SUMMER	Dfb
			SHORT, COOL SUMMER	Dfc
		COLD: DRY WINTER (MONSOONAL)		Dw
POLAR	E	TUNDRA		ET
		PERPETUAL FROST		EF

Climate and Process

Towards the end of the nineteenth century German geomorphology was unique in the heavy concentration which was placed on the study of process. Whereas British, French and American workers had come to view the study of landforms in a broadly historical framework, Central European scholars were especially concerned with process (exogenetic) studies which seemed to form a vital part of a causative chain linking climate with resulting landforms. As we have seen, a fourth element in this causative chain was, for many workers, the effects of tectonic (endogenetic) processes. German concern with the links between climate, process and landforms was especially generated by a very short but intense burst of African colonial activity between the 1880s and the First World War. In 1884 the Society for German Colonization sent an expedition to Tanganyika, in 1888 Dr Goering (the father of the Nazi Reichsmarshal) told the population of South-West Africa that they were to consider themselves annexed by Germany and in 1895 German penetration of the Cameroons began. Economic development began in South-West Africa in 1897, being given great impetus after a boy picked up an alluvial diamond on Lüderitz in 1908, and in East Africa and the Cameroons after 1907. Although scientific work terminated after the First World War, German geomorphologists had ample opportunity to explore landforms associated with tropical desert, savanna and summer-wet tropical conditions. In addition, there was a German expedition to Antarctica (Wilhelm Land) in 1902 and occupation of part of equatorial New Guinea before 1914.

As Mortensen (1943–4) has pointed out, German geomorphology at the end of the nineteenth century was particularly concerned with the study of exogenetic processes and this is nowhere more clearly illustrated than in the two monumental books – Richthofen's *Führer für Forschungsreisende* (1886) and Albrecht Penck's *Morphologie der Erdoberfläche* (1894). Richthofen's guidebook, which was widely used as a text, contained little on landforms, with only about one-tenth of the text devoted to the subject and the remainder dominantly concerned with processes and their sequential effects (figure 12.3), classifications and technical terms. Richthofen did not differentiate geomorphology from geology and, for example, devoted twenty-one pages to the dynamics of aeolian processes (assumed by him to be the dominant desert process) and only three pages to desert landforms. Although only about one-half of Penck's classic two-part work is specifically devoted to the study of processes, it is clear that he believed that present (exogenetic) processes could be used to explain existing landforms and much of the space devoted to landforms contained important speculations regarding the relations of form to process (e.g volume II, plains, constructional hills, valleys, fluvially dissected landforms, basins, coastlines, etc). More than half of volume I concerns weathering, mass movements, aeolian action, fluvial action and glacial action, the remainder being devoted to the mathematical description of the major morphological features of the earth.

As has been suggested, work in exotic environments gave impetus to studies of processes and to the establishment of links between climate and landforms. In his work on China, Richthofen (1877) distinguished between landforms in areas of interior drainage and those in peripheral humid regions, and his

Figure 12.3 Examples of Richthofen's erosional sequences

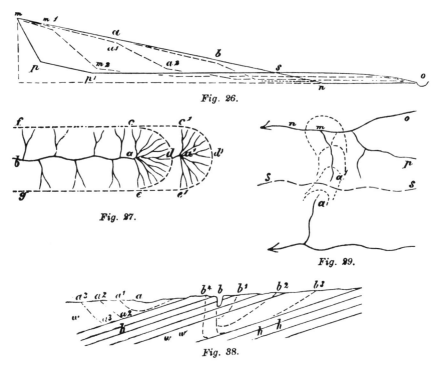

Fig. 26.

Fig. 27.

Fig. 29.

Fig. 38.

Source: From Richthofen (1886), Figures 26, 27, 29 and 38, pp. 139, 143, 146 and 166

researches on loess and wind action led him to ascribe desert landforms dominantly to aeolian processes. Prior to this, Schweinfurth (1865) proposed that some Egyptian wadis had been cut by wind action and, more important, Walther (1890, 1900) advocated the paramount nature of wind action in deserts, despite American work to the contrary. In 1904 (a) Passarge proposed that the desert plains studded with residuals in South-West Africa were too flat to be entirely the work of running water. This notion was opposed by Davis (1905L), and Passarge in later work on Algeria (1910) gave greater credit to fluvial action, a view supported by subsequent American work on pediments (Lawson 1915, Bryan 1922, who introduced this term). It is interesting that the influential Walther never changed his view as to the importance of wind action and Kaiser (1920a, b, 1923a, b, 1926) relied prominently on wind action in his analysis of the desert landforms of South-West Africa (see volume 4).

A second important geomorphological environment, the distinctive land-forms of which were explored in the early years of the present century, was the seasonally wet/dry tropical savanna regions. Bornhardt (1900) described the typical inselberg landscapes of East Africa consisting of almost flat plains with isolated, dome-shaped residuals (to which Bornhardt gave his name) rising up to 500 m above them like islands in an ocean. Passarge (1904a, b, c) originally ascribed both plains and inselbergs to wind scour but later (1924a) to occasional deep humid weathering punctuated by fluvial and aeolian evacuation of debris. In 1908 the Frenchman Hubert described a youthful inselberg landscape in the hinterland of Dahomey with groups of 300 m high inselberg domes which showed evidence of exfoliated sculpture. Three years later the British worker Falconer (1911) wrote of the more subdued (older?) inselberg landscape in Hausaland, Northern Nigeria, where huge plains are broken by low swells, groups of boulders, knobs of granite and occasional dome-shaped inselbergs. Falconer proposed that the residual inselbergs consist of more resistant rocks than the surrounding plains and this was supported by Passarge (1912). Another British worker, Cushing (1913), similarly thought that the inselbergs near the east coast of Peninsular India were composed of resistant quartzite rising from a metamorphic plain: however Cushing ascribed this landscape to marine planation leaving residual stacks! Further work by Holmes (1919) in Mozambique was of great importance in that, although he believed that the granitic and gneissic structures controlled the form of the inselbergs, he did not entirely ascribe their location and existence to geological differences. He was particularly concerned with the ephemeral stream channels found at the base of inselbergs and believed that areas with larger clusters of higher inselbergs on less flat plains were less eroded and therefore at an earlier stage of development than in other localities. Thorbecke (1921), later to organize and edit the Düsseldorf conference of 1926, gave a detailed description of inselberg landscapes in Tikar, India (syenite inselbergs on granite), and Krenkel (1922) and Obst (1923) described similar features in the classic area of East Africa.

Work on humid tropical landforms had not proceeded much beyond Dana's (1849) graphic description of the steep, straight slopes, knife-edged ridges and amphitheatre valley heads in Hawaii (see volume 1, chapter 20). In 1910 Chamberlin and Chamberlin noted similarities between regions of Mexico, Hawaii, China and Japan where steep slopes (up to 60°) rise from flat floodplains at a fairly sharp basal angle with no basal talus. They noted the effects of rapid weathering, the thin soils and the evidence of sheetflow removal of debris. Behrmann (1921, 1924) described similar features in New Guinea and in 1923 Sapper published his first contribution on tropical landforms which was soon to assume great importance.

Early Morphoclimatic Classifications and the Düsseldorf Conference

Davis regarded fluvial landforms as being 'normal' (1899H; see volume 2, chapter 10) in the sense that he was best able to envisage the geomorphic effects of differing climates, perhaps assisted by tectonic uplift or block faulting, as having been (accidentally!) imposed on previous fluvial landscapes, leading to cyclic modifications characteristic of other climatic 'phases'. Davis' recognition of the effects of climatic accidents (1905J; see volume 2, chapter 14) led him to elaborate in the first decade of this century the concepts of the glacial cycle (1900H, 1906J; see volume 2, pp. 307–18) and of the arid cycle (1905L; see volume 2, pp. 302–7). The former was clearly associated with previously fluvially dissected mountains and the latter with interior drained, enclosed basins produced by block faulting, as in the American Basin and Range province (Stoddart 1969). Thus, to Davis, both the glacial and arid cycles, although distinct from the normal cycle, were in some sense corollaries of it following the imposition of valley glaciation by a fall in temperature or the fragmentation of drainage and baselevels by tectonic activity and a decrease of humidity. It is interesting that in later years Davis (1930E; see volume 2, pp. 669–75) blurred his previous distinction between the sequences of humid and arid slope development and placed even greater emphasis on the fluvial action of desert sheetfloods and streamfloods (Davis 1938).

Albrecht Penck (1905, 1910, 1913) also identified these three morphoclimatic regimes – humid, arid and nival (glacial) – but had a different view as to their significance. He regarded them as equivalent, core morphoclimatic environments wherein the close relation between climatic regions and exogenetic processes led to the development of characteristic and distinct landform assemblages. His basis of division was predominantly hydrological with the snow line and the arid/humid boundary being crucial but so difficult to define as to leave many areas as 'marginal' with complex morphoclimatic relationships.

A most important elaboration and clear statement of the concept of climatic geomorphology was made by Martonne (1913) (Birot 1973). He stressed that magnitude of spatial scale distinguishes climatic from structural geomorphology and expressed the belief that significantly different landforms, or 'topographic facies', could be developed under at least six present climatic conditions, although with many transitional areas:

1 Warm humid facies – characterized by intense chemical decomposition, few bare outcrops, gentle slopes and flat plains, as encountered in the equatorial regions of Africa and South America.
2 Humid temperate facies – with characteristics similar to 1 but with

generally steeper slopes, found for example in western and central Europe and eastern United States.

3 Dry season facies –where mechanical weathering is important giving escarpments and steep slopes, as in Mediterranean, tropical savanna and subtropical with dry season climates.

4 Desert facies – with bare arid surfaces characterized by wadis and dunes.

5 Cold dry facies – with permanently frozen subsoil, polygonal soils and coarse debris.

6 Cold humid facies – glacial environments.

Thus Martonne added to the existing three morphoclimatic regions (2, 4 and 6) three further ones which were to become the humid tropical (equatorial; selva) (1), the tropical wet-dry (savanna) (3) and the periglacial (5). At about the same time Penck (1914a) refined his previous distinction between humid and arid morphoclimates by adding subhumid and semi-arid to make a fivefold classification. He drew particular attention to the fact that humid landforms are less steep, their hilltops more convexly rounded, their slope bases more sweepingly concave and their slope elements less distinct than their more angular arid counterparts.

The most serious attempt in morphoclimatic analysis and classification following the First World War found expression in a conference on the morphology of climatic zones held in Düsseldorf on 22 and 23 September 1926 and published in the following year (Thorbecke 1927a). Contributors consisted of most of the leading German climatic geomorphologists and the result was articles devoted to the following topics:

1 Climate and landforms. An introduction by Franz Thorbecke (1927b) of Cologne justifying the association of distinctive landforms with existing climates.

2 Landforms in hot wet equatorial climates by Walter Behrmann (1927) of Frankfurt, dealing significantly with New Guinea (see also Behrmann 1921, 1924).

3 Landforms in the periodically dry tropics with a dominant wet season by Franz Thorbecke (1927c), treating the savanna landforms of the Cameroons and parts of India.

4 Landforms of the periodically dry tropics with a dominant dry season by Fritz Jaeger (1927) of Berlin, concentrating on the southern belt of what had been German South-West Africa.

5 Landforms in marginal monsoon climates by Heinrich Schmitthenner (1927) of Heidelberg, with particular reference to parts of China.

6 Landforms in regions of winter rainfall by Hans Mortensen (1927c) of Göttingen dealing with the Mediterranean climatic region of northern Chile (see also Mortensen 1927a, b).

7 The production of landforms in hot dry deserts by Siegfried Passarge (1927) of Hamburg, treating Egypt and other areas.

8 Desert landforms, with particular reference to the Namib region of South-West Africa, by Erich Kaiser (1927) of Munich. Kaiser worked in Namibia between 1914 and 1919 and was particularly concerned with the processes of desert formation, especially aeolian deflation (see also Kaiser 1926).

9 Landforms of interior and high deserts by Fritz Machatschek (1927) of Zürich, the author of an important later two-volume work on regional geomorphology (Machatschek 1938–40). His Düsseldorf contribution dealt with the south-west United States, central Asia and Libya.

10 Landforms of the arctic by Fritz Klute (1927) of Giessen, concentrating on Greenland and Spitsbergen.

Despite its authoritative and comprehensive character, the emphasis of the Düsseldorf conference was to some extent dictated by the interests of the personalities involved. Some of the regions treated (i.e. 4, 5, and 6) were not morphoclimatically significant; the four arid contributions (4, 7, 8 and 9) were not morphoclimatically distinguishable, indicating that at that time it was becoming increasingly difficult to distinguish between Penck's (1914a) arid, semi-arid and subhumid processes and landforms. Also the inclusion of the contribution by Passarge (1927) did not subscribe to the underlying morphoclimatic theory on grounds of climatic change (Passarge 1926). Despite this criticism, the conference proceedings embody a robust statement of the philosophy of morphoclimatology, most importantly in the contribution by Thorbecke (1927c), as stressed by Derbyshire (1973). Thorbecke described landforms developed in the savanna zone around the River Mbam in the central Cameroons on granites, syenites and gneisses with some basalt plains. The characteristic geomorphic features were held to be due to the present climatic regime wherein the short dry season was *morphologically effective* (an important concept). During this dry season the granitic cliffs shatter, exfoliate and retreat parallel, and during the longer wet season the fragmented basal material quickly disintegrates chemically and is washed away by sheetfloods. The latter material builds up the floodplains and forces the stream channels against the retreating slope bases. These dominantly lateral processes produce the stepped landscape, sharp basal angles, outliers and inselbergs characteristic of the morphologically effective features of the periodically dry tropics with a dominant wet season (figure 12.4).

The Development of Tropical Geomorphology

The Düsseldorf conference, in contrast to the existing high-latitude, temperate humid and arid subtropical emphasis in the English-speaking worlds, was very much biased in favour of the study of tropical morphoclimates. Of the nine chapters devoted to specific regions by the conference proceedings, three were specifically tropical, two on the subtropics and two

Figure 12.4 The process of inselberg landscape formation in the Cameroons according to Thorbecke (1927c)

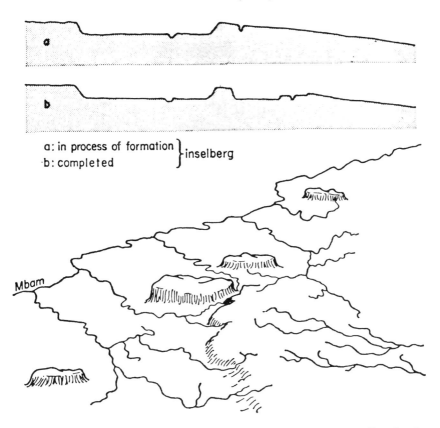

Source: From Derbyshire (1973), Figure 6.1, p. 102. By permission of Macmillan, London and Basingstoke.

on the subtropical margins. One of the contributors, Karl Sapper, gave focus and impetus to tropical climatic geomorphology by his publication in 1935 of the *Geomorphology of the Humid Tropics* (Krynine 1936) following earlier publications (Sapper 1923, Sapper and Geiger 1934). Sapper began work in Central America in the early 1890s but in his book he drew widely on his other experience as well as on German and other research, especially that of Behrmann in New Guinea, Thorbecke in the Cameroons and Freise in Brazil. An introductory chapter describes the geomorphologically significant aspects of tropical climates (e.g. precipitation seasonality and intensity; the chemical effectiveness of tropical rainwater containing excessive carbon dioxide, nitric acid and ammonia), and three short concluding chapters deal with morphologically significant activity of plants and animals, the influence of man on landscape (including deforestation and soil erosion) and the action of the sea in the humid tropics. However, out of a total of 154 pages, 125 pages comprising chapter 2 are devoted to:

1 The constantly humid tropics (i.e. Köppen's Af and Am regions), which include the permanently humid rain forest with no dry season and parts of the rainy monsoonal regions with only a very short dry season.
2 The seasonally humid tropics which, although total annual precipitation may be high, include regions with a marked dry season and savanna-type vegetation (i.e. Köppen's Aw region).
3 The cool elevated tropical mountains, including Central America, Kilimanjaro and the high Andes where processes are compared with those of the temperate zones. This is a short section.

Thus Sapper emphasized within the 'humid tropics' the distinction between the area of all-year rainfall (i.e. selva; rain forest) and those with a marked dry season (i.e. savanna; dry season tropics; periodically dry tropics).

Sapper (1935; see previously 1923, 1933, Sapper and Geiger 1934) gave a very distinctive description of the morphoclimatology of the constantly humid, hot tropical rain forest regions which have sharp boundaries, a dense vegetation cover, little mechanical weathering or unconcentrated soil erosion and accentuated chemical weathering giving deep regolith on flatter areas, often with laterites. The regions especially identified were the Central American rain forest, the West African coastal plain, the New Guinea rain forest, the equatorial mountains of West Africa, the coast ranges of Brazil and the basins of the Amazon and Orinoco. In areas of significant relief erosion is dominantly vertical, valleys and gullies are cut by the torrential runoff, their steeply concave sides are subject to frequent landslides and earthflows producing parallel slope retreat. The upper parts of the slopes are especially steep and terminate in serrated, knife-edged ridges. Rainwash and lateral erosion, dominant in the savanna lands, are virtually absent. Much of the debris produced by valley and gully erosion is partly unweathered bedrock and this produces fans of arkosic material and, finally, the clay and silt on the coastal flatlands.

Sapper proposed a sequence of geomorphic development in these environments:

1 Initial – deep gashes cut by canyons (e.g. Honduras and Guatemala);
2 These develop into steep-sided valleys separated by knife-edged ridges, associated with systems of ravines, the whole region being vegetated except for recent landslide scars (e.g. New Guinea);
3 Headward erosion breaks through the major divides and the landscape becomes a confusion of jagged peaks and deep canyons (e.g. Pacific islands and the West Indies).

4 The residual peaks are reduced and, although still densely vegetated, are slowly reduced by soil creep, rainwash and the beginnings of lateral erosion. Resistant rocks still maintain sharp relief but less-resistant outcrops become gently rolling hills.
5 Finally – aggradation becomes dominant and the region is covered by fluvial silt and clay spread by shifting, meandering rivers (e.g the Amazon and Orinoco basins).

This sequence is not a 'cycle' in the Davisian sense, although Sapper uses the terms 'youth' and 'maturity'. In contrast with the normal cycle, Sapper's tropical sequence begins with erosion being generally slow, due to the effect of the vegetation, then being very fast as the valleys and slopes develop, and finally very slow.

The work of Sapper was paralleled and complemented by that of Freise (1933a, b, 1935, 1936a, b, 1938) who had worked for some 30 years in the humid tropical Brazilian coastal ranges of Rio de Janeiro, Espirito Santo and Minas Geraes, as well as in the savanna area further inland. He studied in particular an area of several hundred square kilometres of virgin forest near Rio, from which he derived the very exaggerated estimated rate of erosion and solution of 30 kg/m²/year. More importantly, he proposed the following sequence in which local denudational events are linked to vegetation:

1 Bare rock slopes of 48–52° on granite and 35–38° on gneiss;
2 Humus formation begins;
3 A layer of decomposed soil is formed favouring accelerated chemical weathering and trees begin to invade;
4 The soil mantle attains 10–11 m in thickness;
5 Earthflows occur beneath the vegetation root mat, the vegetation cover deteriorates and surface crusts form;
6 Earthflows and rainwash destroy the surface vegetation, which is replaced by grass;
7 Complete landslide stripping occurs and bedrock is exposed. This process takes place all over the steep rain forest slopes in a very patchy manner with each patch having a seven-stage timescale of 300,000–400,000 years.

Although the humid tropical landforms of the Brazilian coastal areas behind Rio and Santos, together with their steep slopes and deep weathering, were further described by Martonne (1940), it is interesting that notions regarding such landforms did not find their way into his *Traité de Géographie Physique* before the Second World War. After the war, as will be seen in volume 4, more detailed researches were undertaken, particularly by Rougerie in the Ivory Coast after 1950. Rougerie (1960) stressed the chemical efficacy of humid tropical water, the complexity of the factors controlling the thickness of the regolith and the location of the forest/savanna boundary, and in particular the importance of overland flow. The text by Wilhelmy (1958), who had worked on the Brazilian Pantanal and the Pacific coastal plain of Colombia, stressed the effects of different climates on the landforms produced on similar lithologies.

The Düsseldorf conference laid particular emphasis on the inselberg landforms of the tropical wet–dry savanna lands and subsequent work was particularly concentrated on regions with a long dry season and, therefore, scanty vegetation. It also became clear that such regions are climatically and vegetationally marginal between the perenially humid tropics and the semi-arid lands. At the conference Jaeger (1927) described, among others, areas of southern South-West Africa having (presumably) earlier-stage erosional inselbergs with steep-sided valleys and lacking extensive valley floors. Soon afterwards Passarge (1928, 1929) treated inselbergs with steep lower slopes, a sharp basal nick and little talus in an area of northern Nigeria having semi-arid characteristics. Between 1927 and 1929 Wilhelm Credner (1931, 1935) worked in Malaya and Siam and, in particular, identified inselberg forms on granite in the monsoon region of Malaya covered with equatorial jungle, raising questions regarding the overlap of the two morphoclimatic regions and of the effects of climatic change which we shall treat later (see also Lehmann (1936) on Java). A most important treatment of morphoclimatic marginality was that by Krebs (1933, 1942) on the inselberg landscape on schists, quartzites and basalts in the southern Peninsular Indian regions of Madurai and Tirunelveli. Here inselbergs rise from a low-gradient (2–3 per cent) surface at an elevation of some 200 m in an area exhibiting a range of precipitation regimes. Coining the term 'tropical inselberg' to differentiate it from W. Penck's more general use of the term, Krebs proposed that in southern India humid tropical landforms occur where annual rainfall exceeds 50 inches with no long dry season; that tropical inselbergs are associated either with an annual rainfall of 24–50 inches or with more than 50 inches only where there is a long dry season; and that below 24 inches true arid landforms occur.

Sapper (1933, 1935), expanding on Clozier's (1932) general description of tropical inselbergs, treated the landforms of the seasonally humid (35–80 inches) hot regions of the savannas of Central America, the Cameroons Highlands and the monsoon regions of south-east Asia. His field experience was mainly related to the granite turtleback hills south of the Parima Mountains in Guyana and the Wute Plain of the central Cameroons. Sapper stressed the alternating seasonal dominance of chemical weathering, rilling and sheetflood action and lowland inundation in the wet season and of mechanical weathering and exfoliation in the dry season to produce in the savanna regions a more complex set of geomorphic processes than in the humid tropics. These processes were held to be responsible for the production of isolated residuals (inselbergs) on plains, particularly on granites and gneisses, and, under extreme conditions of sheetflood action and large-scale granite exfoliation, of the sugarloaf inselbergs encountered in such regions as eastern Brazil and northern Ethiopia. On the surrounding piedmonts there are alluvial fans of angular material in places.

Before the Second World War a number of publications were concerned with savanna landforms, the inselberg problem and the distinction between savanna and more arid landforms. Krynine (1935) studied the semi-arid/savanna conditions in the state of Tabasco, southern Mexico, resulting in arkosic sediments being generated by granites and Ackermann (1936) distinguished between savanna lowlands in Northern Rhodesia and true arid pediments by showing that savanna slopes (20–30°) are bordered by a zone of sheetflooding leading down to seasonally flooded valley floors which he termed 'dambos'. These, characteristic of savanna regions, were held to differ from the arid playas later described by Jaeger (1939) in South-West Africa. Jessen (1936, 1938), working in the Planalto of Angola, postulated the cause of parallel cliff retreat as due to the basal debris retaining moisture and promoting year-round accentuated chemical weathering. Although parts of the area consist of less dissected inselberg landscapes with restricted lowlands, he described a dominant rolling upland surface formed up to 1,700 m on gneisses, schists, other metamorphics and intrusive igneous rocks. The American geomorphologist Bailey Willis (1936), working in East Africa, believed that the more resistant rocks of the inselbergs allowed them to be very ancient geological features (possibly Palaeozoic) which had long resisted erosion, while the huge surrounding plains were eroded down, possibly following deep weathering (see Büdel 1957a, later). Willis termed such savanna residuals 'bornhardts' after the German worker who in 1900 made the gneissic inselberg landscape in Madjedje, East Africa, a classic. Gillman (1937), also working in East Africa, did not believe that a more resistant geological outcrop was necessary for a typical inselberg to form. In eastern Brazil Freise (1938) found similar inselberg forms to those of Mozambique on granites and gneisses in rain forest regions, whereas Martonne (1940) contrasted the humid landforms of the coastal mountains with the inselbergs and less accentuated relief of the dry season hinterlands of Rio de Janeiro and Santos. Bishopp (1941) worked on the inselberg landscape in the Rio Branco valley of northern Brazil.

By the Second World War climatic geomorphology was making a significant impact on the English-speaking world, as evidenced by Cotton's (1942) excellent résumé. Although carried out along Davisian lines, as the title *Climatic Accidents* suggests, Cotton was firmly committed to the morpho-climatic approach but was particularly concerned with the overlap between what he believed to be the true inselberg landscapes of the savanna climates and semi-arid landscapes. For him, the distinction resided in the origin of the plains, with semi-arid inselbergs rising from pediment surfaces and

Figure 12.5 The double surface of savanna erosion: A, the surface of the tropical savanna plain (*Spül-Oberfläche*); B, the buried weathering surface (weathering front) (*Verwitterungs-Basisfläche*) with marginally extended bedrock surfaces (*Spülpedimente*) and an inselberg; C, reduction of the surface and lowering of the weathering front to give stepped marginal bedrock surfaces (*Rand-Spülpedimente*); D, further removal of debris exhuming shield inselbergs (*Schildinselberge*)

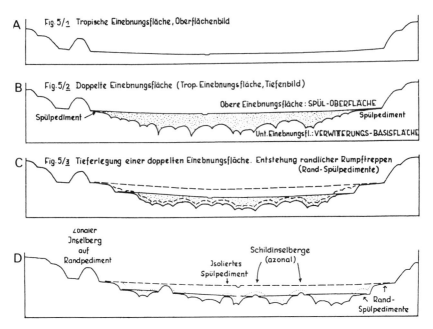

Source: From Büdel (1957a), Figures 5 and 7, pp. 209 and 213

bornhardts from flatter, regularly inundated savanna plains. This view of the morphological integrity of current savanna processes was later supported by Clayton (1956) who, working in the Wute Plain of the central Cameroons backed by a 400–600 m scarp bordering the Ndomme Plateau (see Thorbecke 1927c), was impressed by the current enlargement of the plain by scarp recession in which he believed lateral stream cutting (held to be so important under semi-arid conditions) played only a minor part.

As will be seen, after the Second World War there was an increasing tendency to view more and more landforms everywhere as climatogenetic – the products of complex climatic changes. It is therefore interesting that the most recent dominantly morphogenetic view that savanna landforms are formed under a continuous wet–dry seasonal regime was given by Büdel (1957a, b, 1961) whose name is now so associated with landform theories based on climatic change. He proposed that in more seasonally humid savanna and monsoon regions 'double sufaces of levelling' are developed associated with savanna plains and inselbergs (figure 12.5A). The subaerial or wash-surface (*Spül Oberfläche*) is dominated by sheetwash erosion and by marginal slope retreat producing wash-pediments (*Spülpedimente*) and subaerial inselbergs often of considerable size (figure 12.5B). The second surface of levelling is a buried 'weathering baselevel' (*Verwitterungs-Basisfläche*) or 'weathering front' some 30–60 m or more beneath *in situ* weathered material. This front is the zone of maximum chemical weathering and is highly irregular in form depending on the local susceptibility of the bedrock to chemical weathering (figure 12.5B). Continued lowering of the plain surface under continued processes or following uplift or, possibly, climatic change may exhume low central inselbergs (*Schildinselberge*) and produce a succession of marginal (*Rand*) *Spülpedimente* (figure 12.5C). Continued surface degradation plus deep weathering produces a flat landscape with a combination of subaerial and exhumed inselbergs (figure 12.5D).

The Later Rationalization of Morphoclimatic Regions

Following the Düsseldorf conference interest in morphoclimatic classific-ations continued unabated (Cotton 1942, Büdel 1944, 1948a, b; Martonne 1946, Peltier 1950) and Visher (1941, 1945) presented thirty maps of geomorphologically relevant climatic variables to an American audience which, following Davis, had consistently underplayed the effect of climate on landforms. Between the beginning of the twentieth century and its middle the number of generally recognized morphoclimatic regions had risen from three to about nine, although many overlaps and ambiguities continued to raise problems. One was the view that radically different climates are capable of producing similar landforms (Mortensen 1930, 1933, 1943–4, Birot 1949) and, for example, King's (1951) view that inselberg and pediment production is not limited to desert or savanna conditions. It was characteristic of this period that repeated efforts were necessary to place landform differences, rather than purely climatic ones, at the centre of the stage (Mortensen 1930). It is significant that since the rise of climatic geomorphology the number of proposed morphoclimatic regions has been significantly less than the number of major climatic regions proposed by climatic classifiers. This suggests that greater climatic differences are necessary, operating over a longer timespan, to generate distinctive landform assemblages than to have biogeographical significance. For example, a recent simplified morphoclimatic classification based on a number of sources (Büdel 1948b, Peltier 1950, Troll 1958, Tanner 1961, Wilson 1969) involved taking the seventeen major climatic divisions of Köppen (figure 12.2), dividing into two each of the four which appeared morphoclimatically ambiguous (i.e. Cs, Cfa, Dfc and Dw) to produce twenty-one divisions which were regrouped to generate eight morphoclimatic regions (figure 12.6). These eight regions fall into two groups wherein the major geomorphic processes are either non-seasonal or seasonal:

Figure 12.6 The major present morphoclimatic regions of the world roughly classified according to mean annual temperature (°C), mean annual precipitation (mm), mean number of wet months (above 50 mm) and mean temperature of the warmest month (°C)

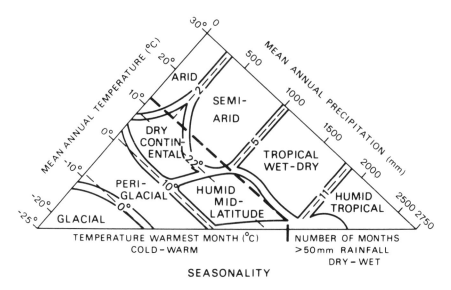

Source: After many sources, including Peltier, Tanner, Wilson, Büdel and Troll. From Chorley *et al*, 1985, Figure 18.2, p. 469.

A   Non-seasonal, including the glacial, arid and humid tropical. These have non-seasonal processes, generally low average erosion rates, highly infrequent and episodic erosional activity (e.g. glacial surges, desert rainstorms and slope mass failures, respectively), and a tendency for the location of their cores to persist latitudinally (at 90° 25° and 0°, respectively) during climatic changes, even if the climatic type is completely obliterated on occasions.

B   Seasonal, including tropical wet–dry, semi-arid, dry continental, humid mid-latitude and periglacial. These have processes which are more distinctly seasonal in their operation; in places high average erosion rates; erosional activity which, although it may be episodic, shows some consistency over a period of years; and a tendency for considerable changes of their size and location accompanying global climatic changes. Such regions can be divided into two groups:

(1) warmer climates (tropical wet–dry and semi-arid) where geomorphic processes differ most significantly in terms of the length of the wet season:

(2) cooler climates (dry continental, humid mid-latitude and periglacial) whose geomorphic processes differ mainly in respect of summer temperatures, as well as with some regard to precipitation amounts.

Figure 12.7 The development of morphoclimatic regionalization in the twentieth century (see figure 12.6 for explanation)

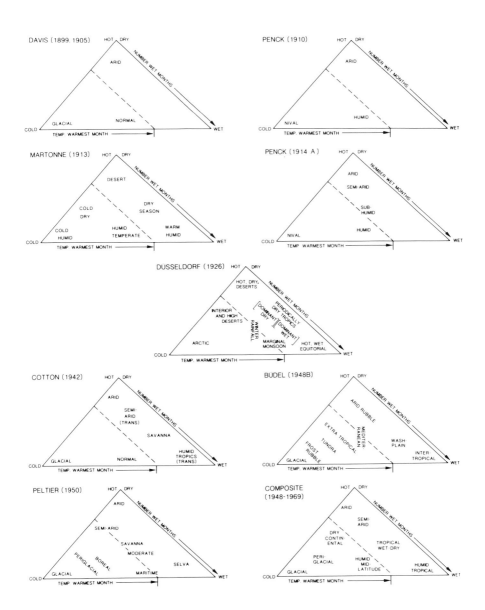

Using this simple triangular basis it is possible to demonstrate graphically the development of morphoclimatic regionalization during the first half of the twentieth century (figure 12.7). As we shall see, after about 1950 the emphasis in climatic geomorphology shifted towards ideas involving climatic change which will be taken up in detail in volume 4 (e.g. Büdel 1963, Tricart and Cailleux 1965, and earlier). Major reviews of climatic geomorphology have been provided by Maull (1938, 1958), Cotton (1942), Peltier (1950), Tanner (1961), Tricart (1953, 1961, 1965), Tricart and Cailleux (1965), Schou (1965), Holzner and Weaver (1965), Stoddart (1969), Wilson (1969), Rathjens (1971) and Derbyshire (1973).

Using the classic Davisian cyclic basis, C. A. Cotton (1942) identified six morphoclimatic regions including four main types and two transition types, each with characteristic mature landforms (figure 12.8):

1 Normal (main) – employing the term in the Davisian sense as applied to humid temperate landforms.
2 Glacial (main) – including the Davisian glacial cycle but excluding periglacial landforms.
3 Humid tropical (transitional) – following the forms and processes set out by Sapper and Freise. Believed to be transitional between 1 and 6.
4 Arid (main) – landforms dominated by interior drainage (Mortensen 1927b), bajada and pediment extension (Bryan 1922, Berkey and Morris 1927, Sauer 1930), basin capture, slope retreat and the replacement of desert mountains by low domes. Fluvial (sheetflood and streamflood) processes dominate but are increasingly assisted through time by aeolian processes. A sharp break appears to exist between the hillslope (30–35° or more), on the one hand, and the pediments (5–7° in the higher parts and 3–4° in the lower ones) and the virtually flat bajadas, on the other.
5 Semi-arid (transitional) – difficult to distinguish from arid conditions in terms of processes but with the upper pediment slopes being steeper and breaks of slopes less abrupt. Early erosional stages may be similar to those of the normal cycle but lateral planation produces steep pediments which may be incipient or terrace-like in places. The semi-arid morphoclimatic region is often difficult to distinguish from the arid and the savanna regions.
6 Savanna (main) – exhibiting the flat plains and abrupt inselbergs already described.

A more sophisticated morphoclimatic classification was proposed by Büdel (1944, 1948b). Stressing the subtlety of the climatic processes controlling landforms (e.g. number of frost–thaw cycles), Büdel pointed out that most major landforms change more slowly than climate (and are thus relics of numerous past climates) and that any attempt to link present-day climate and landforms must concentrate on those forms which are small scale and rapidly

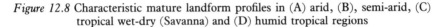

Figure 12.8 Characteristic mature landform profiles in (A) arid, (B), semi-arid, (C) tropical wet-dry (Savanna) and (D) humid tropical regions

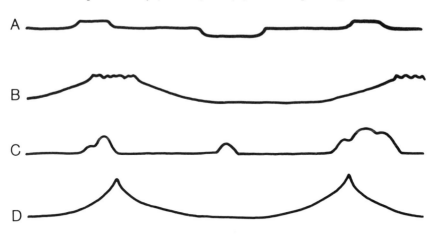

Source: From Cotton (1942), Figure 37, p. 100

evolving. He proposed the following morphoclimatic classification involving eight terrestrial regions and ten subregions (figure 12.9):

1 Glacier zone
2 Frost-rubble zone (*Frostschuttzone*)
3 Tundra zone
4 Extra-tropical (mature: *in situ*) soil zone (*Nichttropische Ortsbödenzone*)
   (a) maritime temperate zone
   (b) subpolar tjäle-free (subpolar without permafrost)
   (c) tjäle zone (subpolar with permafrost)
   (d) continental zone
   (e) steppe zone
5 Mediterranean transition zone (*Etesische Ubergangszone*)
6 Arid-rubble zone (*Trockenschuttzone*)
   (a) tropical inselberg-pediment desert zone
   (b) extra-tropical desert zone
   (c) high-altitude (cool) desert zone
7 Wash-plain zone (humid savanna) (*Flächenspülzone*)
   (a) tropical wash-plain (sheetwash) zone
   (b) sub-tropical wash-plain (sheetwash) zone
8 Inter-tropical mature equatorial soil zone (*Innertropische Ortsbödenzone*)

This classification generated wide interest but was criticized for its azonal organization; its subdivision of permafrost between classes 2, 3 and 4(c); its grouping of taiga and permafrost with maritime forest and steppe (4); and its

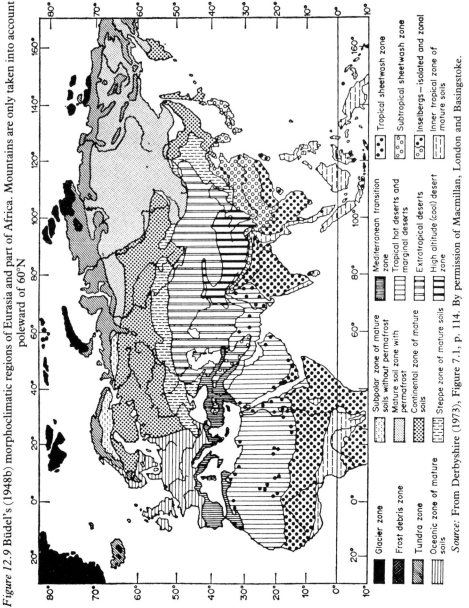

Figure 12.9 Büdel's (1948b) morphoclimatic regions of Eurasia and part of Africa. Mountains are only taken into account poleward of 60°N

Glacier zone

Frost debris zone

Tundra zone

Oceanic zone of mature soils

Subpolar zone of mature soils without permafrost

Mature soil zone with permafrost

Continental zone of mature soils

Steppe zone of mature soils

Mediterranean transition zone

Tropical hot deserts and marginal deserts

Extratropical deserts

High altitude (cool) desert zone

Tropical sheetwash zone

Subtropical sheetwash zone

Inselbergs – isolated and zonal

Inner tropical zone of mature soils

Source: From Derbyshire (1973), Figure 7.1, p. 114. By permission of Macmillan, London and Basingstoke.

separate identification of the Mediterranean landforms (5) being not worthy of distinction (despite the works of Mortensen (1927c) and Birot (1937)).

In 1950 Peltier produced a process-based classification of so-called 'morphogenetic regions' (figure 12.10) based on an analysis of the ranges of temperature and precipitation within which six major geomorphic processes operate (figure 12.11). Peltier identified the following nine regions (Peltier 1950, Table 1, p. 215):

1 Glacial – average annual temperature range 0–20°F; average annual rainfall range 0–45 inches. Dominant processes–glacial erosion, nivation and wind action. Based on the work of Davis (1900H, 1906J).

2 Periglacial – average annual temperature range 5–30°F; average annual rainfall range 5–55 inches. Dominant processes – strong mass movements, moderate to strong wind action and weak fluvial action. Peltier, developing the work of Troll (1944, 1948), identifies a distinct periglacial cycle (figure 12.12) in cold, humid, subarctic regions associated with the production of three coexisting erosion surfaces:
   (a) A surface of downwasting produced by congeliturbation;
   (b) A surface of lateral planation produced where the water table and the zone of frequent nivation coincide;
   (c) A stream graded, or aggraded, surface.

Davis regarded periglacial action as a climatic accident but Peltier followed Troll in believing its characteristics to be sufficiently significant and persistent to be separately categorized. Peltier considered each cycle (including the periglacial (Troll 1948) to be normal within its own regime and an accident only when one morphoclimate temporarily encroaches on another regime. The question remained, however, as to whether periglacial conditions have persisted long enough in any one locality to produce a distinctive set of landforms, as distinct from 'mere embroidery' of the landscape.

3 Boreal – average annual temperature range 15–38°F; average annual rainfall range 10–60 inches. Dominant processes – moderate frost action, moderate to slight wind action and moderate fluvial action. Essentially Köppen's Dfc region.

4 Maritime – average annual temperature range 35–70°F; average annual rainfall range 50–75 inches. Dominant processes – strong mass movements and moderate to strong fluvial action. Köppen's Cf b (maritime west coast) region. It has been pointed out that Peltier's regions 3 and 4 have no dominant geomorphic characteristics to distinguish them from regions 2 and 6.

5 Selva – average annual temperature range 60–85°F; average annual rainfall range 55–90 inches. Dominant processes – strong mass movements, slight slope wash and no wind action. This humid tropical morphoclimate was based on the work of Bornhardt, Sapper, Freise and Cotton.

6 Moderate – average annual temperature range 35–85°F; average annual

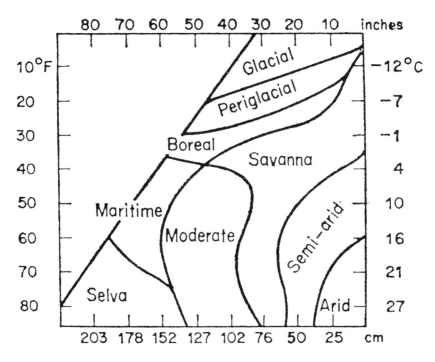

Figure 12.10 Morphogenetic regions by Peltier

Source: From Peltier (1950), Figure 7, p. 222

rainfall range 35–60 inches. Dominant processes – strong fluvial action, moderate mass movements, slight frost action and no significant wind action. This approximated to Davis' (1899H) 'normal' cycle.

7 Savanna – average annual temperature range 10–85°F; average annual rainfall range 25–50 inches. Dominant processes – strong to weak fluvial action and moderate wind action (see Cotton 1942, 1961).

8 Semi-arid average annual temperature range 35–85°F; average annual rainfall range 10–25 inches. Dominant processes – strong wind action and moderate to strong fluvial processes. Includes the dry continental regions (Cotton 1942).

9 Arid – average annual temperature range 55–85°F; average annual rainfall range 0–15 inches. Dominant processes – strong wind action and slight fluvial and mass movement processes (Davis 1905L).

Just after the middle of the present century morphoclimatic classification reached its peak and it was possible to identify regions in somewhat the following manner (see figure 12.6) (Chorley *et al* 1984).

Figure 12.11 Classification of geomorphic process regions by Peltier

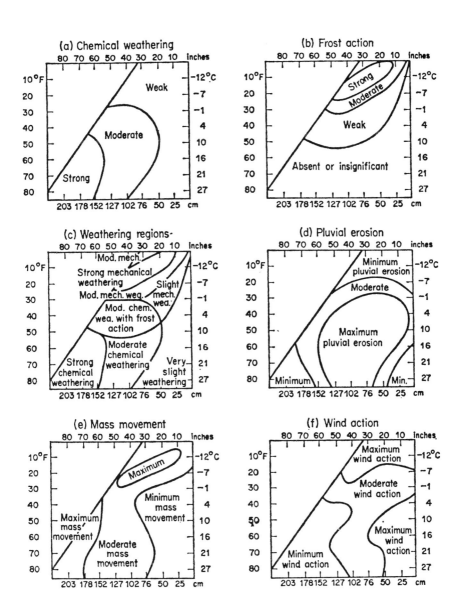

Source: From Peltier (1950), Figures 1–6, p. 219

Figure 12.12 The periglacial cycle

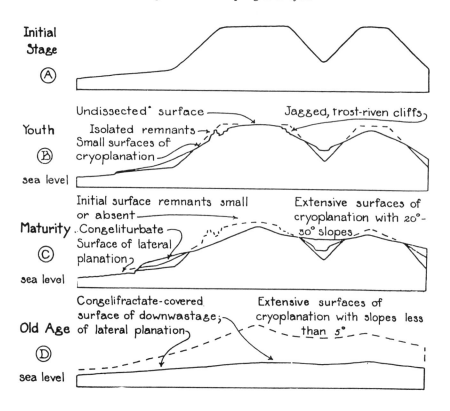

Source: From Peltier (1950), Figure 8, p. 225

Glacial

Other names given to whole or part of this region – subglacial, nival, cold humid, arctic; Köppen region EF.

Geomorphic processes – frost weathering (max.); mechanical weathering (mod.); chemical weathering (min.); mass wasting (min.); fluvial processes (min., except for seasonal meltwater); glacial scour (max.); wind action (max.).

Morphological features – alpine topography; abrasion surfaces; kames; eskers; till forms; fluvio-glacial features.

Arid

Other names – desert; true desert; tropical and subtropical desert, hot dry desert, arid rubble; Köppen region BWh.

Geomorphic processes – frost weathering (min., except at high latitudes); mechanical weathering (max., especially salt and thermal); chemical weathering (min.); mass wasting (min.); fluvial processes (min. at present; commonly evidence of previous action, especially gullying and sheetwash); glacial scour (nil); wind action (max., perhaps overestimated by past workers).

Morphological features – dunes; playas; deflation basins; angular, debris-covered slopes; fossil fluvial forms (e.g. fans and arroyos).

Humid tropical

Other names – selva; rain forest; inner tropical zone, intertropical; Köppen regions Af and Am.

Geomorphic processes – frost weathering (nil); mechanical weathering (min.); chemical weathering (max., giving thick, fine-grained debris even on steep slopes covered with thick vegetation); mass wasting (max., but highly episodic and irregular on steep slopes, especially in the form of earthflows and slips); fluvial processes (mod.– min. slopewash and rainbeat; min. stream corrasion due to lack of coarse debris; max. chemical and suspended-load transport); glacial scour (nil); wind action (nil).

Morphological features – low-gradient rivers; wide, flat or gently un-dulating floodplain floors up to several kilometres wide; steep slopes (frequently greater than 40°) rising abruptly from valley floors, stabilized by dense vegetation but susceptible to infrequent landslides; knife-edged ridges maintained by parallel retreat of steep slopes.

Tropical wet-dry

Other names – savanna; tropical sheetwash zone; moist and dry savanna; tropical savanna, wash plain, periodically dry tropics (part), dry season; Köppen region Aw and Cw.

Geomorphic processes – frost weathering (nil); mechanical weathering (min.– mod.); chemical weathering (seasonal max.– accelerated by standing water, giving deep weathering and two weathering zones, at the surface and at the weathering front; surface crusts of Al and Fe generated during dry seasons); mass wasting (mod.– max.); fluvial processes (mod.– max. – highly seasonal sheetfloods, rill wash and channel flow); glacial scour (nil); wind action (min. – mod.).

Morphological features – steep irregular slopes of coarse debris retreating parallel by basal chemical weathering and sheetflood removal; wide planation surfaces (less than 3.5°) bounded by steep slopes and inselbergs; irregular bedrock surfaces broken by inselbergs and tor-like features probably pro-duced by stripping along the weathering front; badlands on shales and other impermeable rocks.

The morphological features of this region may not be greatly different from those of the semi-arid region *in appearance*, but may differ in terms of their generating processes. This introduces an important matter of principle in morphogenetic classification.

Semi-arid

Other names – peripheral or marginal hot deserts; desert margins; thorn savanna; semi-desert; semi-arid steppes; Mediterranean or summer-dry subtropical zone (lower-latitude parts only), dry season, arid rubble (part); Köppen regions BSh and Cs (part).

Geomorphic processes – frost weathering (min., except at higher latitudes); mechanical weathering (min. – mod., especially thermal and salt); chemical weathering (min. – mod.); mass wasting (mod., but infrequent); fluvial processes (max. but episodic in the form of sheetwash, gullying and ephemeral stream action giving high overall erosion rates); glacial scour (nil); wind action (mod. – max.).

Morphological features – pediments (1–4°) backed by cliffs and angular talus slopes (25–35°) and inselbergs; integrated ephemeral stream systems; arroyos; badlands; alluvial fans; local dunes.

Dry continental

Other names – steppe zone; mid-latitude grasslands; semi-arid steppes; degraded steppes; extratropical deserts (part); savanna (part), cold, dry interior and high deserts, extratropical (part), semi-arid (part); Köppen regions BSk, BWk and Cfa (part).

Geomorphic processes – frost weathering (min. – mod., but highly seasonal with bedrock affected at times); mechanical weathering (min.– mod.); chemical weathering (min. – mod.); mass wasting (mod., but seasonal activity important); fluvial processes (mod.– max., especially in the form of spring and summer sheetwash and gullying); glacial scour (nil); wind action (mod., fossil loess in places).

Morphological features – pediments flanked by steep, scree-covered slopes; badlands; alluvial fans; arroyos; like tropical wet–dry landforms, these may not appear to differ greatly from the semi-arid ones, but processes may do so (especially those associated with frost and spring thaw).

Humid mid-latitude

Other names – oceanic and continental zones of mature soils; temperate marine and continental regions; moist subtropical; humid temperate; maritime; mid-latitude forests; Mediterranean zone and summer-dry subtropical (higher-latitude parts only); moderate (part); forest on Pleistocene permafrost (part), normal, extratropical (part), marginal monsoon (part), humid; Köppen regions Cfa (part), Cfb, Cs (part), Dw (part), Dfa (part), Dfb and Dfc (part).

Geomorphic processes – frost weathering (min. – max., especially in higher-latitude and continental location where this region is transitional with the periglacial region); mechanical weathering (min. – mod.); chemical weathering (mod.); mass wasting (mod. – max., creep dominant); fluvial processes (mod.); glacial scour (nil, except on mountains and as fossil evidence); wind action (min.).

Morphological features – smooth, soil-covered slopes; ridges and valleys; streams with wide range of grain sizes.

Periglacial

Other names – frost debris zone; tundra; subpolar zone; high arctic barrens; forest with permafrost; humid microthermal; boreal (part); mature soil zone with permafrost frost rubble zone, boreal (part); Köppen regions ET, Dw (part), Dfa (part) and Dfc (part).

Geomorphic processes – frost weathering (max., permafrost conditions or subsoil freezing during significant parts of the year; many months under snow cover; spring thaw phenomena); mechanical weathering (max., especially nivation); chemical weathering (min.); mass wasting (max., talus creep and solifluction everywhere during the spring and in northern parts during the summer); fluvial processes (mod., slopewash and valley cutting concentrated in limited thaw season); glacial scour (min., except on mountains and recent fossil evidence); wind action (mod. – max.).

Morphological features – scree slope (about 25° sometimes gullied), solifluction slopes (15–20°) and cryoplanation surfaces (less than 5°) with residuals rising above them; solifluction lakes and terraces; outwash plains; patterned ground; loess and dunes.

As we shall see, after about 1950 work on climatic geomorphology was directed almost exclusively towards the morphogenetic aspects of climatic change, rather than to simple morphoclimatology (Stoddart 1969).

Climatic Change and Morphogenetic Regions

After about 1950 rather more complex classifications of morphoclimatic significance were proposed (e.g. Birot 1960, Tricart and Cailleux 1955, 1965, Tanner 1961, Common 1966, Wilson 1969) but they were increasingly directed to the morphogenetic aspects of the effects of past climatic changes on landforms, particularly in the influential work of Büdel (1951, 1953, 1955, 1957b and, especially, 1963). It is interesting that the now obvious objections to the simple morphoclimatic approach, based on an awareness of the importance and relative rapidity of climatic changes, did not command wider support until after the Second World War. This is not to say, of course, that previous workers ignored the geomorphic effects of climatic change. The effects of Quaternary climatic changes (see volume 1, chapter 13 and this volume, chapter 2) had been increasingly recognized for more than a century and the longer-term effects of Tertiary climates (particularly Pliocene and Miocene) were also a subject of increasing speculation.

As long ago as 1865 Schweinfurth speculated on the more pluvial conditions which possibly resulted in the cutting of Egyptian wadis but it was late-nineteenth-century and early-twentieth-century Quaternary glacial research (see chapter 2) which highlighted the full geomorphic implications of past climatic changes. At the turn of the century no scholar was more responsible for promoting these implications than Albrecht Penck. Penck's work with Brückner (1901–9) on the Alpine ice ages stressed the reality and significance of recent climatic changes and, incidentally, tended to confirm Davis in his views on the significance of climatic 'accidents'. Penck's approach was quite different, however:

> We see on the earth's surface not only the features of the present climate but also those of a past climate. Very extended areas, formerly covered by ice, are now exposed to river action . . . . We meet with proofs of climatic changes also in desert regions . . . . (Old shorelines and extended river valleys) indicate a former moister climate; the desert with its surface features extends here over a region of former water action in the same way as, further north, river action is now displayed on a surface formerly covered by glaciers.
>
> (Penck 1905: 169–70)

> If there are oscillations in the situation of the climatic belts of the earth, it must be asked if they are connected with the disappearance of existing climatic zones and the appearance of new ones. . . . There is also much interest in the study of all border regions of climatic belts for every movement of climatic zones of the earth would here produce changes.
>
> (Penck 1905: 173)

Penck (1905, 1910, 1914a, b) believed that extensive climatic changes (e.g. glacial/fluvial and pluvial/arid) had affected the marginal areas between core morphoclimatic regions and that these marginal areas exhibited landforms owing their origins to a range of different climates. This view of climatic change was to some extent shared by Martonne (1913), although not in such a clear manner. In contrast with Albrecht Penck, Siegfried Passarge was never an advocate for simple morphoclimatic mechanisms. As early as 1904 (a, b, c) he recognized that alternations between arid and savanna conditions had taken place in South-West Africa and in 1912 he opposed the idea of stressing the effects of present exogenetic processes in the production of landforms, as distinct from past ones. In 1926 Passarge distinguished between climatic belts and 'landscape belts', pointing out that climate is not the only control over landforms and that vegetation, rock type and regolith also determine the effect of the formative processes. In the tropics he drew attention to three major shifting morphoclimatic belts – equatorial hot–wet, savanna with a predominant wet season and savanna with a predominant dry season – but suggested that the first two had similar present landscape features (vertical downcutting, dominant landslips, bowl-shaped valleys and sharp ridges), whereas the third is dominated by lateral erosional processes and the formation of iron-rich crusts. Clearly there had been a shift in climatic belts with block-debris slopes occurring under present equatorial hot-wet rain forest and evidence of past pluvial conditions in currently drier areas. Passarge proposed that 'present-day surface landforms are for the most part not the result of present climates, but the product of Pleistocene processes'; that present-day agencies are in the process of modifying Pleistocene forms towards a state in equilibrium with existing climates; but that *major* landforms do not reflect present processes.

Before the Second World War these ideas had not found wide application in regional landform classifications, although shortly afterwards Martonne (1946) stressed the effect of multiple climatic change on tropical landforms. King (1951) suggested that present semi-arid pediplains may have been originally formed as savanna surfaces and Dresch (1953) postulated that the savanna inselbergs of the Sudan may be of polyclimatic origin and of great antiquity. After the mid-1950s virtually all morphoclimatic regionalization was accorded a significant palaeoclimatic (morphogenetic) element. In his 1956 Rio de Janeiro lectures, Birot (1960: 57) stated that in 'almost all areas of the world the frequency of climatic change has meant that hardly anywhere has an area developed under a single climate throughout the whole of an erosion cycle'. In all zones from the arctic to the equator Birot acknowledged the influence of Tertiary and Quaternary climatic changes, pointing out that in the humid tropics the geomorphic effects of past climates are more rapidly removed than in more temperate regions. Tricart and Cailleux (1955, 1965) (plate 12.1) proposed the following classification of morphoclimatic zones (i.e. those showing evidence of processes characteristic of a given latitudinal belt) with inbuilt evidence of past climates:

1 The cold zone.
   (a) Glacial
   (b) Periglacial (later subdivided into five by Tricart).
2 The mid-latitude forest zone (affected by past climates, particularly Pleistocene, and by human activity).
   (a) Maritime (relict Pleistocene glacial and periglacial forms survive).
   (b) Continental (Pleistocene permafrost may survive).
   (c) Mediterranean (relict Pleistocene periglacial forms least important).

Plate 12.1 André Cailleux (centre) with students on an erratic block transported by floating ice at Valenton, Seine

Source: Courtesy A. Cailleux

3 The dry zone. Subdivided on bases of:
   (a) Water deficiency into steppe, xerophytic bush and desert;
   (b) Winter temperatures into cold and warm.
4 The humid tropical zone.
   (a) Savannas, affected by earlier drier conditions, evidences of semi-arid pediplanation and of climatic changes in the form of 'cuirasses' (i.e. laterites, calcretes and silcretes).
   (b) Tropical rain forests.

From these broad morphoclimatic zones Tricart and Cailleux (1965 and 1972, figure 8, pp. 172–3) developed a classification of world morphoclimatic regions (figure 12.13) in which morphoclimatic and morphogenetic (i.e. relict) influences are not clearly distinguished:

1 Glacial regions;
2 Periglacial regions with permafrost;
3 Periglacial regions without permafrost;
4 Forest on Quaternary permafrost;
5 Maritime forest zone of mid-latitudes with mild winters;
6 Maritime forest zone of mid-latitudes with severe winters;
7 Mid-latitude forest zone of Mediterranean type;
8 Semi-desert steppes:
   (a) Semi-desert steppes with severe winters;
9 Deserts and degraded steppes without severe winters;
10 Deserts and degraded steppes with severe winters;
11 Savannas;
12 Intertropical forests;
13 Azonal mountain areas.

Great stress was placed on extra-zonal features resulting from climatic residual forms and it is clear that the above classification included such extra-zonal features, particularly under 2–4, 5–6 (see 2(a) and (b) in their morphoclimatic zones), 8, 9–10 (the term degraded steppe implies climatic change) and 11 (see 4(a) in their morphoclimatic zones).

No work more clearly marked the ascendancy of morphogenetic geomorphology than the climatogenetic studies of Julius Büdel. Following his earlier morphoclimatic studies and those of Passarge (1919), Büdel in the early 1950s turned to the geomorphic evidence for past climatic changes. In 1951 and 1953 he suggested that the main effects of Pleistocene climatic changes were experienced in the latitudinal belt 36–77°N with relatively small changes in the tropical belts, a view later challenged (particularly by Tricart and Cailleux (1955, 1965)). Büdel (1955) later clarified his view on the extent of Pleistocene tropical climatic changes by proposing that at the beginning of the Quaternary the savanna and rain-forest areas, previously more extensive,

Figure 12.13 World morphoclimatic regions by Tricart and Cailleux

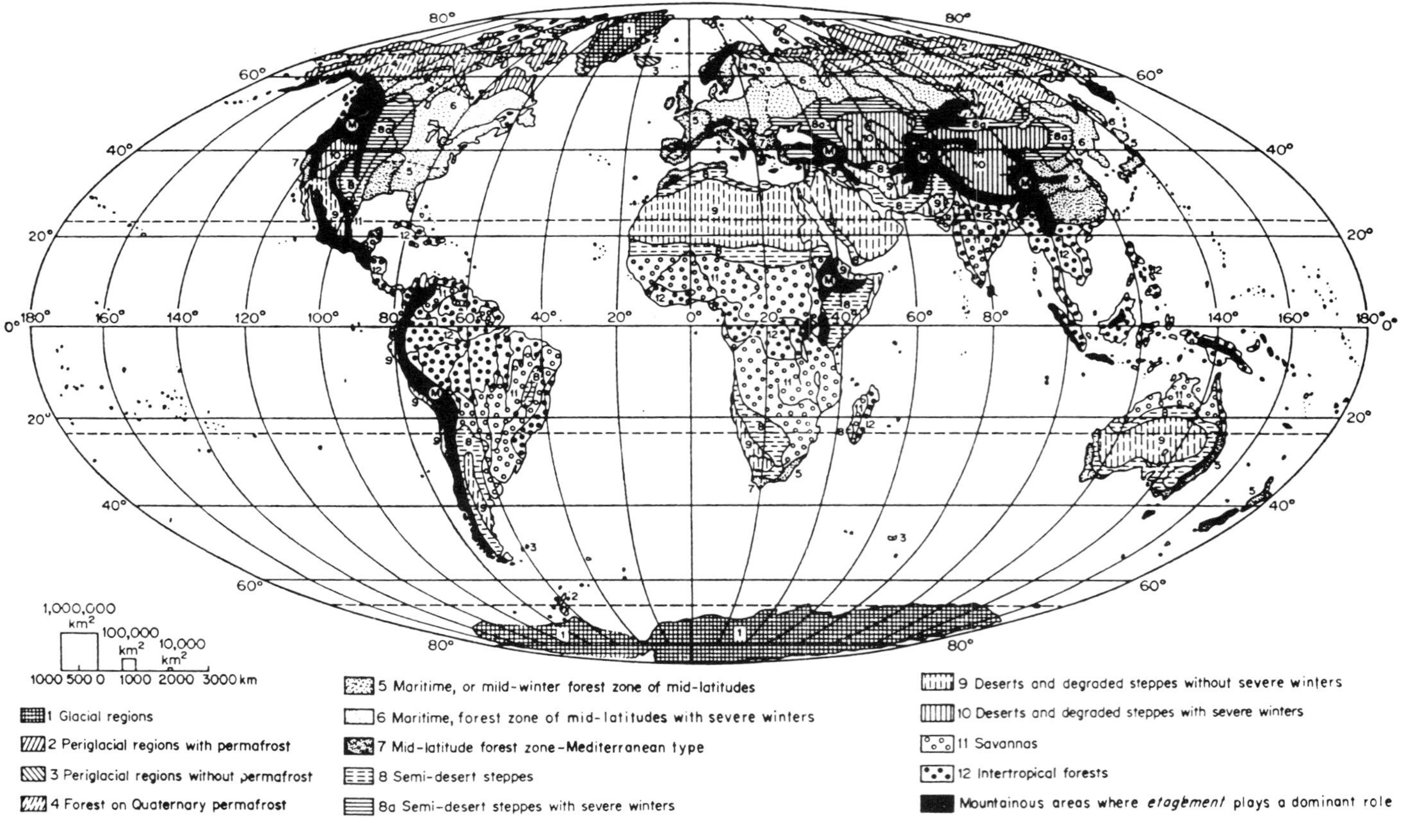

Legend:

1 Glacial regions
2 Periglacial regions with permafrost
3 Periglacial regions without permafrost
4 Forest on Quaternary permafrost

5 Maritime, or mild-winter forest zone of mid-latitudes
6 Maritime, forest zone of mid-latitudes with severe winters
7 Mid-latitude forest zone—Mediterranean type
8 Semi-desert steppes
8a Semi-desert steppes with severe winters

9 Deserts and degraded steppes without severe winters
10 Deserts and degraded steppes with severe winters
11 Savannas
12 Intertropical forests
Mountainous areas where *eragement* plays a dominant role

Scale:
1,000,000 km²
100,000 km²
10,000 km²
1000 500 0 1000 2000 3000 km

Source: From Tricart and Cailleux, (1965) and (1972) Figure 8, pp. 172–3. Courtesy SEDES, Paris and Longman Harlow.

contracted to more or less their present extent. However, during glacial phases the rain forest was believed to have expanded to about 15°N over present savanna lands and, in turn, savanna conditions extended north over parts of the present Sahara, much of which had been savanna during pre-Quaternary times. Indeed, Büdel (1957a) followed a suggestion by Jessen (1938) by proposing that during much of the Tertiary until the end of the Pliocene hot and seasonally humid savanna conditions occurred in Central Europe and that the accompanying savanna planation accounts for many of the 'peneplains' and 'piedmont benchlands' identified in temperate latitudes, later to be subjected to vertical valley cutting during the colder and wetter Quaternary. Part of Franconia in Germany. (the *Fränkische Gäufläche*) was presented as a classic example of Late Tertiary savanna planation forming a surface at about 300 m with fossil inselbergs, subsequently dissected by Pleistocene valleys and lying below an earlier (savanna?) Mio–Pliocene erosion surface encountered at about 480 m on the Steigerwald and Spessart (figure 12.14). This thesis was supported by Louis (1957; see also 1959 and 1960) who, noting that savanna planation surfaces appear to have developed at considerable elevations independent of grand baselevel in the tropics, proposed that many higher-level erosion surfaces in the mid-latitudes were originally formed as gently rolling savanna plains (*Flächmuldenlandschaften*) and then subsequently dissected (into Kerbtal types) when conditions became more cold and moist. Such surfaces are not the *Endrumpf* of Walter Penck and his *Piedmontreppen* were postulated by Louis to be in reality the remains of marginal stepped surfaces (*Rumpftreppen*) of the seasonally wet tropics. Büdel's major climatogenetic work (1963) proposed that the major task of the geomorphologist was to identify generations of landforms produced by past (fossil) and contemporary exogenetic processes steered by climate, together with other factors such as rock differences, uplift characteristics, distances from baselevel, human influence, etc. Noting that it usually takes a very long time (i.e. 10^7 years) for a whole landscape in the slowly evolving temperate zone to adjust to present climates, it was clear to Büdel that most landscapes are a complex mixture of evolving (contemporary) and relict (past) forms. He proposed that Tertiary and Quaternary world landform evolution was dominated by the existence and fluctuation of the following five morphoclimatic zones (figure 12.15):

1 Glaciated zone – polar regions and high mountains.
2 Zone of pronounced valley formation – in presently unglaciated polar regions, especially underlain by permafrost.
3 Extratropical zone of valley formation – of minor present importance but, as fossil landform evidence suggests, of much wider significance and extent during Pleistocene glacial stages.

Figure 12.14 A proposed Late Pliocene savanna (tropical planation) surface in Franconia (*Fränkische Gaüfläche*) exhibiting marginal wash pediments (*Spülpedimenten*) at 345 and 390 m, subaerial inselbergs (e.g. *Sodenberg*) and central exhumed inselbergs (e.g. *Rothreisig*). The Late-Pliocene surface is dissected by Pleistocene valleys and lies below a Mio-Pliocene (*Spessart–Steigerwald*) surface at 480 m

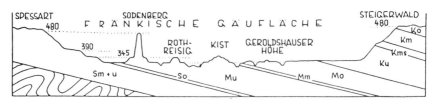

Source: From Büdel (1957a), Figure 9, p. 215

Figure 12.15 Major present morphoclimatic zones according to Büdel

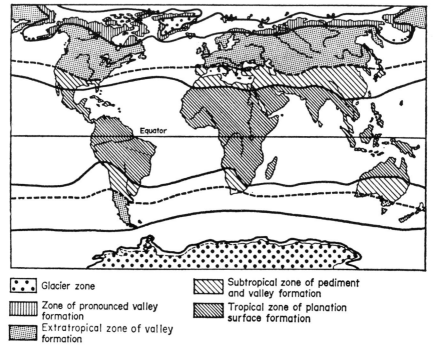

Source: After Büdel (1963), Figure 1; from Derbyshire (1973), Figure 13.1 p. 207. By permission of Macmillan, London and Basingstoke

4 Subtropical zone of pediment and valley formation (transitional between zones 3 and 5).

5 Tropical zone of planation surface formation (i.e. savanna surfaces – *Rumpfflächen*) interrupted by inselbergs. Such surfaces occur now in fossil form in Central Europe and other areas.

Büdel's scheme assumed that since the beginning of the Pliocene zones 1 and 2 (with considerable Pleistocene oscillations) have tended to expand equatorward at the expense of zones 3, 4 and 5. Figure 12.16 gives an example of Büdel's climatogenetic interpretation of a German *Mittelgebirge* valley. A detailed review of Büdel's work has been given by Kiewietdejonge (1984) and we will return to Büdel in Volume 4.

Figure 12.16 Landform generations for an idealized cross-section of a German *Mittelgebirge* valley showing broadly: the Late Pliocene high tropical planation (i.e. savanna) surface at just over 300 m; the oldest (Villafranchian) Pleistocene terraces associated with broad valley cutting; the main Pleistocene valley cut under tundra and frost-rubble conditions; the Holocene (Late Glacial) floodplain and present floodplain.

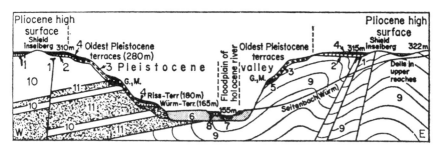

Source: After Büdel (1963) Figure 5. From Derbyshire, 1973, Figure 13.5, p. 223. By permission of Macmillan, London and Basingstoke

As noted in chapter 1, the early 1960s saw a revolution in global tectonic theory whereby the movement of crustal plates since the Triassic (including the Tertiary) was recognized to have caused great non-zonal (i.e. non-latitudinal) variations in climate and associated exogenetic processes. These matters must await consideration in a later volume of this series, as must the detailed consideration of the geomorphic operations of exogenetic processes (including those of Quaternary and earlier times), particularly during the period from about 1890 to the middle of the twentieth century, which will form the basis for volume 4 in this *History of the Study of Landforms*.

References

Ackermann, E. (1936) 'Dambos in Nordrhodesien', *Wissenschaft Veröffentlung Deutschen Museum der Landerkünde zu Leipzig* N.S., 4: 149–57.

Behrmann, W. (1921) 'Die Oberflächenformen der feuchtwarmen Tropen', *Zeitschrift der Gesellschaft für Erdkunde zu Berlin* 56: 44–60.

—— (1924) 'Das westlich Kaiser-Wilhelms-Land in New Guinea', *Zeitschrift der Gesellschaft für Erdkunde zu Berlin* Erganzungschaft 1.

—— (1927) 'Die Oberflächenformen im feuchtheissen Kalmenklima', *Düsseldorfer Geographische Voträge und Erörterungen* part 3, pp. 4–9.

Berkey, C. P. and Morris, F. K. (1927) *Geology of Mongolia*, New York: American Museum of Natural History.

Birot, P. (1937), 'Sur la morphologie de la Sierra Guadarrama occidentale', *Annales de Géographie* 46: 25–42.

—— (1949) *Essai sur quelques problèmes de morphologie générale*, Lisbon: Centro de Estudos Geraficos.

—— (1955) *Les méthodes de la morphologie*, Paris: Presses Universitaires de France.

—— (1960) *Le cycle d'érosion sous les differents climats*, Rio de Janeiro, Centro de Pesquisas de Geografia de Brasil, Universidade de Brasil (English trans. 1968 by C. I. Jackson and K. M. Clayton, London: Batsford).

—— (1973) 'Emmanuel de Martonne, précurseur de la géomorphologie climatique', *Bulletin d'Association de Géographers Français* pp. 408–9, 551–4.

Bishopp, D. W. (1941) 'Geomorphology of British Guiana', *Geological Magazine* 77: 305–29.

Bornhardt, W. (1900) *Zur Oberflächengestaltung und Geologie Deutsch-Ost-Afrikas*, Berlin: Reimer.

Bryan, K. (1922) 'Erosion and sedimentation in the Papago Country, Arizona', *US Geological Survey Bulletin* 730B: 19–90.

Büdel, J. (1944) 'Die morphologischen Wirkungen des Eiszeitklimas im glitocherfreien Gebiet', *Geologische Rundschau* 34: 482–519.

—— (1948a) 'Die klimamorphologischen Zonen der Polarlander', *Erkunde* 2: 22–53.

—— (1948b) 'Das System der klimatischen Geomorphologie', *Verhandlungen der Deutschen Geographentag*, Munich, 27: 69–100 (trans. Roger S. Mays in E. Derbyshire (ed.) (1973) *Climatic Geomorphology*, London: Macmillan, ch. 7).

—— (1951) 'Die klimazonen des Eiszeitalter', *Eiszeitalters und Gegenwart* 1: 16–26 (trans. as 'Climatic zones of the Pleistocene', *International Geology Review* 1: (9): 72–9).

—— (1953) 'Die "periglaziale" morphologische Wirkungen des Eiszeitklimas auf der ganzen Erde', *Erdkunde* 7: 249–65. (trans. as 'The "periglacial" morphological effects of the Pleistocene climate over the entire world', *International Geology Review* 1: (3): 1–16).

—— (1955) 'Reliefgenerationen und plio-pleistozaner Klimawandel im Hoggargebirge', *Erdkunde* 9: 100–15.

—— (1957a) 'Die doppelten Einebrungsflächen in den feuchten Tropen', *Zeitschrift für Geomorphologie* N.F., 1: 201–28.

—— (1957b) 'The Ice Age in the tropics', *Universitas* 1: 183–91.

—— (1961) 'Die Flächenbildung in den feuchten Tropen', *Verhandlungen der Deutschen Geographentag* vol. 33.

—— (1963) 'Klima-genetische Geomorphologie', *Geographische Rundschau* 15: 269–85 (trans. Joyce M. Perry and E. Derbyshire as 'Climatogenetic geomorphology' in E. Derbyshire (ed.) (1973) *Climatic Geomorphology*, London: Macmillan, ch. 13).

Chamberlin, T. C. and Chamberlin, R. T. (1910) 'Certain valley configurations in low latitudes', *Journal of Geology* 18: 117–24.

Cholley, A. (1950) 'Morphologie structurale et morphologie climatique', *Annales de Géographie* 59: 321–35.

Chorley, R. J., Schumm, S. A. and Sugden, D. E. (1984) Geomorphology, London: Methuen.

Clayton, R. W. (1956) 'Linear depressions (*Bergfassniederungen*) in savanna landscapes', *Geographical Studies* 3: 102–26.

Clozier, R. (1932) 'Inselbergs et formes d'érosion tropicales', *Annales de Géographie* 41: 219–20.

Common, R. (1966) 'Slope failure and morphogenetic regions', in G. H. Dury (ed.) *Essays in Geomorphology*, London: Heinemann, pp. 53–81.

Cotton, C. A. (1942) *Climatic Accidents in Landscape Making*, 2nd edn, Christchurch: Whitcombe & Tombs.

—— (1958) 'Alternating Pleistocene morphogenetic systems', *Geological Magazine* 95: 125–36.

—— (1961) 'The theory of savanna planation', *Geography* 46: 89–101.

Credner, W. (1931) 'Das Kräfteverhältnis morphogenetischer Faktoren und ihr Ausdruck in Formenbild Südostasiens', *Bulletin of the Geological Society of China* 11: 13–34.

—— (1935) *Siam, das Land der Thai*, Stuttgart, Reprinted by D. Zeller, Osnabrück, 1966.

Cushing, S. W. (1913) 'The East Coast of India', *Bulletin of the American Geographical Society* 45: 81–92.

Dana, J. D. (1849) In '*United States Exploring Expedition, during the years 1838, 1839, 1840, 1842 and 1843, under the command of Charles Wilkes*, Philadelphia, 10: 380–90.

Derbyshire, E. (ed.) (1973) *Climatic Geomorphology*, London: Macmillan (Intro. pp. 11–8).

Dokuchayev, V. V. (1883) *The Russian Chernozem*, St Petersburg.

Dresch, J. (1953) 'Plaines soudanaises', *Revue de Géomorphologie Dynamique* 4: 39–44.

Falconer, J. D. (1911) *Geology and Geography of Northern Nigeria*, London: Macmillan.

Freise, F. W. (1933a) 'Beobachtungen über Erosion an Urwaldgebirgsflüssen des brasilianischen Staates Rio de Janeiro', *Zeitschrift für Geomorphologie* 8: 1–9.

—— (1933b) 'Brasilianische Zuckerhutberge', *Zeitschrift für Geomorphologie* 8: 49–66.

—— (1935) 'Erscheinungen des Erdfliessens in Tropenurwalde; Beobachtungen aus brasilianischen Küstenwalder', *Zeitschrift für Geomorphologie* 9: 88–98.

—— (1936a) 'Das Binnenklima von Urwalden im subtropischen Brasilien', *Petermanns Geographische Mitteilungen* 82: 301–4, 346–8.

—— (1936b) 'Bodenverkrustungen in Brasilien', *Zeitschrift für Geomorphologie* 9(6): 233–48.

—— (1938) Inselberge und Inselberg-Landschaften in Granit und Gneissgebiete Brasiliens', *Zeitschrift für Geomorphologie* 10: 137–68.

Gillman, C. (1937) 'Zum Inselbergproblem in Ostafrika', *Geologische Rundschau* 28: 296–7.

Glinka, K. D. (1924) *Differents types d'après lesquels se forment les sols et la classification de ces derniers*, Comité International de Pedologie, IV, Commission no. 20.

—— (1927) *Treatise on Soil Science*, trans. C. F. Marbut (4th edn 1963, Israel Program for Scientific Translation, Jerusalem).

Herbertson, A. J. (1905) 'The major natural regions: An essay on systematic geography', *Geographical Journal* 25: 300–12.

Holmes, A. (1919) 'The Pre-Cambrian and associated rocks of the District of Mozambique', *Quarterly Journal of the Geological Society of London* 74 (for 1918): 31–97.

Holzner, L. and Weaver, G. D. (1965) 'Geographic evaluation of climatic and climato-genetic geomorphology', *Annals of the Association of American Geographers* 55: 592–602.

Hubert, H. (1908) *Mission Scientifique au Dahomey*, Paris: Larose.

Jaeger, F. (1927) 'Die Oberflächenformen im periodisch trockenen Tropenklima mit überwiegender Trockenzeit', *Düsseldorfer Geographische Vorträge und Erörterungen* part 3, 18–25.

—— (1939) 'Die Trockenseen die Erde', *Petermanns Geographische Mitteilungen*, 1–60.

Jessen, O. (1936) *Reisen und Forschungen in Angola*, Berlin: Reimer.

—— (1938) 'Tertiärklima und Mittelgebergsmorphologie', *Zeitschrift der Gesellschaft für Erdkunde zu Berlin* 36–49.

Kaiser, E. (1920a) 'Studien während des Krieges in Südwestafrika', *Zeitschrift der Deutschen Geologischen Gesellschaft* 72: 50–76.

—— (1920b) 'Bericht über geologische Studien während des Krieges in Sudwestafrika', *Abhandlungen Giessener Hochschulgesellschaft* 2: 1–58.

—— (1923a) 'Was ist eine Wüste?' *Mitteilungen der Geographische Gesellschaft in München* vol. 16 (3), 1–20.

—— (1923b) 'Kaolinisierung und Verkieselung als Verwitterungsvorgänge in der Namibwüste Sudwestafrikas', *Zeitschrift für Krystallographie* 58: 125–46.

—— (ed.) (1926) *Die Diamentenwüste Sudwestafrikas*, Berlin: Reimer.

—— (1927) 'Über Wüstenformen, insbesondere in der Namib Südwestafrikas', *Düsseldorfer Geographische Vorträge und Erörterungen* part 3, 68–78.

Kiewietdejonge, C. J. (1984) 'Büdel's geomorphology', *Progress in Physical Geography*, 8, 218–48 and 365–97.

King, L. C. (1951) *South African Scenery*, Edinburgh: Oliver & Boyd.

Klute, F. (1927) 'Die oberflächenformen der Arktis', *Düsseldorfer Geographische Vorträge und Erörterungen* part 3, 91–100.

Köppen, W. (1901) 'Versuch einer Klassifikation der Klimate vorzugsweise nach ihren Beziehungen zur Pflanzenwelt', *Geographische Zeitschrift* 6: 593–611, 657–79.

—— (1923) *Die Klimat der Erde: Grundriss der Klimakunde*, Berlin: Gruyter.

—— (1936) 'Das geographische System der Klimate', In Köppen, W. and Geiger, R. (eds) *Handbuch der Klimatologie*, Berlin, vol. 1, part C.

Krebs, N. (1933) 'Morphologische Beobachtungen in Südindien', *Sitzungsberichte der Preussischen Akademie der Wissenschaften Math.-Phys. Klasse* 23: 694–721.

—— (1942) 'Uber Wesen und Verbreitung der tropischen Inselberge', *Abhandlungen der Preussischen Akademie der Wissenschaften Math.-Natur. Klasse* no. 6.

Krenkel, E. (1922) *Die Bruchzonen Ostafrikas*, Berlin: Bornträger.

Krynine, P. D. (1935) 'Arkose deposits in the humid tropics: A study in sedimentation in southern Mexico', *American Journal of Science* 5th series, 29: 353–63.

—— (1936) 'Geomorphology and sedimentation in the humid tropics', *American Journal of Science* 5th series, 32: 297–306.

Lawson, A. C. (1915) 'Epigene profiles of the desert', *University of California, Publications of the Bulletin of the Department of Geography* 9: 23–48.

Lehmann, H. (1936) 'Morphologische Studien auf Java', *Geographische Abhandlungen* vol. 9 (3), 1–114.

Louis, H. (1957) 'Rumpflachen, Erosionszyklen und Klimamorphologie', *Petermanns*

Geographische Mitteilungen 262: 9–26 (trans. Roger S. Mays as 'The problem of erosion surfaces, cycles of erosion and climatic geomorphology', in E. Derbyshire (ed.) (1973) *Climatic Geomorphology*, London: Macmillan, ch. 9).

—— (1959) 'The Davisian cycle of erosion and climatic geomorphology', *Proceedings of the International Geographical Union Regional Conference in Japan, 1957*, pp. 164–6.

—— (1960) *Allgemeine Geomorphologie*, Berlin: De Gruyter.

Machatschek, F. (1927) 'Die oberflächenformen der Binnen-und Hochwusten', *Düsseldorfer Geographische Vorträge und Erörterungen* part 3, pp. 79–90.

—— (1938–40) *Das Relief der Erde: Versuch einer regionalen Morphologie der Erdoberfläche*, 2 vols, Berlin: Bornträger.

Martonne, E. de. (1913) 'Le climat facteur du relief', *Scientia* year 7, 13: 339–55 (trans. E. Derbyshire as 'Climate: Factor of relief', in E. Derbyshire (ed.) (1973) *Climatic Geomorphology*, London: Macmillan, ch. 3).

—— (1940) 'Problèmes morphologiques du Brésil tropical Atlantique', *Annales de Géographie* 49: 1–27, 106–29.

—— (1946) 'Géographie zonale', *Annales de Géographie* 55: 1–18.

Maull. O. (1938) *Geomorphologie: Enzyklopädie der Erdkunde*, Leipzig: Deuticke.

—— (1958) *Handbuch der Geomorphologie*, Vienna: Deuticke.

Mortensen, H. (1921) 'Morphologie der samländischen Steilküste', *Veröffentlichungen der Geographische Institut der Albertus-Universität zu Königsberg* vol. 3, 1–40.

—— (1927a) 'Geographische Forchungsreise in Chile 1925', *Forschungen und Fortschritte* 3: 166–8.

—— (1927b) 'Der Formenschatz der nordchilenischen Wüste', *Abhandlungen der Gesellschaft der Wissenschaft zu Göttingen, Math.-Phys. Klasse* N.F. XII, vol. 1, 1–192.

—— (1927c) 'Die Oberflächenformen der Winterregengebiete', *Düsseldorfer Geographische Vorträge und Erörterungen*, Part 3, 37–53.

—— (1930) 'Einige Oberflächenformen in Chile und auf Spitzbergen im Rahman einer vergleichenden Morphologie der Klimazonen', *Petermanns Geographische Mitteilungen* 209: 147–56.

—— (1933) 'Die "Salzsprengung" und ihre Bedeutung für die regionalklimatische Gliederung der Wüsten', *Petermanns Geographische Mitteilungen* year 79, 130–5.

—— (1943–4) 'Sechzig Jahre moderne geographische Morphologie, *Offentlichen Sitzung der Akademie der Wissenschaften zu Göttingen, Jahrbuch* 33–77.

Obst, E. (1923) 'Das abflusslose Rumpfochollenland in N-O Deutsch-Ostafrika, Part II', *Mitteilungen der Geographische Gesellschaft in Hamburg* vol. 35, 1–16.

Passarge, S. (1904a) *Die Kalahari: Versuch einer physisch-geographischen Darstellung der Sandfelden des Südafrikanischen Beckens*, Berlin: Reiner.

—— (1904b) 'Rumpflächen und Inselberge', *Zeitschrift der Deutschen Geologischen Gesellschaft* 56: 193–209.

—— (1904c) 'Die Inselberglandschaften in tropischen Afrika', *Naturwissenschaft Wochenschrift* N.F., 3: 657–65.

—— (1910) 'Verwitterung und Abtragung in der Steppen und Wüsten Algeriens', *Verhandlungen des Deutschen Geographentages zu Lübeck*, 7, 102–24.

—— (1912) 'Physiologische Morphologie', *Mitteilungen der Geographischen Gesellschaft in Hamburg* 26, 133–337.

—— (1919) 'Die Vorzeitformen der deutschen Mittelgebergelandschaften', *Petermanns Geographische Mitteilungen* year 65, 41–6.

—— (1920–9) *Die Grundlagen der Landschaftskunde*, 4 vols, Hamburg: Friederichsen, de Gruyter and Co. (see especially (1920) *Die Oberflächengestalt der Erde*, vol. III.)

—— (1921–4) *Vergleichende Landschaftskunde*, 4 vols, Berlin: Reimer (See especially (1921) *Aufgaben und Methoden*, vol. I).

—— (1924). Das Problem afrikanischer Inselberglandschaften', *Petermanns Geographische Mitteilungen* 70: 66–70, 117–20.

—— (1926) 'Morphologie der Klimazonen oder Morphologie der Landschaftsgurtel?' *Petermanns Geographische Mitteilungen* 72: 173–5 (trans. Roger S. Mays as 'Morphology of climatic zones or morphology of climatic belts?' in E. Derbyshire (ed.) (1973) *Climatic Geomorphology*, London: Macmillan, ch. 5).

—— (1927) 'Die Ausgestaltung der Trockenwüsten im heissen Gürtel', *Düsseldorfer Geographische Vorträge und Erörterungen* part 3, 54–67.

—— (1928) *Panoramen afrikanischer Inselberglandschaften*, Berlin: Reimer.

—— (1929) 'Das Problem der Inselberglandschaften', *Zeitschrift für Geomorphologie* 4: 109–22.

Peltier, L. C. (1950) 'The geographic cycle in periglacial regions as it is related to climatic geomorphology', *Annals of the Association of American Geographers* 40: 214–36.

Penck, A. (1894) *Morphologie der Erdoberfläche* 2 parts, Stuttgart: Engelhorn.

—— (1905) 'Climatic features in the land surface', *American Journal of Science* 4th series, 19: 165–74.

—— (1910) 'Versuch einer Klimaklassifikation auf physiographischer Grundlage', *Sitzungsberichte der Preussischen Akademie der Wissenschaften, Math.-Phys. Klasse* 12: 236–46 (trans. Roger S. Mays as 'Attempt at a classification of climate on physiographic basis', in E. Derbyshire (ed.) (1973) *Climatic Geomorphology*, London: Macmillan, ch. 2).

—— (1913) 'Versuch einer Klimaklassifikation auf physiographischer Grundlage', *Sitzungsberichte der Preussischen Akademie der Wissenschaften, Math.-Phys. Klasse* 15: 77–97.

—— (1914a) 'Die Formen der Landoberfläche und Verschiebungen der Klimagurtel', *Sitzungsberichte der Preussischen Akademie der Wissenschaften, Math.-Phys. Klasse* 4: 77–97.

—— (1914b) 'The shifting of the climatic belts', *Scottish Geographical Magazine* 30: 281–93.

Penck, A. and Brückner, E. (1901–9) *Die Alpen im Eiszeitalter*, 3 vols, Leipzig: Tauchnitz.

Rathjens, C. (ed.) (1971) *Klimatischen Geomorphologie*, Darmstadt: Wissenschaftliche Buchgesellschaft.

Richthofen, F. von (1877) *China*, vol. 1, Berlin; Reimer.

—— (1886) *Führer für Forschungsreisende: Anleitung zu Beobachtungen über Gegenstände der physischen Geographie und Geologie*, Berlin.

Rougerie, G. (1960) 'Le façonnement actuel des modelés en Côte d'Ivoire forestière', *Mémoires de l'Institut Français d'Afrique Noire* no. 58.

Sapper, K. (1923) *Die Tropen*, Stuttgart.

—— (1935) 'Geomorphologie der feuchten Tropen', *Geographische Schriften* vol. 7.

Sapper, K, and Geiger, R. (1934) 'Die dauernd frostfreien Räume der Erde und ihre Begrenzung, *Meterologische Zeitschrift*, 465–8.

Sauer, C. O. (1930) 'Basin and range forms in the Chiricahua Area', *University of California Publications in Geography* no. 3(6), 339–414.

Schmitthenner, H. (1927) 'Die Oberflächengestaltung im aussertropischen Monsumklima', *Düsseldorfer Geographische Vorträge und Erörterungen, Part 3* pp. 26–36.

Schou, A, (1965) 'Klimatisk geomorfologie', *Geografisk Tidskrift*, 64: 129–61.

Schweinfurth, G. A. (1865) 'Der Nil und das Baer'sch Gesetz der Uferbildung', *Petermanns Geographische Mitteilungen*, 126–8.

Stoddart, D. R. (1969) 'Climatic geomorphology: Review and re-assessment', *Progress in Geography* 1: 159–222.

Supan, A. G. (1884) *Grundzuge der Physischen Erdkunde*, Leipzig: Von Veit (2nd edn 1899, 5th edn 1911).

Tanner, W. F. (1961) 'An alternative approach to morphogenetic climates', *Southeastern Geologist* 2:(4) 113–26.

Thorbecke, F. (1921) 'Die Inselberglandschaft von Nord-Tikar', *Festschrift A. Hettner*, Breslau, pp. 215–42.

—— (ed.) (1927a) 'Morphologie der Klimazonen', *Düsseldorfer Geographische Vorträge und Erörterungen, Part 3*, 1–100.

—— (1927b) 'Klima und oberflächenformen: Die stellung des problems', *Düsseldorfer Geographische Vorträge und Erörterungen, Part 3*, 1–3.

—— (1927c) 'Der Formenschatz im periodisch trocknen Tropenklima mit uberwiegender Regenzeit', *Düsseldorfer Geographische Vorträge und Erörterungen, Part 3*, 10–18 (trans. Roger S. Mays and E. Derbyshire as 'Landforms of the Savanna Zone with a short dry season', in E. Derbyshire (ed.) (1973) *Climatic Geomorphology*, London: Macmillan, ch. 6).

Tricart, J (1953) 'Climat et géomorphologie', *Cahiers Information Géographique* no. 2, 39–51.

—— (1961) 'Les charactéristiques fondamentales du système morphogénetique des pays tropicaux humides', *Information Géographique* 25: 155–69.

—— (1965) *The Landforms of the Humid Tropics, Forests and Savannas*, Paris: SEDES (English trans. 1972 by J. Kiewiet de Jonge, London: Longman).

Tricart, J. and Cailleux, A. (1965) *Introduction à la Géomorphologie Climatique*, Paris: SEDES (earlier publication 1955).

—— (1972) *Introduction to Climatic Geomorphology*, London: Longman.

Troll, C. (1944) 'Structurboden, frostklimate und solifluktion der Erde', *Geologische Rundschau* 34: 545–694.

—— (1948) 'Der subnivale oder periglaziale Zyklus der Denudation', *Erdkunde* 2: 1–21.

—— (1958) 'Climatic seasons and climatic classification', *Oriental Geographer* 2: 141–65.

Visher, S. S. (1937) 'Regional contrasts in erosion in Indiana, with special attention to the climatic factor in causation', *Bulletin of the Geological Society of America* 48: 897–929.

—— (1941) 'Climate and geomorphology: some comparisons between regions', *Journal of Geomorphology* 4: 54–64.

—— (1945) 'Climate maps of geologic interest', *Bulletin of the Geological Society of America* 56: 713–36.

Voyeykov, A. (1885) 'Flüsse und Landseen als Produkte des Klimas', *Zeitschrift der Gesellschaft für Erdkunde zu Berlin* 20: 92–110.

—— (1887) *Die Klimate der Erde*, 2 vols, Jena: Costenoble.

Walther, J. (1890) 'Die Denudation in der Wüste', *Abhandlungen der Königlich Sächsischen Gesellschaft der Wissenschaften Math.-Phys. Klasse* 16.

—— (1900) *Das Gesetz der Wüstenbildung* Berlin: Reimer.

Wilhelmy, H. (1958) *Klimamorphologie der Massengesteine*, Braunschweig: Westermann.

Willis, B. (1936) *East African Plateaus and Rift Valleys*, Washington D.C.: Carnegie Institution.

Wilson, L. (1969) 'Relationship between geomorphic processes and modern climates as a method in paleoclimatology', *Révue de Géographie Physique et de Géologie Dynamique* 11: 303–14.

Indexes

Subject and Place Index
Index of Persons

Subject and Place Index

As stated in Volume 2, in the study of landforms principle and place are symbiotic and cannot safely be divorced. But since the present volume covers most of the surface of the globe, terrain features are usually given as examples to illustrate a type of landform and the broad formative process associated with it. Place is subordinated to principle and places on which no useful information is added other than the name are omitted in an effort to diminish a welter of names of only local significance.

The subject aspect of the text has been equally or more tantalizing. Geomorphology has attracted, and continues to attract, numerous terminologists whose ability to coin new words seems limitless. For English, the *Supplement to the Oxford English Dictionary*, 4 volumes, 1972–1986 (Clarendon Press, Oxford) enlightens many of the generally accepted new landform terms. In the indexes below the development of concepts and theories holds sway.

References to figures are added in italics at the end of relevant entries.

D. M. Beckinsale

Index of Persons

by
Robert and Monica Beckinsale
and
Richard and Rosemary Chorley

This biographical information has been obtained from a large number of sources and individuals, the most important being listed below.

References to illustrations are given in italics at the end of the individual entries concerned.

American Men of Science (1906–   ) (Later *American Men and Women of Science*).

Association of American Geographers Directory (1987), New York.

Dictionary of Scientific Biography (1970–78), C. C. Gillispie (Ed.), 15 Vols., New York: Charles Scribners' Sons.

Directory of Soviet Geographers (1967, 1977 and 1988), published in *Soviet Geography*, Vol. 8, pp. 513–610; Vol. 18, pp. 436–538 and Vol. 29, pp. 152–366; V. H. Winston and Son Inc. for the American Geographical Society.

Geographers: Bibliographical Studies (1977–   ), London: Mansell, for the International Geographical Union.

Geologists and the History of Geology (1980), William A. S. Sarjeant, Vols. 2 and 3, London: The Macmillan Press Ltd.

Geologists and the History of Geology (1987), William A. S. Sarjeant, Supplement I, Malabar, Florida: Robert E. Krieger Pub. Co.

A Historical Catalogue of Scientists and Scientific Books (1984), R. M. Gascoigne, New York and London: Garland.

International Geomorphology: Directory 1989, H. Jesse Walker (Ed.).

Kratkaya geograficheskaya entsiklopediya (1966), A. A. Grigor'yev (Ed.), Vol. 5, Izdatel'stvo 'Sovietskaya Entsiklopediya', Moscow.

Library of the Geological Society of America, Boulder, Colorado.

Library of the Geological Society of London, Burlington House, London.

Library of the Royal Geographical Society, Kensington Gore, London.

Library of the Scott Polar Research Institute, Cambridge (Courtesy of Miss Rosemary Graham).

Orbis Geographicus (1960–   ), Franz Steiner: Stuttgart.

Individuals: Profs. H. Bremer and M. Schwarzbach of Cologne; Prof. S. A. Schumm of Ft Collins; Prof. A. Rapp of Lund; Prof. L. Starkel of Krakow; Chancellor O. Granö of Turku; Dr S. Kozarski of Poznán; Dr P. S. Savill of Oxford; Dr D. Turnock of Leicester; Dr S. A. Lukyanova of Moscow; Prof. B. G. Thom of Sydney; Prof. S. Dragomirescu of Bucarest.